PENGUIN BOOKS

THE ESSENTIAL URBAN FARMER

NOVELLA CARPENTER is the author of the bestselling *Farm City: The Education of an Urban Farmer.* She lives and farms in Oakland with her partner Billy, one cat, three ducks, five rabbits, two goats, and fifty thousand honeybees.

WILLOW ROSENTHAL is the founder of City Slicker Farms in Oakland, California, and teaches adults and children about urban gardening and consults with various groups to implement urban farming strategies. She lives in North Berkeley with her husband where they tend a very small veggie garden of their own.

ALSO BY NOVELLA CARPENTER

Farm City: The Education of an Urban Farmer

THE ESSENTIAL URBAN FARMER

Novella Carpenter

and

Willow Rosenthal

PENGUIN BOOKS

PENGUIN BOOKS

Published by the Penguin Group

Penguin Group (USA) Inc., 375 Hudson Street, New York, New York 10014, U.S.A.

Penguin Group (Canada), 90 Eglinton Avenue East, Suite 700,
    Toronto, Ontario, Canada M4P 2Y3 (a division of Pearson Penguin Canada Inc.)

Penguin Books Ltd, 80 Strand, London WC2R 0RL, England

Penguin Ireland, 25 St Stephen's Green, Dublin 2, Ireland
    (a division of Penguin Books Ltd)

Penguin Group (Australia), 250 Camberwell Road, Camberwell, Victoria 3124, Australia
    (a division of Pearson Australia Group Pty Ltd)

Penguin Books India Pvt Ltd, 11 Community Centre,
    Panchsheel Park, New Delhi - 110 017, India

Penguin Group (NZ), 67 Apollo Drive, Rosedale, Auckland 0632, New Zealand
    (a division of Pearson New Zealand Ltd)

Penguin Books (South Africa) (Pty) Ltd, 24 Sturdee Avenue,
    Rosebank, Johannesburg 2196, South Africa

Penguin Books Ltd, Registered Offices:
80 Strand, London WC2R 0RL, England

First published in Penguin Books 2011

10 9 8 7 6 5 4 3 2

Copyright © Novella Carpenter and Willow Rosenthal, 2011
All rights reserved

Photographs: © Sophia Wang

Hand drawn illustrations: Bronwyn Barry

Computer generated illustrations: Arianna Rosenthal

LIBRARY OF CONGRESS CATALOGING IN PUBLICATION DATA
Carpenter, Novella, 1972–
    The essential urban farmer / Novella Carpenter and Willow Rosenthal.
        p. cm.
    Includes index.
    ISBN 978-0-14-311871-8
    1. Urban agriculture.   I. Rosenthal, Willow.   II. Title.
    S494.5.U72C36 2012
    635.977—dc23
                                    2011039313

Printed in the United States of America
Set in Linoletter Std
Designed by Sabrina Bowers

Dedicated to fellow urban farmers, past, present, and future.

CONTENTS

Please note that illustrations throughout are not necessarily to scale.

INTRODUCTION

Willow and I first bonded over urban farming. We were both growing vegetables, beekeeping, and raising chickens and ducks in the middle of the city of Oakland, California. When we met, at the turn of the century, I had recently started Ghost-Town Farm, one-tenth of an acre farm on squatted land near downtown. Willow had founded City Slicker Farms a few years before, as a nonprofit urban farming organization devoted to making affordable, urban-grown organic produce available to low-income residents in West Oakland. This is the book that we wished we had when we first started out, a how-to manual that speaks directly to farmers trying to grow food and raise animals in the city.

We became passionate about urban farming for a variety of reasons. One is the way urban agriculture connects urban people to the food they are eating. The lettuce someone seeds, waters, and then harvests for dinner makes the freshest, most delicious salad they have ever had. Backyard chicken eggs are a revelation, partially because they are so fresh and partially because you raised the hen who laid this special gift. We realized that many city folks don't think they can produce their own food, and so they miss out on these connections. By growing even a little food in the city, these experiences become accessible.

Speaking of accessibility, urban farming is a way for people of all income levels to eat fresh, local, organic food. I knew that I didn't have enough money to buy organic produce or meat, and so I decided to raise it myself. An average urban backyard (25 feet by 40 feet), if cultivated intensively, has the potential to grow all of the fruits and vegetables for one person. Growing edibles in the city—even on a deck or small backyard—makes economic sense for people who have more time than money. Due to low incomes and lack of access to grocery stores, urban people fail to get the healthy nutrition they need. A few packets of seeds costing less than twenty dollars can produce enough vegetables for a year's worth of eating. If government regulations were changed and financial support given, many of the fruits and vegetables consumed in a city could be grown within the city itself, through a combination of backyard gardening, community gardening, school gardens, commercial gardens, and increasing urban agriculture on currently unused municipal land. This would mean everyone would have access to healthy organic food!

I say organic because this is the farming method that we encourage everyone to use. Organic means that you don't use chemically synthesized fertilizers or pesticides—two things that your neighbors in the city do not need to be exposed to. Other aspects of organic farming that we encourage, and explain in this book, are: building soil fertility through crop rotation; proper application of compost and green manures; and controlling weeds and pests by mulching, picking by hand, or using natural sprays or mixtures.

Rural organic farms do not necessarily follow practices that are sustainable for the earth, animals, or human beings. Growing in the city also means that you can go a bit beyond organic by growing a variety of crops on one site (instead of growing a single crop [monoculture]), using water efficiently, integrating livestock, and using city wastes to create a more closed-loop nutrient system. This avoids the use of factory-made fertilizers, using fossil fuels for operating farm machinery, or shipping produce and inputs—and, unlike commercial farms, ensures fair and safe labor practices.

Another method of urban farming is intensive farming, which enables growers to achieve high food yields in small spaces. Crops are spaced as tightly as possible and soil fertility is continuously built to support their growth. To achieve the highest yields possible while also maintaining the overall health of the plants and animals, we focus on the soil; because of this, the crops will thrive.

We have taken principles from intensive farming (sometimes called French intensive), biodynamic farming, permaculture, and edible landscaping, rolled them up in a ball and called it urban farming. As you begin to build your urban farm, you'll no doubt encounter other useful methods and ideas. There is no perfect way: raising food is a constantly changing dialogue between you the farmer and the land-scape, animals, community members, and political and social circum-stances.

There is no better time to start urban farms. We're entering a golden era, as farms spring up on rooftops in Brooklyn and Chicago; in abandoned lots in Detroit and San Francisco; in community gardens and in backyards. The thing is, none of this is new! The historical record shows that up until only recently, growing food in cities all over the world was the norm rather than an oddball fad. Think back to an-cient Mesopotamia or the Incan empire, both highly urban societies; the people relied on urban and periurban farms for their food. Learn about the French market gardens built during the Paris Commune. Read up on Detroit mayor Hazen Pinagree, who developed a system of farms in Detroit in the 1890s. Remember the victory gardens grown on the grounds of the White House, in New York's Bryant Park, on the grounds of businesses and people's backyards during both world wars.

It wasn't until the 1950s, when the highway systems were built and the era of cheap fossil fuel began, that the strict division between rural and urban began to take shape. In the 1970s, during the oil crisis, an ecological movement started to grow that encouraged self-sufficiency in cities. Books on sustainable living, such as *The Integral Urban House,* demonstrated that one could grow vegetables and raise honeybees and chickens in the city.

There is a strong element of social justice in this latest wave of interest in growing food in the city. There is often vacant land in so-called blighted areas—empty lots that could be used by the people who live there to produce food. This is why Willow founded City Slicker Farms. With neighborhood support, she started a few gardens on empty lots and set up a farm stand to sell the produce at affordable prices; no one was turned away for lack of funds. Neighbor interest in self-sufficient food production led to the addition a few years later of a backyard garden program for low-income residents. Today City Slicker Farms has built two hundred backyard gardens that produce tens of thousands of pounds of food per year in West Oakland. Urban farming is empowering: It can create self-sufficiency in communities who need it most.

A move toward more food production in the city is a way to combat other threats to humanity: climate change, the irresponsible use of fossil fuels, and a ballooning urban waste stream. Small urban farms don't have to use fossil fuel–derived fertilizers or pesticides. Urban-grown produce doesn't need to be shipped in refrigerated trucks or by airplane. We can rechannel the wastes away from landfills and toward productive uses, such as fertilizer. Restaurant food scraps can feed urban chickens and rabbits; busy cabinetry shops can supply shavings for garden paths; coffee grounds from the local café can be used to make great compost. By recycling wastes we reintroduce city people to that great American ideal of thrift.

At the time Willow and I started talking about this book I had just helped launch a store in Berkeley that sells organic locally made animal feeds, farm equipment, and books and tools for self-sufficient

SUSTAINABILITY BULL'S-EYE: FARM INPUTS AND OUTPUTS

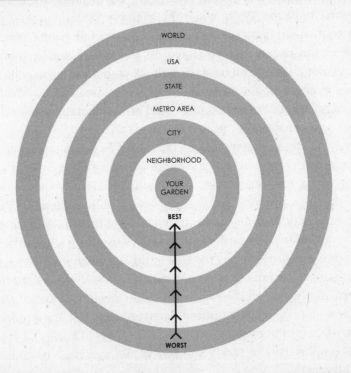

When sourcing materials keep in mind that local is more sustainable.

living, and that also provides classes about urban chickens, goats, and rabbits. The classes often sold out in a few days, so I knew there was a great hunger to learn these skills. We noticed that there was no definitive how-to manual for setting up, maintaining, and expanding an urban farm and began talking about collecting all the how-to knowledge we'd gathered over the previous ten years into a manual. Our goal then was to write that book, to create a one-stop resource for both beginning and seasoned urban farmers. Our years-long efforts to distill our field experience has resulted in *The Essential Urban Farmer*, the book you hold in your hands.

The Essential Urban Farmer is divided into three parts and reflects our philosophy: First observe; then take action; then sustain the results; and finally, expand your efforts. The first part, Designing Your Urban Farm, can be read as a beginner's guide to advance planning. It covers choosing a site, testing the soil, and creating a layout that uses the space you have to grow the most food possible. Though many people want to just dive right in, it's often best to first take a step back, assess what you have, and build a strategy. We encourage baby urban farmers to start with small steps, like growing some lettuce or carrots. Then move on to bees or chickens. You'll build on your knowledge base and stay sane by moving slowly.

Part II, Raising City Vegetables and Fruits, involves getting your hands dirty in order to grow vegetables and fruits. You'll read about how to build soil fertility, set up growing beds, start seedlings, plant seeds and plants, and irrigate your crops. If you live in an apartment or have a small amount of growing space, go to page 113. After you've planted, you'll need to learn how to defend your garden (organically, of course) from pests and disease. And then, finally, you'll find out about best practices for harvesting your homegrown bounty.

If you've already tried your hand at growing vegetables, or have chickens but are looking to expand your farm animal repertoire, Part III, Raising City Animals, is for you. This section covers six animals that will thrive on your city farm: honeybees, chickens, ducks, turkeys, rabbits, and dairy goats. There are in-depth directions for setting up housing, feeding, and maintaining your menagerie. We've also included tips for saving money on your small-scale animal operation by using the city's waste for food, bedding, and building materials. Finally, we also include step-by-step instructions for processing animals for the dinner table.

At the back you will find an elaborate resource section that lists materials and tools, places from which to order farm equipment and livestock, listings of urban farms all across the country, tables of information for seed savers, and sample seeding and planting calendars. Since we are based in California, we consulted and visited seasoned practitioners throughout the country to learn about their climate zones. Their best practices are touched on throughout. We are indebted to them, and welcome input from you, our readers, and fellow urban farmers.

Whether you're just getting into urban agriculture or at it for a while, we're sure you'll find something of use here, even if it's primarily a sense of solidarity. So please send ideas and other resources to essentialurbanfarmer@gmail.com.

Finally, though we often shy away from the R word, we daily see more evidence that urban farming is becoming a real revolution. The most important things that Willow and I have learned haven't been about chickens or vegetables, rabbits or fruit trees; it has to do with the people in our communities. We've seen firsthand the way animals and plants bring our neighbors together. Cultivating life allows people to feel a connection to the earth, and to each other. We encourage you to not just sit with this book but to go out and seek experienced mentors, to gather and share knowledge and resources. We all eat, and the two of us think that if we all grow a little bit of what we eat, the world would be a better place.

—NOVELLA CARPENTER

PART I

DESIGNING YOUR URBAN FARM

Whether you have an urban farm or are just starting out, this section is an invaluable primer. Although it may be tempting to just start and ask questions later, this section will give you a deeper understanding of what you are doing and how it fits into a variety of different farming philosophies. It will also help you ensure that you're growing food that is safe to eat—a huge consideration for the urban agriculturist.

The biggest hurdle is finding land that is appropriate for farming. The soil must be tested for heavy metals. Local ordinances and neighbors should be consulted. If you have a landlord, he must be convinced that it is a good idea. Before beginning the fun stuff—planning where to place your veggie beds and animal housing, collecting tools and materials for building—you must secure a safe location.

W ➤ E

CHAPTER 1

CHOOSING YOUR URBAN FARM SITE

First off, where will you be farming? Selecting a site for many of you will be easy. It'll be built on the yard, deck, or roof of the house you own. If this is the case and you already have land access, skip to chapter 2, to get started. For the rest, who are looking for land—apartment dwellers, ambitious backyard gardeners ready to expand, renters, or owners whose land isn't suitable—we offer ways to secure some. If you're a renter with the perfect farming situation, first get permission from your landlord.

HOW TO SWEET-TALK YOUR LANDLORD AND BE A RESPONSIBLE STEWARD

In the interest of getting your rental deposit back, and of using the space you live in responsibly, it's important for renters to get clear, written permission from their landlord for planned changes to the landscape. It is especially important with deck and rooftop gardens to discuss what you want to do and to explain how your methods will protect built structures. It works best to try to think like an owner when approaching your landlord. It's likely their main concerns will be: increased liability exposure; property value reduction; higher water bills; messes; and neighbor complaints.

When getting permission, address these concerns up front in a written, proactive proposal. Let your landlord know if you want community members to be involved. Ask if their property liability insurance would cover such participation or if they would need you to take out an additional policy. List everything you want to do, make a map, and show that you have discussed the proposal with any neighbors who might be affected and that you have a plan, if they want, for how to return the property to its original state when you eventually move. If the water bill isn't in your name, let your landlord know you will handle increased water bills, for instance, by offering to pay an extra amount or transfering the bill into your name. If you approach the subject this way, your landlord might just see that you are responsible and agree to your plan. Be open to their suggestions and concerns, and be willing to compromise, and you'll likely come to an agreement.

Being a renter might also affect the type of farm you design. You may want to keep your trees in movable containers, and skip building wood boxes for your veggie beds, so you don't lose your investment when you move.

FINDING VACANT LAND

In most cities there are hundreds or thousands of acres of unused land that could be producing food: empty lots; utility company right of ways; neglected backyards; median strips; parks. Vast sunny lawns at institutions such as churches, universities, hospitals, prisons, and senior homes eat up water that could be used to grow produce. Once your eyes are trained to see such possibilities it can be hard to turn off the habit of fantasizing beautiful veggie gardens growing in these spaces or imagining a herd of goats grazing. The first step in turning the fantasy into a reality is finding out who owns or controls the land and contacting them. Vacant land generally falls into three categories:

+ privately owned by individuals, companies, or religious groups;
+ owned by the city, county, or state; for example, county tax assessor offices or park districts;
+ owned by a public authority or district; for example, transportation authority; utility district.

If you spot a vacant lot with potential, first write down the street addresses on either side of the site and the cross streets of the block. Then check with your county tax assessor's office. You will find maps showing each block and street address in your city either online or at their office. Its parcel number will be listed on the map. Carefully check to see if your site comprises more than one lot, or if it's the yard of a nearby house or business.

With the parcel number in hand you can look up the ownership history, which will be listed in a separate online section or written source material filed at the office. The most recent owner will be listed along with their mailing address. Knowing the historic use of the property will tell you its previous use; for example, if you simply see a list of individual names, most likely it was a residence or vacant lot for a long time. But names such as Acme Chemical Company or Billy's Junk Yard may indicate problems you don't want to touch with a ten-foot dig bar.

BASIC SITE CONSIDERATIONS

So, you're out cruising around looking at various places to farm: maybe one is an abandoned lot; maybe one is the backyard of a friend; maybe one is a piece of land owned by a church. For a baseline, research the following:

- Who owns or controls the land?
- Is there access to water? If you have to install municipal water service or a well it could cost you thousands of dollars.
- Is there vehicle or cart access for moving in materials?
- Is there good sunlight for most of the day? If not, do you have permission to trim or remove trees, bushes, or vines?
- Is the site protected from extreme wind?
- Does the slope of the land make terracing essential?
- Is the site prone to flooding or standing water?
- Is there garden-ready soil over most of the site, or is it covered with blacktop, concrete, or gravel?
- Could the soil be contaminated with toxins? (See chapter 2.)
- Can you use the land long enough to merit the significant time and energy you'd have to put in?

Survey the land and locate the municipal water valve, a concrete lid somewhere in the sidewalk or road adjacent to the plot; if it's a friend's backyard, find the hose bib (the outdoor water source). Locate walk-in and drive-in access. Hang out to observe the sun and wind patterns. It's also important to get onto the site with a dig bar and check for gravel, concrete, and blacktop. Concrete removal can be prohibitively expensive, so consider a new site if the entire lot is paved. Before beginning negotiations, you will want to take soil samples and test for contamination. (See chapter 2.)

CONTACTING THE LANDOWNER AND CREATING AN AGREEMENT

Unfortunately, in most counties, the owner's phone number is not provided at the tax assessor's office. If it's a governmental agency, you can look it up easily, but those of private owners are more difficult to track: They may not live at the address provided in the records. Try calling information for their number. But often your only option is to send written correspondence. Sadly, you may find a *perfect* site that has sat unused for years—and send off letters never to receive a reply.

Local realtors are also a good source of information on ownership and land history. They may know the contact information for an owner you can't otherwise find, and they often know the use history of land. Realtors can be very friendly to the idea of community beautification through gardening. Hey, greening raises property values and sales commissions!

Once you have made contact, the owner may turn out to be amenable, and even enthusiastic, about having their land used productively. Government-owned land is a different ball of wax. Government bureaucracy can result in a maddening circle: a long list of blasé government employees that ends back with the first buck passer. Being tenacious and having an unfailingly friendly, compassionate attitude can pay off for you in the end. Government agencies rarely have systems in place for dealing with such inquiries, which means that yours is unexpectedly adding to someone's workload. You may have to blaze the trail and help the agency create a system to allow private citizens to use government land for urban farming.

The following guidelines will improve your chances of a positive outcome when approaching either private or government owners:

First write a one-page letter that includes the following:

- who you are;
- a brief description of your proposal for the land, including any community benefit (e.g., community garden plots; donations of produce to food banks);
- a proposal to lease the land for free or for a comfortable price;
- your willingness to pay for water, obtain any necessary liability insurance, and leave the property as you found it, should you leave;
- an offer to provide a more detailed proposal along with personal references;
- your contact information;
- your gratitude for their consideration.

If they respond, write a letter or e-mail proposing a meeting. Include a proposal that describes the following:

- your ideas for the land, including a layout diagram and any community benefit;
- the minimum amount of time you are willing to use the land; for example, that you would like to use the property for at least three years, with a provision for renewing your agreement on a year-to-year basis (any less time, e.g., three months, isn't worth your while);
- any structures you want to put up, and any existing plants or structures you'd like to remove to facilitate your plan;
- how you will maintain the property, in terms of safety and aesthetics (owners can be cited by the city for blight and will be concerned about this);
- what you will do if and when you leave the land (e.g., cleanup; restoring; finding a qualified successor);
- who will have access to the property;
- who will work the land;
- options for how the water bill would be paid (e.g., either by transferring it to your name or by arranging to pay the owner);
- your willingness to obtain liability insurance, including a possible provider and/or organization name and contact information;
- your intention to obtain the necessary business licenses and agricultural inspections;
- character and/or professional references.

Note: If you want to incorporate animals, like chickens or bees, it might be best to keep it to yourself at this point. After a year of smooth sailing with growing produce, broach the delicate subject of livestock with the land owner.

If the owner agrees to your using the property:

⁜ Negotiate an agreement in writing.

⁜ Ideally meet in person to hammer out the details.

⁜ Create a written document for signing. It's not necessary to involve a lawyer, but you may want to have it reviewed by one.

If you can't meet the owner's demands, it would be better to find another site than to commit to something that's uncomfortable.

Set up the payment system for the utilities:

⁜ There should be easy ways to communicate about and pay the water bill. The ideal solution is to transfer the water service into your name, since busy landlords can be tardy in passing them along.

⁜ If the bill isn't transferred into your name have the landlord send you quarterly copies with your portion noted.

Set up liability insurance:

⁜ Find out if the owner has a liability policy, and if so, if it will cover your activities. If not, offer to pay for increased coverage, or take out another policy.

⁜ Find out if your city has a program to provide it for community gardens, or if there are nonprofit urban farming organizations willing to add your site to their policy. Because rates are calculated based on the total number of acres, if your garden is small enough, adding you may not increase their rates.

⁜ Consider purchasing a personal umbrella insurance policy as well, since few insurance policies will cover everything someone could sue for.

If the owner is a governmental agency:

⁜ Contact your city parks and recreation department to find out if they have a community garden program. If so, ask if they help residents start new ones and what support they offer.

⁜ Contact the following organizations to find out if they know anything about efforts to facilitate the use of government land for urban farming in your area:

  • The Community Food Security Coalition;

  • The American Community Gardening Association;

  • Local nonprofit urban farming and community gardening organizations;

  • Commercial urban farming operations;

Note: Joining forces under the umbrella of an already existing non-profit organization can streamline the process and lend credibility to your project.

* Contact your local city council member and enlist their help in gaining approval for your project. In addition, city council members often have discretionary funds they can devote to pet projects.
* Recruit a list of supporters by pounding the pavement. You will want to reach out to local residents, business associations, community groups, and churches to ask them to sign a letter and/or attend city council meetings in support.

PURCHASING VACANT LAND

Creating productive and beautiful urban farms on loaned land can eat up thousands of dollars and wo/man hours, only to be lost once someone decides to "develop" it. If you own the land though, you don't have to worry about losing it, and it will be truly sustainable. Purchasing land for urban farming can make sense or not, depending on your real estate market.

If you've decided to purchase vacant land, consider your proximity to the site. You will be spending a lot of time there, and if you have to commute your energy may flag. The best scenario is next door, or within a few blocks of your residence. As the owner, you will be responsible for any necessary cleanup of garbage or toxic waste, so be especially careful in your assessment. Purchasing a former auto mechanic's used parts yard, or the former site of an industrial building, is dicey unless you have the funds for a major cleanup project. Sites covered with blacktop or concrete will require a huge amount of time and money to rehabilitate.

In addition to the purchase price, as noted previously, you will need to buy liability insurance and pay yearly property taxes. Loans for purchasing vacant land are structured differently than those for purchasing residences. Banks often won't make loans for it, and if they do, the interest rate will be higher. Private loans can be obtained, again at higher rates than home loans. The best scenario is obviously to save up your money, find a site that's cheap, and purchase it outright. In rare cases agricultural or open space preservation land trusts may be able to help finance or protect your land from future development. Having your land designated as protected open or agricultural space by such an entity can also reduce your property tax liability.

Another way to purchase land is through county property tax default auctions. Counties auction properties when the owner has failed to pay their annual property tax for more than five to ten years, depending on the county. Success stories using this strategy abound: empty lots in the high-end Bay Area real estate market purchased for under ten thousand dollars; entire lots in Detroit purchased for a few hundred dollars. Focusing your search on less desirable neighborhoods helps, but this isn't a rule of thumb. Know that auctions aren't always logical: It just depends on who shows up or logs in (many are conducted online now) to bid on the day. It's important to know that full payment must be made upon your successful bid. Private auctions often operate on a similar basis.

W   E

CHAPTER 2

URBAN SOIL, WATER, AND AIR CONTAMINATION

After you've figured out where you're going to farm, the next step is to test for toxins in the soil. It's a downside to growing in the city but don't be bummed by this reality: Since farming encourages testing and cleanup of toxic soil, we think that the possibility of contamination in urban areas should be cause *for* not against it. You have an incentive to find out if you are currently unknowingly being exposed to toxins. If you are, farming or gardening can reduce it by containing and covering the soil and by the cleaning actions of plants and soil organisms.

For example, the main pathway for lead poisoning, especially in children, is ingesting contaminated dirt. When a garden is planted and mulched with compost, wood chips, etc., the dirt is less exposed and is contained by roots. Adding compost dilutes the soil, lowering contamination levels. Compost also binds to heavy metals, making them less bioavailable to plants and people. This is perhaps the most effective method for remediating soil containing lead. (For more information, see "Lead in Urban-grown Vegetables," Cornell University Cooperative Extension.) As a defense, plants often concentrate more of the heavy metals they absorb in their roots, which are not usually the edible portion of your produce. Also, the amount in the fruits and vegetables grown in these environments is much lower than that in the soil itself, so the chance of being poisoned by eating produce is usually quite low.

At the bare minimum, everyone should test their soil for heavy metals, no matter what. It's relatively cheap, and it will give you peace of mind. If you need further prodding, take the survey below. If you answer yes to any of the following questions, you will definitely need to test your soil for heavy metals.

- Is it possible your land is the former site of a factory or business that might have used chemicals, motor oil, etc.?
- Was your house built before the use of leaded paints was discontinued in 1940?
- If so, was its upkeep neglected at any time and the paint allowed to peel onto the ground?
- Was the paint scraped or sandblasted without proper containment?
- Was your house built prior to 1945? (That's when the burning and burying of garbage was finally prohibited in most cities. Though this practice probably continued into the seventies in many low-income communities.)
- Did a historic house or structure burn down on the property?
- Is the property near a freeway, busy road, or airport? (Tire dust contains toxic heavy metals, and prior to 1975 leaded fuel was used; also, jet fuel still contains lead.)

Heavy Metals

The most likely way to be poisoned by heavy metals is by actually eating or breathing in the dirt itself, not from eating produce grown in soils containing them. Still, it's good to know if heavy metals are present in dangerously high quantities and take appropriate precautions.

The reason they are so harmful to humans is that the body mistakes them for essential nutrients and stores them in tissue. Lead is the most common heavy metal contaminant in urban soils and the primary cause of heavy metal poisoning in children. Most of what is in garden soils was left from the days of leaded plumbing, paint, and fuel. It binds to cells in the body and is absorbed readily instead of calcium and iron. Health problems from lead exposure range from kidney, nervous system, and thyroid damage, especially in pregnant women, to the stunting of brain development in children.

Though urban industry can result in quite a cocktail of heavy metal elements, the most common elemental contaminants after lead are arsenic, mercury, and cadmium. Arsenic is released into the environment through metal smelting and galvanizing, by power plants, and by the manufacture of chemicals and pesticides; it is also common in paint, rat poison, fungicides, and wood preservatives. It is a primary ingredient in pressure-treated lumber commonly used in outdoor structures. Look for the telltale greenish tint to the wood. Like lead, arsenic is likely to remain in soil unless removed. Former orchards are common sites of high arsenic levels, and since many cities sprawled into surrounding farmland, you may be living on the site of what used to be an orchard.

We are also unfortunately surrounded by unsafe mercury levels. Mining operations and various industrial processes release mercury, toxifying urban soils. Most concerning of all, mercury was also added to paint as a fungicide until 1990.

Cadmium is a by-product of mining, smelting, and metal plating, and is used in batteries, Polyvinyl chloride (PVC) plastics, and paint pigments. It can be found in soils because insecticides, fungicides, sludge, motor oil, and commercial fertilizers using it are still on the market. Cigarettes also contain cadmium, so the old practice of burying trash can be a source of elevated levels.

We hope you're scared now—really, really scared. Since testing for heavy metals is cheap, and the results of exposure can be extremely serious, unless you are certain there is no risk you should just go ahead and test the soil. Luckily, soil-testing labs will test for both the dangerous and the useful nutrients and soil texture all in one fell swoop. Depending on the lab you choose, the hazardous elements tested for can include arsenic, cadmium, chromium, copper, lead, manganese, mercury, molybdenum, nickel, selenium, and/or zinc. We recommend using a lab associated with a university, as their tests are reliable and prices are usually low. Make sure to specify which elements you want included in the test in addition to lead.

Acceptable levels of heavy metals in soil

There are currently no national standards for safe amounts of heavy metals in agricultural soils. In 1996, the U.S. Environmental Protection Agency (EPA) set soil screening levels (SSL) to be used as an evaluation tool for cleaning up contaminated residential properties and Superfund sites.

ENVIRONMENTAL PROTECTION AGENCY SOIL SCREENING LEVELS FOR HEAVY METALS IN SOIL

Heavy Metal Element	EPA Soil Screening Level (in milligrams per kilogram)	Acceptable Parts per Million in Soil
arsenic	1.0	1.0
cadmium	0.4	0.4
chromium	2.0	2.0
lead	400	400
nickel	7.0	7.0
selenium	0.3	0.3
zinc	620	620

Heavy Metal Testing Procedure

Here's how to prepare your soil samples before having them tested by a lab. See page 501 for a resources list of soil-testing facilities.

* Draw a diagram of your land and break it up into 5 to 20 quadrants, depending on your budget, by drawing horizontal and vertical lines at regular intervals. Be sure to note where any structures are and where north is. You will take a mixed soil sample from each quandrant.

* Number your quadrants on your diagram.

* Make sure you plan to take samples from at least two areas that you find suitable sun for growing. That way, if one is contaminated and the other is clean, you can go ahead with the second area.

* Get plastic or paper bags—two for each quadrant—and label them with the quadrant sample number. You will send one bag to the lab and keep one in case of loss, or if you want to do other types of testing later.

* Bring a shovel, trowel, bucket, permanent marker, and the bags outside. Your tools must be stainless steel, since galvanized or aluminum tools can influence the test results.

* Starting with quadrant number one on your map, dig approximately four holes randomly scattered within the quadrant. The holes should be between 12 inches and 18 inches deep. Put the soil you dug out to the side of each hole.

* Shave off some soil from one side of each hole, making sure to get some from its entire depth, and put it in your mixing bucket. Stir the soil samples in the bucket. From this mixture, put about one cup in each of the two bags.

* Fill in the holes well to prevent tripping, replacing any remaining soil from the mixing bucket as you go.

- ✿ Clean out your mixing bucket.
- ✿ Repeat for the remainder of the quadrants.
- ✿ Send one set of samples to a lab; make sure to note that they are to test for heavy metals.

Guidelines for Farming and Gardening in Heavy Metal–Contaminated Soils

After you receive your test results you can create a garden plan. Along with them you will receive information on whether the levels are low, medium, or high. Below we offer a number of options to guide your decision making.

Low-level Farming Options

- ✿ Use lime to change the pH balance of the soil if necessary (a pH balance of 6.5 reduces lead availability to plants).
- ✿ Farm directly into the existing soil.
- ✿ It's fine to allow critters such as chickens, ducks, turkeys, goats, and rabbits to eat scraps and weeds from the garden.

Medium-level Farming Options

- ✿ Farm in constructed planter boxes (12-inch boxes will reduce the lead levels; 24-inch boxes, or boxes lined with a root-impermeable barrier, will virtually eliminate exposure).
- ✿ Dilute the soil by adding compost—and high-quality compost is important. Since the legal level for lead in commercially produced compost is 300 parts per million, it's best to use some from a site that provides heavy metal test results for their materials.
- ✿ Lime the soil if necessary (a pH balance of 6.5 reduces lead availability to plants).
- ✿ Make sure paths and beds are mulched to create a barrier to human exposure to dirt.
- ✿ Wear gloves and wash hands after gardening.
- ✿ Plant fruiting crops (they don't concentrate lead in the edible portion of the plant).
- ✿ If you grow root vegetables, peel them before eating.
- ✿ If you plan to keep animals on this land, mulch the soil with at least 6 inches of wood chips. Chickens will dig into the ground and may inadvertently ingest lead, which studies show will be passed on into its eggs. Rabbits, ducks, turkeys, and goats should not be allowed to eat leafy greens and weeds grown directly in the contaminated soil, as their meat and milk will bioaccumulate the lead. Keeping honeybees will be fine, especially since they can forage as far as 2 miles away.

- If you grow leafy greens, soak them in a 1 percent vinegar solution for at least ten minutes, and then wash well to clean off surface dirt.
- Keep the lead where it is by mulching paths in the medium lead section with the plant refuse from that section rather than adding it to compost piles that could be used in other areas.
- A more drastic option is to remove 12 inches to 18 inches of soil and take it to a landfill; replace it with imported topsoil and/or compost. But be aware that your local landfill may not accept contaminated soil. In addition, this just gives the problem to people who live near it, so you have to think about where you stand on that.

HEAVY METAL SOIL REMEDIATION TECHNIQUES

An effective strategy for transforming lead into a harmless form is to add colloidal phosphate, usually derived from the mined remains of ancient marine animals or fish bonemeal, to your soil. It will create lead phosphate, which is immobile in the soil and not absorbable by plants, as well as provide a balanced source of phosphorus and calcium for your plants. Follow the recommendations provided with the colloidal phosphate or bonemeal to know how much to add to your soil. Retest your soil six months later, and add the recommended amount again if necessary.

As of the writing of this book, this method is being tested in Oakland and New Orleans. See appendix 1 for more information and resources.

A word of caution about remediation, i.e., cleaning heavy metals from the soil: Many tout the ability of special plants to absorb and remove heavy metals from soils. Unfortunately, to remove significant levels of lead from your land you would need to grow successive crops of these plants for many years, sending the plant material to a toxic-waste dump. This strategy, called phytoremediation, uses hyperaccumulator plants—such as sunflowers, various *Brassicas*, geraniums, amaranth, and nettles. While researchers have found plants that concentrate amazingly high levels of some metals, unfortunately lead, the most common heavy metal in urban soils, is not among them. A chemical called EDTA (ethylenediaminetetraacetic acid) can be added to the soil to concentrate hyperaccumulator plant uptake of lead, but the soil must then be remediated (cleaned) of the EDTA, an expensive undertaking.

High-level Farming Options

⚜ Observe medium guidelines and/or:

⚜ Take produce samples, dry them, send them to the lab for testing, and act according to the danger posed by the levels found in the results.

⚜ Grow only ornamentals.

⚜ Don't raise animals except bees, which can forage up to 2 miles away.

⚜ Cap the entire area with root-impermeable material and use raised boxes and pathway mulch like wood chips, straw, rocks, etc., that cover the contaminated soil.

⚜ Cap the entire area with root-impermeable material and use the space for other activities, such as for a greenhouse, animal housing on top of the capped soil, or a community gathering space.

⚜ Remove 12 inches to 18 inches of soil to the landfill and replace with imported topsoil and/or compost. (See questions about disposal of contaminated landfill under Medium-level Farming Options.)

⚜ In the end, if you get a high heavy-metal test result, it's up to you to decide whether you want to be ultrasafe (and pave over your yard) or whether you feel okay with growing crops using careful safety guidelines. If you decide to farm, you can send dried plant tissues to the lab to test for heavy metals. These results will tell you if the plants are absorbing them, making them unsafe to eat.

CHEMICAL CONTAMINATION

Unlike the handful of heavy metal contaminants, the list of possible chemical contaminants in urban soils is vast. Health problems from exposure to chemical contaminants are much more varied than those from heavy metals, but typically they include cancer, liver, and nerve damage. You don't want the stuff in you or your food! Because testing for organic compounds (chemicals) can be extremely expensive compared to heavy metals, we encourage you to do some sleuthing instead.

Is My Site at Risk for Chemical Contamination?

Answers to the following questions can help you discover if your garden is at risk for chemical contamination:

1. Is your land the former site of a business that used toxic chemicals?

2. Are or were there any nearby factories that may have released chemicals into the ground or the water table?

You will need to make the rounds of a number of helpful offices to find out the answer to the first one. The zoning designation of your neighborhood will tell you if businesses are now or were allowed there in the past. The city planning and zoning departments will most likely have a map that shows the designations of your neighborhood. They will also have a record of any building permits. For example, if someone applied to build an auto shop, it will be noted in the records. Most cities also have historic preservation departments. The staff at these offices can be very helpful, often having encyclopedic knowledge that extends to many individual properties. They will most likely be able to help you find out the history through various records, including Sanborn maps that show historical land uses. These also often can be found at your local university.

If you find out that a business operated on your land, especially industries such as auto shops, manufacturing plants, dry cleaners, metal smelters, or chemical or galvanizing plants, don't grow food unless you have the funds to test for a wide range of contaminants (a test for one can cost hundreds of dollars) and intend to embark on a major cleanup project. If you are in this situation, contact the United States Environmental Protection Agency for more information.

WATER CONTAMINATION

It would be a good idea to find out about your water table, specifically how high it is. If it is permeable to groundwater runoff and high enough for plant roots to reach it (about ten to fifteen feet below the soil surface), your plants could absorb toxins in this water. While many chemicals are broken down into harmless substances as water soaks through the soil, some can remain in a toxic state in groundwater. Contact your local water district office to find out about the water table in your area. They will have maps showing the aquifers and how close they are to the surface.

If the water table is high, ask around in your neighborhood, especially among longtime residents, to find out if there used to be major industrial or factory activity that could have caused groundwater contamination. Contact the U.S. Environmental Protection Agency to find out if there are any problems with groundwater toxicity in your area. Of special concern are sites near current or former chemical-manufacturing companies. Many of them used unsafe storage practices or simply dumped chemicals on the ground.

If your water is supplied by a regional or city water company, data on heavy metal and other types of water contamination are published quarterly or annually. Often your water company will send this to you automatically, but if not, give them a call. Some outside research on heavy metal levels in municipal drinking water will show you that *no* level of lead is acceptable (this is equally true for food). The U.S. EPA's "action level," i.e., the level at which they take action against a water company, is 15 parts per billion.

Another source of heavy metals in water are plumbing systems within homes. Unfortunately, this does not apply only to old plumbing— new solder also contains lead. Brass, copper, and lead pipes, fittings, and solder containing lead can result in higher water concentrations of lead. Hot water tends to release the lead from such pipes, so using only cold water in the garden is a good practice. Flushing your faucet into a drain for five to thirty seconds releases water that may have been sitting in your pipes, carrying any lead residues away. If you suspect high levels, have your water tested by a lab.

AIR CONTAMINATION

Although urban particulate air contamination causes a lot of health problems in people, it isn't as much of a concern for garden-grown food. Airborne heavy metals and particulates fall onto the soil, especially near industrial plants and freeways, but aren't likely to be absorbed into plants through their leaves. As long as you wash your produce, this contamination shouldn't pose much of a risk. An exception is if your site is very close to a freeway. Freeway driving produces very toxic tire dust that falls within about a hundred yards and can contaminate your soil, resulting in airborne residue on leaves and concentrations of heavy metals in the soil that can be absorbed by your crops.

IS IGNORANCE REALLY BLISS?

This chapter may seem very discouraging, but knowledge is the best protection against the health dangers of living with toxins. The fact is, we are all living in a toxic soup caused by the industrial society we live in, and the more we know about it, the less we will tolerate such production of goods at the cost of our health. If we begin to take responsibility for making our personal environments safe—and one of the ways to do this is through gardening—the collective result will be a safer environment for everyone.

CHAPTER 3

CREATING YOUR FARM SITE LAYOUT

You have a blank slate. Now it's time to create your urban farm. Before breaking ground, you will want to create your garden in your imagination. This will enable you to see how everything will fit together, how the work will flow, and what plants and animals will be the most pleasing to you. Make diagrams. Draw pictures. Make a collage. Move things around. Play. Use your imagination before making the financial investment. The following pages will help you know enough about urban farming to make a preliminary layout plan, but remember: These topics will be covered in-depth in subsequent chapters.

GETTING TO KNOW YOUR SITE

The first step in designing your urban farm is to evaluate the qualities of your site and the best use of each area. Because this is an *urban* and not a rural farm, you'll want to take the realities of the environment into account in your thinking.

Before you start planting and building, you should get to know your land really well. Spend some time on your land observing the sun and wind patterns. Be aware that in North America, if you are observing from late fall through early spring, the sun pattern will be more northerly than it will be in the summer. Sketch the rough dimensions of your land on a piece of paper. Note down the following:

* ❋ existing buildings, fences, entry gates, and sidewalks;
* ❋ water spigots;
* ❋ existing perennial plants (trees, shrubs, etc.);
* ❋ partial and full shade areas;
* ❋ windy areas, and the prevailing wind direction;
* ❋ south-facing walls (these areas will be warmer);
* ❋ slopes and notable land contours;
* ❋ the compass directions. If you don't have an innate sense of this, you're going to have to find a compass!
* ❋ If you plan on keeping animals, note the approximate distance of your neighbors' houses from your fence line.

Often every square foot of ground on urban farms needs to be hand dug to sift out crabgrass, garbage, and concrete. You will need to take stock and be informed about what you're getting into. Knowing about major cleanup needs up front can be inspiring, but finding out about them halfway into the project can be discouraging.

Walk the entire area with a shovel and a dig bar (an eight foot steel bar with a pointed end). Evaluate your soil by noting any areas of gravel, concrete, stones, or garbage that may need to be removed. Every foot or so investigate whether there are buried concrete slabs, old foundations, or other junk below the soil surface by digging holes or poking the dig bar into the ground. If you hit something hard, excavate the edges to see how big it is. Old concrete walkways or low walls can be broken up fairly easily with a sledgehammer and pried out with a crowbar (be careful not to break water and sewer lines). But if you find a large continuous slab you will need to decide whether to remove it or build your garden on top of it. The cost of removing large areas of gravel, blacktop, or concrete can be prohibitively expensive. A cheaper option is to build taller boxed beds over the slab. For more information on what to do if your site is covered with a concrete slab or blacktop, refer to chapter 6.

Next is an evaluation of the weed situation. Note the types and extent of the weeds. If you discover that you have what are called pernicious weeds, you will need to incorporate major weed removal into your plan. Pernicious weeds spread by underground roots. If even one piece of the root is left in the soil, in most cases it will resprout into a new plant. Simply trying to yank them out of the ground isn't very effective on these weeds, and using a tiller just multiplies their growth, so prep yourself for a very labor-intensive weed removal process. Chapter 13 describes how to best tackle the problem.

You may also want to make a plan to remove existing perennial trees and shrubs that create too much shade, acidify your soil (in the case of conifers), or are just plain unattractive. If you have permission, don't be afraid to cut down and remove ornamental shrubs and trees in order to plant fruiting trees or other crops. Unwanted ornamental plants can also be dug up and offered for free on the Internet.

Your site investigation will let you know how much up-front cleanup work you're in for. If you have found any conditions that need to be dealt with, such as buried garbage, concrete, brickwork, blacktop, pernicious weeds, or perennials to be removed, make a list and allot time for the cleanup stage.

CREATING A FARM SITE DRAWING

Now comes the fun part. It's time to make a plan. One of the most enjoyable tasks for a farmer is daydreaming with seed catalogs and gardening books. This is the time to note down specific types of fruiting shrubs and trees you want. It's the time to decide where to locate annual and perennial plantings and what materials you will use to build structures. Of course there is a dynamic between theory and practice. You should think of your design as a set of guidelines that will change as you experiment and find out what works best for your site conditions and your life.

Draw a scale map on a piece of grid paper or poster board. To measure your site, lay a tape measure out on the ground and walk naturally for twenty feet, counting your paces. You can then calculate each pace's feet and inches. Pace your land. Write down the measurements. Decide on a scale that will fit your paper; for example ¼ inch = 1 foot.

One of the best ways to plan your farm layout is to cut out pieces of paper scaled to your diagram and the size of the beds, compost bins, etc., and move them around. Consider the work flow: It can be pretty annoying to have to walk clear around your house to the side yard each time you want a trowel or rake, or to have to move manure into your garden by bucket because your compost piles aren't accessible by wheelbarrow. You will find sample layouts on the next few pages. If your "land" is a deck, a patio, or a rooftop, see page 115 for a layout.

IDEAL FRONT YARD FARM (22' X 50')

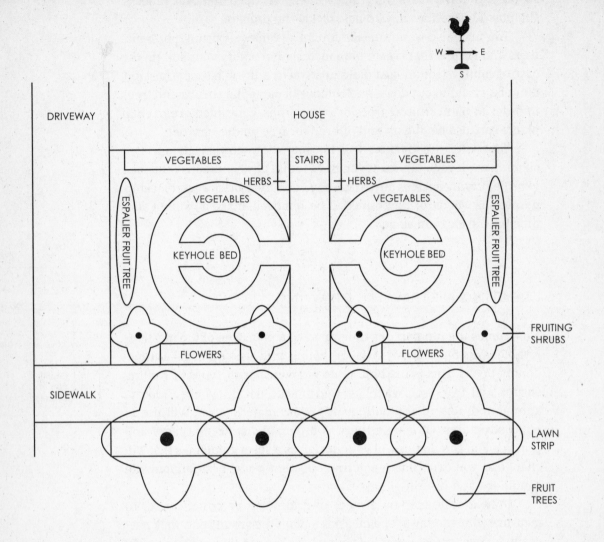

POSSIBLE YIELDS

Tree Fruit	± 300 lbs per year
Shrub Fruit	± 40 lbs per year
Vegetables/Herbs	± 260 – 780 lbs per year

DIMENSIONS

Rectangular Beds	2´ wide
Keyhole Beds	3´ wide beds; 2´ wide paths
Espalier Trees	2´ wide
Flower Beds	2´ wide
Herb Beds	1´ wide
Shrubs	4´ – 5´ wide
Growing Space	260 sq ft (for annuals)

BEDS

Bender board, bricks, or no edging

PATHS

White Dutch clover, wood chips, or decomposed granite

IDEAL BACKYARD FARM (40' X 50')

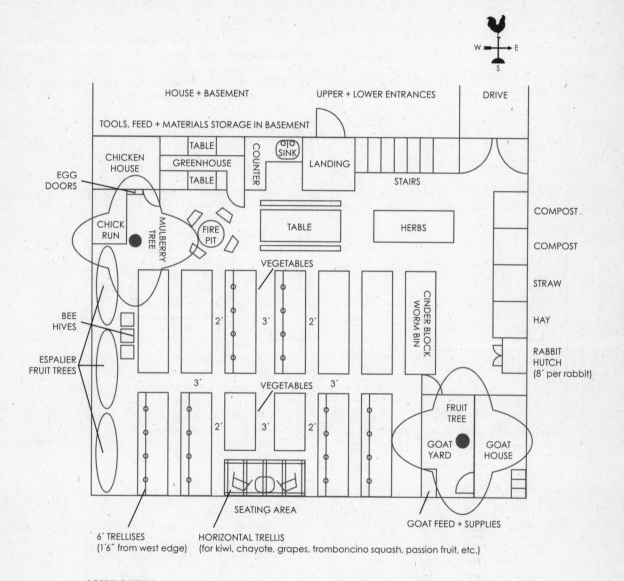

W N E S

HOUSE + BASEMENT
UPPER + LOWER ENTRANCES
DRIVE

TOOLS, FEED + MATERIALS STORAGE IN BASEMENT

CHICKEN HOUSE
TABLE
GREENHOUSE
TABLE
COUNTER
SINK
LANDING
STAIRS

EGG DOORS

CHICK RUN
MULBERRY TREE
FIRE PIT
TABLE
HERBS

COMPOST
COMPOST
STRAW
HAY

VEGETABLES

2′   3′   2′

CINDER BLOCK WORM BIN

RABBIT HUTCH
(8′ per rabbit)

BEE HIVES

ESPALIER FRUIT TREES

3′
VEGETABLES
3′

2′   3′   2′

FRUIT TREE
GOAT YARD
GOAT HOUSE

SEATING AREA

GOAT FEED + SUPPLIES

6′ TRELLISES
(1′6″ from west edge)

HORIZONTAL TRELLIS
(for kiwi, chayote, grapes, tromboncino squash, passion fruit, etc.)

POSSIBLE YIELDS

Milk	± 2 quarts per day (except when nursing)
Eggs	4 dozen per week (10 hens; decreasing in winter)
Rabbit Meat	72 lbs max (bred 3x per year = 24 rabbits @ 3 lbs each)
Tree Fruit	250 lbs per year
Vegetables/Herbs	± 491–1475 per year
Honey	48 quarts per year

BEDS

3 ½′ wide
Constructed of recycled wood
or other material or no edging

PATHS

Minor 2′
Major 3′
White Dutch clover, wood chips,
or decomposed granite

IDEAL EMPTY LOT FARM (50' X 125')

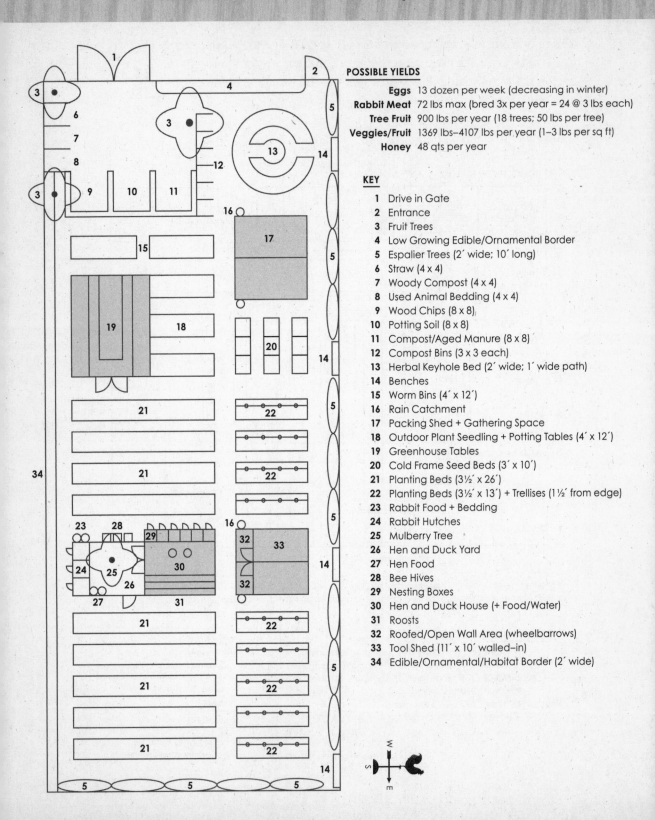

POSSIBLE YIELDS

Eggs	13 dozen per week (decreasing in winter)
Rabbit Meat	72 lbs max (bred 3x per year = 24 @ 3 lbs each)
Tree Fruit	900 lbs per year (18 trees; 50 lbs per tree)
Veggies/Fruit	1369 lbs–4107 lbs per year (1–3 lbs per sq ft)
Honey	48 qts per year

KEY

1. Drive in Gate
2. Entrance
3. Fruit Trees
4. Low Growing Edible/Ornamental Border
5. Espalier Trees (2′ wide; 10′ long)
6. Straw (4 x 4)
7. Woody Compost (4 x 4)
8. Used Animal Bedding (4 x 4)
9. Wood Chips (8 x 8)
10. Potting Soil (8 x 8)
11. Compost/Aged Manure (8 x 8)
12. Compost Bins (3 x 3 each)
13. Herbal Keyhole Bed (2′ wide; 1′ wide path)
14. Benches
15. Worm Bins (4′ x 12′)
16. Rain Catchment
17. Packing Shed + Gathering Space
18. Outdoor Plant Seedling + Potting Tables (4′ x 12′)
19. Greenhouse Tables
20. Cold Frame Seed Beds (3′ x 10′)
21. Planting Beds (3½′ x 26′)
22. Planting Beds (3½′ x 13′) + Trellises (1½′ from edge)
23. Rabbit Food + Bedding
24. Rabbit Hutches
25. Mulberry Tree
26. Hen and Duck Yard
27. Hen Food
28. Bee Hives
29. Nesting Boxes
30. Hen and Duck House (+ Food/Water)
31. Roosts
32. Roofed/Open Wall Area (wheelbarrows)
33. Tool Shed (11′ x 10′ walled-in)
34. Edible/Ornamental/Habitat Border (2′ wide)

You may want to skip ahead to parts II and III at this point to learn more about the various crop and farm animal options so you can decide what to include. Below are some of the practical elements you also might want:

Urban Farm Components

- crop-planting beds (either inground or boxed)
- children's garden beds
- garden beds with access for the handicapped
- vertical growing structures (trellises, grow tubes, grow bags, etc.)
- fruit trees
- fruiting shrubs and vines
- seedling propagation area
- hoop house or cold frames (see page 235 for description) for cold weather growing
- composting area (compost bins, worm composting containers, manure aging bins, brush pile, leaf pile)
- bulk materials storage (mulch, compost, straw, sawdust)
- animal housing and yards
- beehives
- animal supplies storage
- ornamental plantings
- vehicle access
- tool storage
- produce washing and packing setup
- shaded group seating/work/outdoor eating and hanging out areas
- benches
- outdoor cooking area
- rainwater catchment
- grey-water system for diverting used household water to the garden
- water storage cistern

CLIMATE AND ORIENTATION

In most geographical locations the most ideal garden site would be sunny all day, slightly sloped, facing south or southwest, and wind free. In practice it is unlikely that this is what you'll get. The good news is that there are ways to improve less than ideal situations. You can plant food-producing shrubs and trees to shield your farm from wind. Shady areas can be used for animal housing, toolsheds, compost bins, and work or gathering space. If your site is facing east or north you can use season-extending cold frames, tunnels, or hoop houses to heat up your site.

Your climate will also influence your garden layout. If you live in a coastal or northern city, you will need to take advantage of warm spots. Locate seedbeds, planting beds, hoop houses, and cold frames in sunny spots and near heat-storing, south-facing walls. If you live in the South, reverse this to protect summer-grown crops with shade.

DIMENSIONS AND ORIENTATIONS
OF URBAN FARM COMPONENTS

In our years of urban farming we have learned that certain dimensions and orientations of farm components work best in the city.

Rectangular Beds

While they may not be very creative aesthetically, rectangular beds truly are space and time saving. Whether inground, boxed, or edged with some other material, your beds should be no wider than four feet if surrounded by paths. If they are any wider it is too difficult to reach the center without stepping into the bed. To learn how to build garden beds please refer to chapter 6.

Two feet is a good width for beds accessible from only one side. Make your beds as long as you can, but no more than about thirty feet. Beds are best oriented from north to south, allowing for the same light exposure in all parts. However, if the land slope or some other factor makes this difficult, you can run them east to west or at an angle.

Keyhole Beds

Keyhole beds are formed in the shape of a horseshoe. They have more surface area and less path space, making them space efficient. It's hard to build them as boxed beds, since the sides are curved, but you can easily edge them with bender board or various found materials, such as logs, hunks of concrete, or bricks to define the space. The width of its ring(s) should be three and a half feet wide, with inner paths that are one to two feet wide. Using keyhole beds is a great way to achieve a more naturalistic, curvy appearance without sacrificing efficient use of space.

Boxes

Boxed beds can be from four to twenty-four inches high. Eighteen to twenty-one inches is an ideal upper limit ergonomically, unless wheelchair access is required, in which case they need to be thirty inches high to allow a wheelchair to fit underneath the top board.

KEYHOLE BEDS

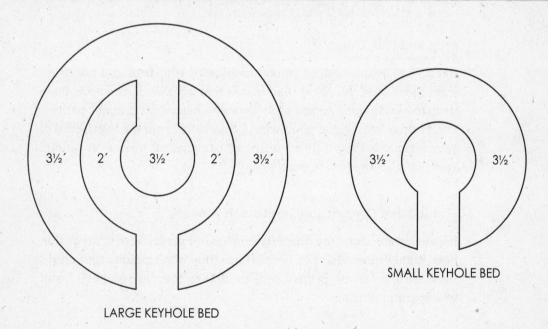

LARGE KEYHOLE BED

SMALL KEYHOLE BED

Paths

A great way to design your paths and beds in a rectangular plot is as follows. Make two paths, each between three and five feet wide, with one centered down the length and the other down the width of the space. They will cross in the center. Create a two-foot-wide planting border around the periphery of the site with another three-to-five-foot-wide path alongside it. These will be your major paths—they are wide enough to accommodate carts and provide disabled access should you need it.

Paths between the beds throughout the rest of the site should be between one and two-and-a-half feet wide—not wider—to maximize growing space. Irregularly shaped beds and paths eat up space that could be used for growing, and just make more weeding work.

If wheelchair access is required, paths must be three feet wide, and there must be periodic five-foot-diameter wheelchair turnaround spaces. A good compromise is to create a wheelchair-access section of the farm.

Fruit and Nut Trees

In order to avoid shading annual crops with your fruit and nut trees, train them to be no more than seven feet high. To save space, train trees to be flat along fences or to grow as a hedge along major paths—or site them where you need wind protection, or on the north side of your farm site. Plant taller fruiting or ornamental trees over animal yards or seating areas to create shade.

Fruiting and Ornamental Shrubs and Vines

Locate shrubs along the property line of your urban farm in areas that need wind protection or in beds where they won't shade other crops. Most shrubs can be pruned and trained to the desired height and width within reason.

Vertical Growing Structures

Trellises—essentially any form of flat growing frame—can be planned along fences and down the length of growing beds. Vertical trellises are preferable to tomato and cucumber cages. If your bed is three and a half feet wide, run the trellis slightly off center, one and a half feet

OFFSET TRELLIS PLAN VIEW

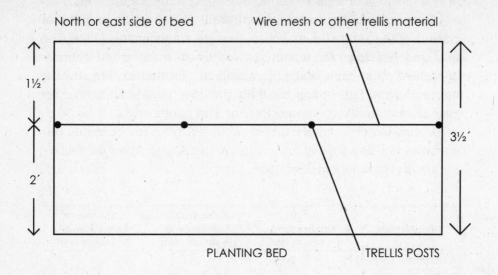

North or east side of bed   Wire mesh or other trellis material

1½′

3½′

2′

PLANTING BED   TRELLIS POSTS

from one side. This will allow for more efficient use of space, and will create a shadier side that is narrower, with more of the shade falling on the adjacent path. This narrow, shadier strip can be used to grow salad greens, radishes, and cool-season *Brassica*s in the summer.

Grow tubes, grow bags, and living walls can be used to increase growing square footage and are ideal for shallow-rooting crops, such as onions and celery. Grow tubes can be built suspended above growing beds or patios. The layout will be about one and a half to two feet wide, and can be as long as you like. The height of the top tier of tubes shouldn't be higher than you can reach, about five feet. Vertically oriented grow bags can be suspended above beds or patios or hung from existing walls and railings. Living walls are shallow receptacles that are suspended from the sides of buildings. Each line of bags or living walls should be afforded about two feet of width and be of an accessible height. See page 115 for illustrations.

Animal Housing and Yards

Local planning and zoning codes are important for the layout of animal structures, since many cities require a certain distance from your and/or your neighbors' house or other buildings. Passersby can pose a risk to your animals through taunting and rock throwing. Ideally, locate animal housing deep within the center of the property or closer to the safety of your house. Provide ample shade. Make sure food and bedding can be stored close by. In many world cultures household meat, milk, and egg animals are located underneath the house or deck. This option naturally provides shade and shelter for your animals and keeps them close for easy caretaking.

Animals need a space that is protected from the elements (indoors) as well as a yard to run around in (outdoors). Allow the following amount of indoor and outdoor space:

Per Animal	Indoor Requirements (in square feet)	Outdoor Requirements (in square feet)	Confinement Requirements* (in square feet)
Chickens	1.5–2	8	4
Ducks	3	15	8
Rabbits	NA	NA	6–9
Goats	10–15	100–200†	50
Beehive	NA	6	NA

*Additional indoor space should be provided in cold climates since outdoor time isn't possible for part of the year.

†Assuming they are given regular walks; more if they are not.

Beehives should also be hidden from possible molestation, and should be out of *your* way as much as possible, with the entrance pointed away from heavily trafficked areas. Bees leave and return to their hives all day long, in pursuit of pollen and nectar—you don't want to be caught in the middle of a bee highway. Allow a two-by-three-foot space for each hive.

Tool Storage

Use space for tool storage that is less than ideal for gardening. Depending on your climate, it may not be necessary to shield tools from the weather on all sides as long as you keep your seeds inside the house. Use the spaces under staircases or decks for tool storage by building a sloped roof of corrugated material suspended from the floor beams above and a Peg-Board wall organizer attached to the post beams holding up the stairs. Peg-Board is one of the top ten most wonderful inventions of all time—use it to hang and organize tools.

Compost Bins, Worm Bins, Bulk Materials Storage Bins, and Animal Supply Bins

To properly decompose organic matter, such as kitchen scraps, weeds, garden trimmings, and animal manure, compost piles must heat up. Compost bins or piles must be at least three feet cubed to achieve this. Each one needs a few feet of access space in front and a three-square-foot space next to it for turning the pile. Be sure your water source can reach your bins, since proper moisture levels (like that of a wrung-out sponge) is a key condition for composting. In areas with a lot of rain or snow, a roof over your compost area is ideal because a compost pile can get too boggy for proper breakdown.

Small, plastic, manufactured worm bins can be kept close to or inside the house. If you have a large volume of food scraps, you may want to construct larger ones, however. If left outside, worm bins should be shaded and protected from wind, rain, and snow. They also can be built either of wood, cinder block, or poured concrete. Three and a half to four feet wide and two feet high are good dimensions; they can be as long as you like. See appendix 11 for sources of worm bin plans.

Warning: Don't put any decomposing material up against a wooden fence. You can imagine the result: a rotten fence! If you have a brick wall or metal fence along the perimeter, this would be an ideal back wall for compost, brush, and material storage bins. Also, if you're in an area prone to rat infestations, know that brush and compost piles will become habitats for rodents (see page 301 for control methods).

In intensive urban farming we don't always let beds lie fallow under cover crop to regain soil nutrients; therefore, large amounts of homemade compost, free municipal or county compost, and locally obtainable animal manures should be applied to beds two to four times a year. These materials must be stored. Sawdust and straw for animal bedding also need a place to live. One of the most important design considerations on the urban farm can be where these materials are stored.

Try to avoid having to unload your materials from your truck into a wheelbarrow and then into a pile or bin; ideally, a driveway would allow you to back right up and shovel the compost right off. If you don't have vehicle access, put the bins or piles along a fence from which you have sidewalk access. You can back your truck up onto the sidewalk and shovel the materials over the fence.

If you plan to have materials delivered by a dump truck, the driveway needs to be eight feet wide. In this case, the storage bins should be six to eight feet wide (the widest dump truck bed will be eight feet), with an overhead space of about fourteen feet clear of tree limbs and power lines.

Propagation Area

You can propagate seedlings in- or outdoors; see chapter 9 for instructions. Indoor seedling propagation involves installing fluorescent lights over a table. You will need a protected space for much of the year to propagate outdoors, such as a greenhouse, a cold frame, or a homemade greenhouse box. Make sure there is a water supply nearby, and find a spot that gets sun, won't shade your annual growing beds, and has good drainage. Building a greenhouse off of a house or garage can save you the effort of constructing the fourth wall. It's best to site seedbeds, greenhouses, and greenhouse boxes near the house or entry to the garden, so that you will be more likely to water and monitor them.

After seedlings are transplanted from open trays to pots, they will need a place to be "hardened off." One or more tables should be set up in the shade or under a shade cloth (a synthetic material that can block varied percentages of sunlight) near the greenhouse. It's great to have another table of a comfortable work height there for the tasks of seeding and transplanting. If you plan to purchase seeding mix in bulk, also include a three-sided storage bin, ideally with a roof, in your plan.

Hoop Houses and Cold Frames for Cold-weather Growing

If you live in an area with hard freezes or long winters, you may want to extend the harvest with greenhouses, hoop houses, or cold frames (see page 235 for an explanation of these structures). Locate such structures in sheltered but sunny locations. If you have differences in elevation on your land, be sure not to site them in cool hollows.

Produce-washing and -packing Setup

Whether it's a homemade plywood-topped table with a hole cut in it for a mixing bowl washtub, a cabinet with a fully plumbed sink, or an old bathtub and table, a place to clean, bunch, and package your produce is indispensable.

The produce-washing and -packing area should be set up in the shade with access to running water. Added bonus: Pipe the wash water from the outdoor sink to growing beds or trees, thus recycling the water.

Group Seating, Working, and Outdoor-dining Area

From the simple—a seating circle of straw bales—to the grand—a gas-fired barbecue on a tiled patio—a place to rest and enjoy the fruits of your labor is essential. Favorites include picnic tables, fire pits, wood-fired cob ovens, brick patios, low-slung decks, and small lawns. Shade provided by trees, arbors, or umbrellas is a must. It's especially pleasant to locate this area in the center of your garden for good views of all the bountiful beauty.

Children's Garden Beds

With your help, children will love to plant, observe, explore, and taste. Building a few small, low box beds and surrounding them with a low white-picket fence and gate will let children know that it is their own special place. Having this space will also help you if your children are getting into *your* beds and digging half the soil out into the path. You can gently guide them to their space and work with them there, where you don't have any expectations aside from their own exploration. Children love hidden nooks and special things all their own. Creating bean or pea tepees where they can hide and wild patches they can tunnel through are great ways to build on their innate love of nature.

Rainwater Catchment and Grey Water Reclamation Systems

Collecting rainwater and household wastewater can cut down on your water bills and conserve a precious resource. Rain barrels and storage racks should be sited close to the gutter downspouts that fill them.

Grey-water systems require constructed wetlands or other methods of cleaning kitchen and laundry wastewater for use in the garden. Allow for a five feet by fifteen feet strip of land near the sewer pipe you will be tapping into for grey-water processing and storage. See chapter 11 for more information on constructing rainwater and grey-water systems.

LAYING OUT YOUR FARM PLAN

After you've evaluated your site, decided what elements to include, and drawn up a site plan, it's time to see how it works in the actual space by laying out your plan with stakes and string. The first preparation you must do is cut down or weed-whack any brush that will get in the way (be sure to wear goggles!). Depending on the size of your farm, you will need to collect, or make, around a hundred two-foot-long stakes. Next, gather together twine, a measuring tape, a hammer, some scissors, and your plan and head outside.

Using your diagram and the measuring tape, begin staking out the various elements you plan to include. Lining things up accurately can be difficult, so try to start with something that seems straight and relatively square, such as a fence or house. Pound stakes securely into the ground at intervals around the perimeter of beds or structures, and then delineate the edges with string. Tie it to one stake and wind it around the rest until you return to your first stake. Stake and label the locations of perennial trees and shrubs.

After staking out your plan, sit with it for a few days or weeks. Observe it from various angles. Walk around and pretend you're gardening. Imagine bees zooming in and out of the beehive: Is it out of the way of human traffic? Pretend you have to muck out the rabbit cages: Are they situated close by a compost area? Ask for opinions. Follow your intuition.

If you are working with a larger site, you may want to break up the work into phases, so it doesn't seem too daunting—physically and monetarily. Don't become overwhelmed. With a consistent effort, even as little as a few hours of work per week, your dream farm *will* become a reality.

W E

CHAPTER 4

CHOOSING FARM-BUILDING MATERIALS, TOOLS, AND SUPPLIES

Now that you have created a plan for your urban farm, you will need to decide which materials to use for your infrastructure (compost bins, chicken coops, hoop house), acquire the proper tools (shovels, aphid spray nozzles, and manure forks), and load up on supplies (beekeeping gear, seeds, and straw) before you can start growing. This chapter will share the information you need to salvage or purchase them.

We're in favor of reusing discarded materials from the voluminous urban-waste stream—it's a great way to reduce your ecological footprint and save money. To find free or affordably priced second-hand materials, some good places to start looking are salvage yards, second-hand stores, big trash pickup days, garbage transfer stations and landfills (sometimes employees won't allow you to take garbage from the dump), and garbage areas behind businesses. Other great resources are freecycle.com, craigslist.com, and state reuse programs that match larger companies and institutions with people who can use what they no longer need (see appendix 11).

When considering free or salvaged materials, look at them with an eye toward the amount of clean-up work that will be necessary to make the materials usable. Removing nails and screws can take significant time and effort. It may be more worth it to pay for wood at your local salvage yard that has already been cleaned than to use free wood full of nails. Seriously cracked, warped, or rotted materials are not worth your time, even if free.

To keep neighbor relations on the sweet side you'd best avoid turning your yard into a repository of possibly, someday useful materials. In cities you can find old windows, pallets, and any number of other free materials any day of the week. An indispensable rule of thumb is to only allow yourself to get materials when you are ready to use them. Of course, there's an occasional gold mine you won't want to pass up, but sticking to this rule will keep you and your neighbors in a land of beauty, not junk.

BUILDING MATERIAL NO-NO'S

Although plywood and pressboard can be used to build inner parts of structures, outdoors they degrade quickly. They should also be avoided for outdoor use due to the toxic nature of the glues used to bind them. A toxic situation will result if soil, plants, or animals come in contact with any of the following materials:

- Pressure-treated wood. You can recognize such wood by its slightly green hue and/or "stapled" look where the insertion of chemicals makes tiny indented lines.
- Plywood or other glued-together pressboards or laminates (if you can see layers on the end grain, it's glued).
- Any material that may have been painted with leaded paint.

The following pages give suggestions for many good garden building materials.

WOOD AND FAKE WOOD

Douglas Fir, Cedar, or Redwood Lumber

If you want it built to last, most experts recommend redwood for all garden structures. However, redwood trees are endangered—it's a bit of a sin to buy new redwood. In addition, new-growth redwood is often much more porous than old-growth wood of any kind, so, practically, using new redwood doesn't make much sense.

While it is very expensive, salvaged redwood and cedar are obtainable. If your budget can handle it, go for it. Salvaged Douglas fir is much easier to find and is usually inexpensive. Salvaged Douglas fir is of two types: old-growth full-dimension lumber and second-growth common-dimension lumber. Old-growth full-dimension lumber will last a lot longer than newer wood and is larger and sturdier. To tell the difference, first look at the size of the boards. Full dimension is just that: for a 2 × 4, it's 2 inches by 4 inches. Common-dimension lumber starts out at the full dimension and is surfaced to be smaller. A 2 × 4 is actually 1½ inches by 3½ inches. Next look at the grain of the wood. The closer together the rings are, the older growth it is, and the longer it will last. (See page 105 for a table of lumber dimensions.)

Old-growth redwood and cedar last the longest outdoors, followed by old-growth Douglas fir, new-growth redwood, and new-growth cedar, New-growth Douglas fir has the least staying power, but salvaged new-growth Douglas fir has the advantage over brand-new wood of having been fully cured and hardened inside of the building it used to be a part of, so it will last longer than new Douglas fir.

When you're at the salvage yard evaluate salvaged boards for rot and straightness. "Sight" your boards to see if they are warped. If you're not sure how to do this, just ask a staff member at the yard—they know all about it and will be happy to initiate you.

You can use logs to edge your beds

Branches and Logs

While they will eventually decompose, tree branches and logs, especially when cut to short lengths, are quite handy for creating low-edged beds. The wood can be oriented horizontally or vertically, depending on the length. To make a stable bed border it's helpful to dig a shallow trench for the bottom of the wood to sit in.

Pallets

It's easy to get both softwood and hardwood pallets for free, and they are great for making three-sided compost and material-storage bins, boxed beds, and container garden boxes. Hardwood pallets last much longer than softwood pallets. Try driving a nail into the wood. If it's really tough it's probably hardwood and will make a very long-lasting building material, but you'll need to predrill holes for the screws. The easiest way to build structures such as planter boxes, storage bins, or

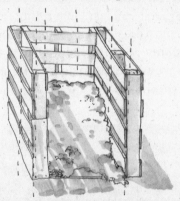

Making a compost or storage bin out of pallets supported by rebar stakes

compost bins out of pallets is to pound rebar or metal T-stakes into the soil every two feet or so and thread the pallet over them perpendicular to the ground. You may want to line the inside with burlap, another fabric, or layers of cardboard to keep materials or soil from falling out. You can also deconstruct pallets to use the lumber.

Milled Lumber from Tree Care and Removal Companies

Some tree-care companies mill the wood they salvage from downed urban trees. This is a great local resource of lumber. Since these are local companies they might even be willing to donate lumber to your cause.

Fiber-cement Board

Fiber-cement siding comes in 4 inch by 8 inch sheets as well as various siding "board" styles. A mixture of fiberglass and cement, it certainly fits the bill for durability. Fiber-cement-board siding is an obvious choice for siding any garden structures you build, such as sheds, animal housing, etc. Its one failing as a boxed-bed building material is that it is a bit too flexible, but with a well-built structure on the inside (use plastic lumber for this), it works well and will never rot. While cement is a nonrenewable mined resource, unlike with ones made of wood, you'll probably never have to rebuild your planter boxes.

Recycled Plastic Lumber and Landscape Edging

Recycling technology has improved by leaps and bounds of late. Scientists have worked hard to find just the right mixture of various types of plastic garbage to make building materials that are strong and durable. For a long time most plastic lumber was problematic as a planter-box building material due to its flexibility. When the beds were filled with soil, the boards would bulge. In addition, it is often composed of recycled materials *and* wood, defeating your goal of using 100 percent recycled materials.

A new type of plastic lumber, made of recycled milk jugs for strength combined with recycled bumpers for rigidity, has finally solved the problem. We left a phone message for Thomas Nosker of Rutgers University Advanced Materials via Immiscible Polymer Processing Center to get the skinny on whether we could build planter boxes out of the stuff. He called us back right away from the U.S. Army base in Fort Bragg, North Carolina, where he was working on the construction of a tank bridge. "Well, if it can support a tank, I think you can use it," he said. He also added that the research showed clearly that nothing toxic leaches out of the material. This is great news for urban farmers. Due to all of its qualities, we judge recycled-milk-jug lumber to be the number-one choice building material for urban farmers who choose to buy new materials. (See appendix 11 for sources.)

Flexible plastic landscape-edging boards, ⅝-inch thick and of various widths and lengths, are great for, yes, edging, and also to create curved raised boxes or terraced beds. They're really bendy, so they can be used to make curved shapes, but if you want a square bed, they're a nightmare. You can also buy stakes of the same material for installation (put the stakes opposite where the pressure is coming from so the fasteners don't pull out, i.e., the outside of the bed). Be sure to find a brand made of 100 percent recycled materials rather than of a mixture of plastic and wood.

Salvaged Solid Wood Doors

Antique solid wood doors, often made of redwood, are a great width for tall boxed beds. The only problem is that they're almost always painted. You must remove the paint to ensure that it doesn't peel into the dirt. To make a solid box, be sure to use 4 inch by 4 inch posts in the corners and fasten with bolts going through.

Sapling Wood

Thin green sapling wood can be woven on sapling wood posts that are at least 1 inch in diameter to make beautiful fences and trellises. It's just like weaving a basket: over, under, over, under, around the posts. As the wood dries it becomes rigid. To find sapling wood, go to a creek or river and look for willow, dogwood, or other river trees. Or, in the city, ask a landscaping company to give you their sapling prunings.

Woven sapling wood fence

Peeler Core Timber Stakes and Bamboo Stakes

Trellises and deer fences can be constructed easily with either of these types of stakes. Peeler Core stakes are usually 4 inches in diameter and have a pointed end, so they can be pounded into the ground. They are often pressure-treated, so look for ones that aren't. Bamboo stakes are a bit trickier to get into the ground without splitting; the thicker the diameter at the bottom the better. To drive bamboo into the ground cut off one end at an angle and the other flat. Both of these types of stakes can be set in soil using a ladder, a baby sledgehammer, a fence-post level, and a tape measure to check for height. What's great is that you can save a lot of time and money (on hardware) by attaching the trellis crosspieces, plastic mesh, or wire mesh with outdoor-rated zip ties. Voilà, you're done.

Tree Bamboo (4 inches–plus Diameter)

Tree bamboo is used in many countries to construct multistory buildings. Because it has only recently gained popularity in the United States, it is scarce but obtainable. It can be used in container gardens

A peeler core stake

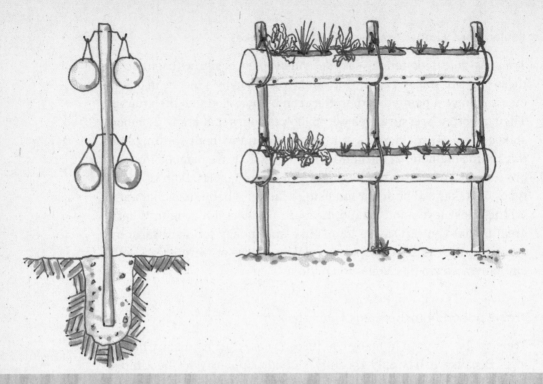

This horizontal tube planter can be made of recycled PVC pipe or bamboo. It can be suspended from a fence or from posts.

to build horizontal trough planters, but it can be used to build structures as well. To build a planter, split the tree bamboo in half, cap the open ends, and drill drainage holes in the bottom. Set it on a patio or suspend it aboveground off fences or posts. Japanese tool companies sell bamboo-splitting tools.

METAL

Corrugated Roofing

Corrugated roofing, which is especially economical when purchased used, can be utilized with the corrugations running either way to make great boxed beds (see page 101 for instructions and illustrations). It is also obviously very useful for roofs.

Hardware Cloth

What a misnomer! This is not cloth at all but woven wire-screen mesh with openings from ½ inch to ⅛ inch. A bit more durable than poultry netting, we often use it sunk into the ground underneath animal housing and fencing to keep critters out. If moles or gophers are a problem, line the bottom of your raised beds with hardware cloth or stucco-reinforcement mesh, both more durable than chicken wire.

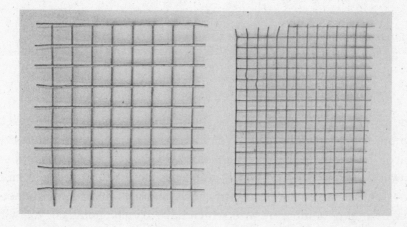

Hardware cloth

Post Anchor Brackets/Post Bases

Post bases are metal brackets used for attaching wood fence posts to a poured concrete fence posthole. Part of the bracket extends into poured concrete and the other part extends upward to support the post. Wood eventually rots. If you've set your fence posts or planter-box supports in cement and they rot out, all you have is a large chunk of unusable cement in the ground. If you've set post bases in cement, however, all you have to do is replace the wooden post to repair the damage.

Poultry and Aviary Netting and Stucco Wire

Poultry and aviary netting, which are also misnomers, are actually made out of metal wire. Typical hexagonal chicken wire, also called

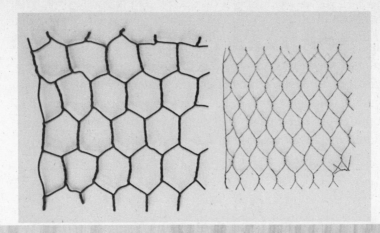

poultry netting, is actually useless in the chicken coop but can be used for other building jobs. Stucco reinforcing wire (exactly the same size but of thicker-gauge wire) is the material of choice to deter gophers, as it won't degrade underground as quickly as poultry netting. For animal housing, use aviary netting. With ⅜-inch openings (unlike poultry netting or stucco-reinforcing wire, which have 1-inch openings), you will avoid infestations of rodents and chicken feed–eating birds.

Rebar

It's quite easy to find salvaged rebar, a multiuse building material, for the urban farm. It can be used to stake plants. Or pound 1½–foot rebar stakes into the ground in the corners of beds to prevent the garden hose from dragging across your plantings as you run it around corners (poke a tennis ball on the top to protect your shins). Rebar can be bent to make a structure for an enclosed chicken run or greenhouse. It can be pounded into the ground to

Hose guide to protect your plants

support hollow building materials, such as salvaged PVC pipe, to make a hoop house. Endless, endless uses. To cut it, use a special metal blade in your power saw of choice.

T-stakes

The t-stake, often used as a basic field-fence post, goes in easy and is hard to get out. Like wooden stakes, they are great supports for trellises and fences, though when pounding in, they're a bit harder to keep straight. They're cheap and can often be found used.

Welded-wire Cement Reinforcement Mesh (Also Called 6-6-6)

Cement reinforcement mesh is the one material we would want on a desert island. Well, as long as we had zip ties, too. Since it's marketed to the commercial construction trade, it's inexpensive compared to

CONCRETE REINFORCEMENT MESH DETAIL

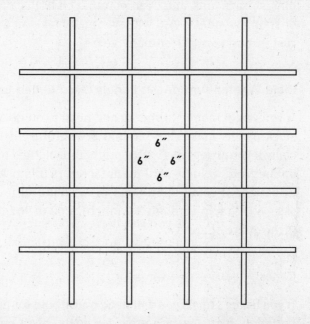

6"
6"   6"
6"

Concrete Reinforcement Mesh aka 6-6-6

trellising materials sold to the home gardener, which are always more expensive. It can be bought in flat sheets or bulk rolls.

This material has holes large enough to allow you to harvest crops and is rigid enough to support the weight of all trellised garden crops. Farmers never mess with those store-bought things called tomato cages, because 6-6-6 is perfect for making cylindrical cage supports for tomatoes, cucumbers, melons, and winter squash (see page 49 for an illustration). You can also use a 6-6-6 cage to make a simple cloche (a glass or plastic case that fits over one plant): Wrap the cylinder in clear polyethylene sheeting or floating row cover (a specially formulated synthetic material) and pop it over a pepper or melon seedling. The cloche will keep the plant warmer in the spring, thus increasing the chance for a good harvest come summer.

Cement reinforcement mesh (6-6-6) can also be bent in an arch over planting beds to make a cucumber trellis, a polyethylene-covered hotbed, a floating row–covered low tunnel, or a shade-cloth frame. Be sure to leave the sharp 6-inch ends of the wire free on each side to anchor it to the ground. Thus, it is self-staking. It can be attached to rebar stakes for animal housing, for example, to form a structure for aviary netting. And last but not least, lined with shade cloth, burlap, or filter fabric and attached to posts, it can form the walls of compost bins, storage bins, and even planting beds. Because it's not galvanized or stainless steel, 6-6-6 will eventually take on a rusted look and will rust through only after many years.

Rigid Welded-wire Mesh Panels (Also Called Livestock Panels)

If you've got money to spend, and need to outdo Martha Stewart, livestock panels make durable and beautiful trellises and fences, especially if you order ones with a bright finish. Rigid panels can be ordered online from welded-wire manufacturers or stock-fencing companies or bought at retail operations in agricultural towns. Look for gauge sizes 6 (³⁄₁₆ inch) through 3 (¼ inch), a 60 to 96 inch height, and 4 to 6 inch square openings.

Cages

If you have animals you will need cages, both for housing and for transportation. Big cages are great for goats, small ones for chickens, and medium-size ones for rabbits. Mostly metal ones have better ventila-

6-6-6 low tunnel with polyethelyne sheeting

tion and don't fall apart like their plastic cousins. Collapsible metal dog crates are the best, as they can be stored flat and are incredibly strong and durable. They are available through craigslist.com, friends, and on curbsides across America.

FABRIC

Repurposed Denim and Burlap

To prolong the life of planter-box wood, staple burlap bags or old blue jeans on the inside. While the wood will still get wet, it will be shielded from the soil decomposition process that rots wood over time.

Floating Row Cover

Also known by the brand names Reemay or Agribon, floating row cover is a white woven fabric essential for protecting crops from cold temperatures and insects. It is often used as a temporary warmth-providing blanket to protect frost-sensitive crops during a cold snap or placed on newly transplanted seedlings to ward off pests that like to feast on young plants.

Deer Netting

Deer are a plague on farmers even in many urban areas. Mark our words: They will eat everything. It's easy to keep them out, though. For an "invisible" deer fence, simply set posts using 6 to 8 foot rebar stakes all around your garden. Attach deer netting to the fence with zip ties or unrottable string, leaving 6 inches of extra material at the bottom. Dig a shallow trench between the stakes. Using irrigation stakes, secure the deer netting at close intervals inside the trench and bury it. Make sure to install a gate so you can get it. If your gate lacks the proper height, attach grape stakes or some other lightweight wood reaching 6 to 8 feet high vertically above the gate.

Plastic Fence Netting

Plastic fence netting, most often seen in use in its orange form during highway construction projects, can be very easy to come by for free and is very useful on the farm. Think temporary poultry enclosure. Why construct a complicated wood and chicken wire "chicken tractor" when you can just enclose the bed you want your chickens to scratch up with this flexible but durable bright orange plastic fencing and a few bamboo stakes.

Woven Shade Cloth

If you're growing crops in an area with scorching heat, or want to grow cool-climate crops such as those for salad mix during the summer, shade cloth is a miracle product. It comes in varied shade-blocking percentages, starting at about 30 percent and going up in increments to 95 percent for almost-full shade. It can be bought by the roll or the foot. Attaching lengths of it over low tunnels will create a cool, shady area underneath. Or cover the greenhouse with shade cloth in the hot summer months. Though it is plastic (boo hoo!), woven shade cloth can last for over a decade.

BLOCKS, BRICKS, AND STONE

Urbanite

The first time someone used the word "urbanite" with us we thought it was some fancy material similar to terrazzo. We found out that it's just a euphemism for broken-up concrete chunks. Urbanite is the quintessential urban farm building material for obvious reasons. The cons for its use are the same as for stackable concrete blocks—it's bulky. The pros are that it's free, plentiful, doesn't rot, and isn't toxic. Standing the chunks on end instead of placing them flat-side down helps avoid wasting space; a shallow trench for the bottom third or so to sit in will be needed for this. The height of the bed will therefore be

Edging a bed with horizontally oriented urbanite

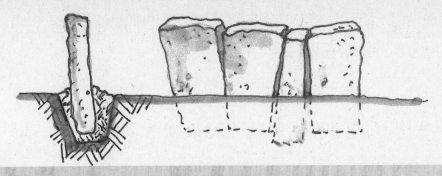

Edging a bed with vertically oriented urbanite

limited to the size of the chunks. If stacking horizontally, angle the wall in slightly to keep it from falling apart.

Stackable Concrete Blocks

Stackable concrete blocks, or cinder blocks, abound in many forms and are a great urban farming building material. Because they are wide, however, the blocks take up a lot of room, which is a drawback to using them to build raised beds. They are expensive when new, so you'll want to look for used or free ones.

ALTERNATIVE MASONRY UNITS

Also know as AMUs, alternative masonry units are an environmentally friendly homemade alternative to cement masonry units (stackable concrete such as cinder blocks). They are made of soil mixed with sand and either straw, rice hulls, or coconut coir. You will need a compressed earth block press to make them, which can be purchased on the Web; it will be quite pricey, but for large-scale projects it could be worth the investment. See appendix 11 for resources.

Bricks

Many urban farm beds are edged with bricks, since they're plentiful at the salvage yard and dump. Clean them well with a chisel and hammer, so they fit snugly, and stack them carefully, so they don't fall out of place. Brick is also a wonderful patio material.

Stone

Whether mortared or dry stacked, stone can be a beautiful material to use for building planter boxes in situ. Traditionally, fieldstone dredged out of a farmer's soil was used to build walls and terraces.

GREENHOUSE AND COLD-FRAME GLAZING

Polyethylene Sheeting

Polyethylene plastic sheeting has many uses, from solarizing weedy ground to covering cold frames to covering homemade greenhouses, hoop houses, or garden-bed low tunnels. Plastic is gross, we know, but it's cheap and useful. To mitigate the environmental harm, order plastic intended for agriculgtural use; you'll be able to reuse it for a number of years. Fasten it securely to cold frames by sandwiching it between rigid materials, such as strips of wood.

Twin Wall Polycarbonate Sheets

These durable if expensive sheets have replaced glass as the farmer's choice for greenhouse glazing. The material lets in the right amount of light and heat but has the added benefit of insulating against cold. The reason? It is formed of two layers of plastic with air-filled channels in between. These somewhat flexible sheets can be attached to a variety of frame shapes to make cold frames, cloches, and greenhouses. Screw down with large-head screws to attach to wood. If attaching to metal, the rigidity allows for punched holes and zip-tie fastening.

Windows

Old windows make great greenhouse boxes, and your animals will probably appreciate the view if you put one in their housing.

OTHER BUILDING MATERIALS

Concrete

We're in favor of garden structures that don't fall down when the tomatoes or cucumbers mature. A little well-placed concrete makes trellis and fence posts stay where you want them. Get a post hole-digger, some post bases from your locally owned hardware store, and get ready. Dig all the holes, run a string between them to keep the posts in line, and have all the posts and tools ready before you start mixing (the concrete won't wait for you). The key to good, hard cement is slow curing. While setting your posts, and for a day or so after, keep the surface of the concrete watered. And be sure to mark anything that's still hardening with caution tape if there are people around.

Cob, Earth Bag, Straw Bale, and Wattle-and-Daub Construction

Cob is an ancient and worldwide building method in which clay soil, carbonaceous material (such as straw), and sand are mixed together to make the natural equivalent of concrete. Used since time immemorial to make the proverbial mud hut, modern green builders have adapted it for all types of urban farm structures, from sheds to benches to wood-fired ovens. We have been lucky enough to participate in some building projects using cob and found it to be a pleasant if time-consuming activity. A cob bench or toolshed is a lovely thing in an urban farm, but it must be built by someone who knows how or is willing to find out and follow through to the last plaster coat and a roof. If not, the structure will gradually melt in the rain, making a heavy, depressing mess.

While we haven't personally done earth bag, straw bale, or wattle-and-daub construction, we've seen projects that look great. Earth-bag construction is the most intriguing, due to its simplicity and affordability. It involves digging up earth, putting it in bags, and stacking and compacting them to make walls. Depending on the height, various simple methods of reinforcement are used. Windows or doors can even be framed out. We can imagine that earth-bag construction could be easily used to make planter boxes of any size and shape.

Wattles

What are wattles? It's fun to say that word. We're sure it has many meanings, but in this instance we are referring to the jute netting–covered straw sausages that are used to prevent erosion on hillsides and freeway verges. We got this idea from our amazing friend John Bela. John was racking his brain to try to figure out how to build planter beds with locally produced sustainable materials. Since there are wattles made from waste rice straw in California, where he lives, he decided to try them. These flexible sausages can be used to edge beds both curved and straight. Nine-inch-diameter wattles can be stacked and kept in place with wood or rebar stakes. They will eventually decompose, but it will take a long time. A company called Earth Saver makes them in California; there may be an equivalent in your state.

GARDEN PATHWAY MATERIALS

Cardboard

We just cannot believe that people still put cardboard into the landfill! It's made out of trees and is a wonderful resource for many reasons. Nonwaxed used cardboard is a great weed blocker for paths and perennial planting areas. After pernicious weeds are removed, paths can be layered with tape-free, broken-down boxes and covered with mulch. This will keep down weeds.

Note: Many landscapers use filter fabric or weed-blocker fabric—synthetic cloth intended to limit weed growth—for paths. However, because these materials are made from petroleum products and don't allow for sufficient drainage and air infiltration, we do not recommend them.

Wood Chips

Local tree-care and removal companies often have large quantities of wood chips to get off their hands. Get on the gravy train by offering to let them dump a truckload in your driveway. Using wood chips on your paths over time creates a great habitat for beneficial micro- and macroorganisms, and eventually builds soil. Because wood chips often smell good and absorb water, they are also lovely to put down in duck runs and goat houses. If you have to buy wood chips, be sure to save money by purchasing in bulk by the truckload instead of in bags. Check before purchasing that they are untreated and undyed.

Crushed Rock, Gravel, and Pathway Fines

Though crushed rock is expensive, if maintained well it is durable and attractive. It can also add a few degrees of heat to cooler garden sites as it stores and then releases the sun's energy. If installing rock, since you won't be able to redig it later, be sure to eliminate weeds thoroughly and lay down as many layers of cardboard or old jeans as will fit.

Smooth gravel or pebbles are easier on the knees than sharp gravel. Light-colored pathway fines, also called decomposed granite, can reflect light up to plants. Pathway fines are compacted by applying water and tamping them down with a tamping tool. Both gravel and pathway fines can end up all over the place if your beds aren't boxed or edged to keep them in their place.

Clover, Drought-tolerant Grasses, or Herb Ground Cover

White Dutch clover is the gold standard for green garden paths. Clover, a legume, absorbs nitrogen from the air and puts it in the soil, where plant roots can use it. Although you may need to edge your paths with boards or some other material to keep the clover from invading your beds, this lovely turf, unlike wood chips, will not need replacing. You'll love being able to garden barefoot. Other options for a living mulch are drought-tolerant grasses and herb ground covers. Just be sure they feel nice on the knees and can withstand foot traffic before planting.

FARMING TOOLS AND SUPPLIES

Like it or not, farming does require using a number of tools. This section is a list of the farm gear you absolutely should own—and crafty methods to acquire it.

Must-have Tools:

- flat shovels
- pointed shovels
- hoes
- leaf rakes
- pruning loppers
- pruning shears
- trowels

In addition to these common tools that don't require explanation, we also recommend the following. Note: These are in alphabetical order, not order of importance.

Aphid Spray Nozzle

A number of companies now make a hose attachment that handily sprays a hard stream of water on the underside of leaves to kill aphids and other soft-bodied insects. This tool is attached to a nozzle arm, quite easy to use, and much more affordable and effective than "natural" bug sprays on the market.

Baskets

Thrift stores abound with cheap baskets. Hang them up by stringing a bungee cord through the handles and attaching the ends to nails or some other handy hanger. Use them for everything that needs carrying excepting liquids. And they're pretty.

Bow Rakes

This hard rake is essential for the final forming of raised beds. Initially the tine side is used to move soil around, then it's flipped for final smoothing with the bow side.

Buckets

These are used for everything from moving compost up and down stairs to collecting weeds, garden waste, and pests, to holding harvested produce. While five-gallon plastic buckets are plentiful and useful, the more attractive and durable galvanized ones can often be purchased inexpensively at junk or antique stores or garage sales.

Cobra-Head Weeder

Unlike other weeders and cultivators, this tool is multipurpose (weeding, furrowing, edging, harvesting) and very sharp.

Dibber

Dib-what? A dibber is an old-fashioned tool used to poke holes for seeds and seedlings that is also useful for digging up roots.

Dig Bar

A long, heavy, steel bar used to penetrate quite deeply into soil without digging a hole. It is used to probe the soil for rocks and concrete and can also be used to pry up what you find.

Flame Weeder

This ingenious device is perfect for pyromaniacs with a lot of weeds. It uses fuel, so you have to deal with that, but if you have annual pernicious weeds, your work will be cut down exponentially. Be very careful not to set wanted plants on fire.

Flat Spades

Unlike those of common flat or pointed shovels, the blades of flat spades are truly flat. These are the best tools for digging, and they can also be used to cut straight edges in lawn or around beds. They often have a D handle.

Harvesting Knives

For harvesting, a good knife is essential. Folding knives are best, but they must have a long enough blade to slice through lettuces or squash stems at a pass.

Harvesting Trays

Although buckets suffice for many vegetables, as well as fruits, those that crush or damage easily, such as tomatoes, summer squash, and peaches, should be harvested into flat containers. Baskets, plastic trays, and cardboard boxes can all fit the bill.

Hay Hooks

You'll need these to move straw bales off the truck for mulching projects or animal bedding. Moving bales of alfalfa and orchard grass for your goats is made easier with this tool.

Japanese Gardening Tools

Japanese hand tools are among the best made and well worth the investment. Put them on your holiday and birthday gift list now. We use the Nejiri weeder, Ika hoe, grass sickle, Hori Hori trowel, hand hatchet, and harvesting knife almost every day. They're ergonomic and stay sharp.

Machete

Machetes are great for chopping up garden clippings for the compost, clearing brush quickly, and cutting down cover crops.

Manure Fork

A manure fork's tines are closer together than those of a pitchfork, which makes it preferable for moving light materials such as straw, hay, manure, wood chips, and sawdust; such material is unlikely to fall through.

Mattock

The mattock resembles a sledgehammer with a hoelike blade instead of a hammerhead. This is one of the best tools hands-down for the initial weeding and loosening of new garden beds. With a few easy swings, work that would take twice as long with a shovel is done.

Pole Fruit-Picker

Instead of using a ladder, use a long pole picker. At the end there's a small metal basket with tines for pulling fruit from high limbs.

Scissors

We keep six or seven pairs on our tool wall and use them all the time for the usual scissor duties plus a lot of our harvesting.

Seeding Tool

A number of simple mechanical devices on the market dispense seeds evenly into garden beds, in the greenhouse and directly into garden, and can be calibrated for various size seeds.

Siphon Mixer

This must-have device mixes concentrated solutions of any kind from a bucket into your hose water so you can spray your garden directly with fish emulsion, or compost or manure tea. If you have drip irrigation, a similar device attaches directly to the drip supply line.

Sledgehammer

We use a "baby sledge" all the time to pound in stakes and posts. The larger version is useful for breaking up concrete or any other demolition projects.

Spading Forks

Unlike pitchforks, which are intended for pitching straw and hay, spading forks are made for digging. They are ideal for loosening compacted soil and are essential for double-digging garden beds.

Stakes and String

Stakes and string are used not only when laying out new beds, but each season when reestablishing and amending them. To make a durable, reusable tool, cut or find wood stakes, drill holes in them, and attach strings of the longest length you will need. In the field, excess can be rolled around one of the stakes to take up slack, and they can be stored tightly rolled.

Standing Weeders

A number of products on the market, classified as twist weeders or tillers, enable the farmer to loosen weeds from a standing position. Very helpful for those with back problems.

Sweep Catcher

A wire frame covered with canvas forms an oversized dustpan, traditionally used by landscapers to sweep up lawn clippings or prunings. This tool is very useful on the farm for its intended purpose as well as for harvesting cut-and-come-again greens. The location of the sweep catcher's handle makes it easy to balance with one hand while sweeping up cut greens. The other hand can be used to cut the greens.

Tarps and Landscaper's Bags

Tarps serve a number of uses in the garden. They can be used to protect things from the elements, to lay down underneath or on top of truckloads of materials, to lay on the ground to gather weeds while you work, or as a temporary sunshade. Landscaper's bags are made out of tarplike material and are great for hauling lighter-weight materials such as weeds, hay, or straw.

Watering Cans and Pitchers

We don't recommend watering your garden with a watering can, but along with plastic pitchers they can be very useful for mixing up fish emulsion, compost, and manure tea and watering plants with used wash water.

Watering Wand

For hand watering, you will need watering wands or spray nozzles with an on/off switch that can be fixed in the on position, so you don't have to keep holding down the button or lever while you water.

Wheelbarrows

Garden carts aren't very helpful in urban gardens, since they require wide paths to maneuver. We recommend you get a traditional 6-cubic-foot contractor's wheelbarrow and a traditional light-duty 4-cubic-foot wheelbarrow. The latter is truly wonderful on the urban farm, because you can easily lift it up to go up steps or to dump soil into tall boxed beds. This is much easier than having to shovel material out of the wheelbarrow. If a larger wheelbarrow is desired, one with a dual front wheel is better than a cart. Avoid plastic wheelbarrows.

For garden construction projects you may need the following tools and hardware:

- bolt cutter (for cutting welded wire fencing and chain)
- cable, cable clamps, and a swage-it tool for crimping the clamps for permanent trellises (for perennials such as grapes or kiwis)
- cat's claw (for removing nails)
- chain and S hooks or carabiners
- chalk snap line
- circular saw and blades (make sure the wrench to change blades is present)

- clamps (large C or bar clamps)
- crowbar (for prying nailed-together wood apart and removing nails)
- drill-and-bit set (corded or battery- or hand-powered)
- exterior-rated bolts, washers, and nuts
- exterior-rated screws and nails
- exterior-rated zip ties
- hammers
- hand screwdrivers
- heavy-duty staple gun and ¼-to-⅜-inch staples
- jigsaw and blades
- ladder
- level
- locking (vice grips) and regular pliers
- measuring tapes
- palm sander or hand sanding block and sandpaper
- pencils and sharpener
- plumb bob
- post level
- sawhorses
- small sledgehammer
- speed square
- three-prong outdoor extension cords and splitters
- t-square
- wood shims

Acquiring Tools

Gardening tools are plentiful and cheap at garage sales, salvage yards, and junk stores. During the last decades almost everyone has at some point had the notion to attempt an ambitious gardening project and then not seen it through. Why not help them clear out their overfull garages and help yourself to a deal. If you do need to buy tools, buy high quality; Japanese- and English-made tools are especially good choices.

Public library branches called "tool libraries" have cropped up in some cities. These are extremely useful, since you don't have to buy or rent one that is used infrequently. If you don't have one in your area, encourage your local public library to start one.

Farming Supplies

If you don't want your farm to be a money sucker, we recommend that you compare prices between your retail gardening center and rural farm-supply stores that have an online or phone-ordering capability. Farm supply companies geared toward the commercial farmer are often the most economical source. Another option is to make biannual pilgrimages to rural feed mills and farm supply stores to load up on what you need.

Hardware stores (especially those serving professional contractors) and restaurant- and irrigation-supply stores (geared toward landscaping businesses) also carry many of the supplies urban farmers need, sometimes at a lower cost than retail garden stores; an example is copper stripping used to deter snails. If you buy the copper materials designed for home gardeners the package will yield enough for one small planter box. For a slightly higher price at the hardware store you can buy a 12-inch by 20-foot roll of copper roof flashing and cut it into three 4-inch-wide strips, and you have enough to cover the lip of four large beds.

Helpful farming supplies available from commercial suppliers:

- baling wire
- beneficial insects
- bird netting
- deer netting
- diatomaceous earth
- dormant tree oil and sprays
- drip irrigation supplies and stakes
- floating-row cover material and hoop stakes
- gopher, mole, and rodent traps
- grafting supplies
- jute twine in bulk rolls
- legume inoculant
- powdered and liquid amendments
- restaurant-size lettuce spinner
- rooting compound
- shade cloth
- sticky insect-trap paint
- wooden plant labels

SEEDS AND PLANTS

With a few exceptions, the highest-quality, freshest seeds can be bought through seed companies that sell to commercial and home gardeners through catalogs and online. Seeds found in the store on the rack can be old, but more important, rack seeds are often just your standard fare. The varieties are most likely not selected for your climate, and they're definitely not chosen to be the early and space-saving varieties best for urban farmers. You won't find cool things like pelleted carrot seeds (they're coated with clay, which makes sowing tiny seeds easier and economizes seed). In addition, depending on the size of your farm, you can save a lot of money buying seeds in larger quantities than what's in the usual off-the-rack packet, especially for expensive items like seed potatoes and onion and garlic sets. If you purchase seeds or plants retail, shop only at nurseries that take the time to know their business and stock appropriate varieties for your climate.

HAY AND STRAW

If you plan on integrating livestock into your urban farm, prepare to get intimate with hay. Hay is baled grasses and legumes, which can include orchard grass, Timothy hay, alfalfa, and oat hay. It's full of nutrients and is the primary feed for cows, horses, and goats. You'll see that chickens enjoy eating alfalfa and oat hayseed heads, as do rabbits. Straw, on the other hand, is nutrient-free grain straw used in hennesting boxes and to line goat pens and rabbit cages. Straw is also used as a mulch and compost ingredient in the garden. Many urban farmers drive long distances for a hay-and-straw source. We discovered that our local racetrack has a hay supplier that sells directly to the public; racetrack hay tends to be very high quality.

Another source for hay and straw is to grow and cut your own. What epitomizes farming more than a waving field of fodder grass? Seeding and harvesting hay in a few empty lots can more than satisfy your needs, especially in climates with ample spring and summer rain (arid areas will require irrigation, which may be difficult to arrange). Empty lots with wild grasses can be cut for use as compost or mulch material. Be sure you harvest before the seeds mature fully, or you'll have imported a monster into your garden. A well-sharpened scythe is a wonderful tool to learn to use.

BEE SUPPLIES

More and more cities are playing host to beekeeping-supply stores that sell veils, supers, extractors, smokers, foundations—you name it. It's a great thing to support these businesses, especially if you need an extra super pronto during the main nectar season. But you'll save money if you buy supplies directly through a reputable manufacturer. You'll definitely save more money if you are willing to buy parts and then assemble the frames and boxes yourself. Shipping costs can get pretty high, though, so make sure it is worth it before ordering online. (See appendix 11 for a list of bee supply companies.)

The bees themselves can be ordered online and sent through the mail in something called a "bee package." Alternately, you can find a local beekeeper who might sell you a full colony, or a nuc, which is a minicolony. See page 361 for more information about keeping bees in the city.

POULTRY GEAR

For the hens, ducks, and turkeys you'll need items such as waterers, feeders, and nesting boxes. You'll also need inputs such as oyster shells, pelleted feed, scratch, kelp meal, and diatomeceous earth for controlling flies. Since bags of pellets and oyster shells are heavy, we don't recommend buying these online. Luckily, most pet stores now carry poultry feed, and they can special order pretty much anything you need. If this seems too costly, a drive out to the country will probably yield a "hay and grain" store that will sell bags of poultry chow and other supplies at a bargain price. Even better is to find a feed mill and buy directly from them. See page 384 for more information about keeping poultry in the city.

RABBIT SUPPLIES

Rabbit cages, feeders, and waterers are widely available from pet stores. The pet store rabbit cages are often too small for keeping meat rabbits, and the expensive hutches they sell are made out of wood, which isn't good for you or the rabbits. So we buy large-size rabbit cages (see appendix 11 for sources). In a pinch, metal dog crates can be used. Rabbit feed—pelleted mixes of alfalfa meal, barley, and wheat—is sold at pet stores as well. We find that the rabbit feed at farm supply stores tends to be higher quality. If possible, going to the feed mill where the pellets are made is the best scenario—the feed will be less expensive and fresher. See page 439 for more information about keeping rabbits in the city.

GOAT SUPPLIES

Some urban goat owners drive long distances to load up on dairy pellets, grain mixtures, and bales of hay at a rural feed store. Even at these stores it is difficult to find some essential items, such as milking pails, hoof trimming shears, teat dip, and disbudding irons (see page 494). Luckily the Internet provides. Online suppliers carry everything you could ever want for your goats, including stanchions (stands on which you milk your goat), kidding kits, and all manner of goat health merchandise. They also carry cheese-making supplies, including cheese molds, cultures, rennet, and books. (See appendix 11 for suppliers.) See page 462 for more information about keeping goats in the city.

W ▸ E

CHAPTER 5

PLANNING WHAT TO GROW AND RAISE

It's time to get personal and do some food-focused research and planning to decide what crops you'd like to grow and which animals make the most sense to raise.

VEGETABLES

Traditionally, planning a home vegetable garden meant planning a summer garden: tomatoes, beans, squash, peppers, eggplant, and corn. The self-sufficiency-oriented urban farmer will want to expand his or her horizons. While summer veggies are a special treat, if you really want to make a dent in your food bills, you'll need to shift your focus to those vegetables we eat every week; for many of us that includes salad greens, cooking greens, carrots, beets, potatoes, broccoli, and herbs. The good news is that not only can these crops be grown in the summer along with your tomatoes, but depending on your region, they often thrive from early spring through late fall.

You'll also want to save your precious growing space for crops that are expensive to buy in the store. Dry beans, bulb onions, and cabbage are so inexpensive to buy, it makes more sense to use your precious space to grow parsley, chard, and edible pod peas—expensive items. Additionally, you may not want to grow things that take years to start producing, like asparagus, or take hours to harvest, such as raspberries. These crops are luxury items that should be incorporated into your garden plan with some thought.

To use your small space effectively, you should focus your efforts on growing crops that:

- can be harvested from spring through fall or, better yet, in winter;
- mature quickly;
- are high yielding for the amount of space they take up;
- are high in nutrients;
- are easy to grow;
- are the most expensive to buy in the store; and
- thrive in your climate.

Which vegetables most closely fit the above criteria? If you focus on fast-maturing varieties, the following crops are the most efficient.

	Harvest Spring Through Fall	Mature Quickly	High Yielding	High in Nutrients	Easy to Grow	Expensive in the Store
Salad greens	✔	✔	✔	✔	✔	✔
Cooking greens	✔	✔	✔	✔	✔	✔
Root vegetables, including potatoes	✔	Depends on variety	✔	✔	Depends on conditions	✔
Brassicas, except for cabbage and Brussels sprouts	✔	Depends on variety	✔	✔	Depends on conditions	✔
Herbs	✔	✔	✔	✔	✔	✔
Early onions (see page 70)	✔	✔	✔	✘	✔	✔

Focusing on growing the above means growing fewer summer fruiting crops, especially the lower-yielding ones, such as corn and melons. The higher-yielding summer crops are cucumbers, green beans, summer squash, and tomatoes. If you live in a hot climate, also feel free to try peppers, eggplant, and okra. It almost always makes more sense to buy corn and melons from your local rural organic grower than tie up half your garden beds in such low-yielding crops.

A lot of space and time can be lost growing the wrong variety or "cultivar" of a particular crop. Check with your county cooperative extension service or locally owned plant nursery, ask around among other gardeners, or ask your farmers' market vendors for the best crops and varieties for your region. Since specific cultivars are bred for different seasons, make sure that the one you are growing and the time of year are a good match. Don't grow giant winter spinach in the summer. You may watch it bolt (go to seed) a month after planting. When choosing cultivars, urban farmers should focus on early varieties to maximize yields. It's also a good idea to check out those bred for smaller spaces, such as bush summer squash.

Decide who your produce is destined for and grow what you eat, what your family eats, or what your community wants. There's nothing more unmotivating than realizing that despite all the planning, amending, and planting, you aren't eating out of your garden. Maybe you got excited about some new vegetable you had never heard of but then found you didn't know how to cook it. Maybe you thought you would be a self-improvement miracle and eat kale at every meal from now on. Start from where you are in your taste preferences rather than an idealized version of yourself.

Do an inventory of what you actually eat in a week. Perhaps you find you love mashed potatoes—guess what, you can grow potatoes! If you eat a lot of fried eggs, laying hens should be a prominent part of your design. Radishes may be easy to grow, but do you eat that many radishes? Children generally love broccoli, so if you have kids, you might want to consider it. Don't fall prey to too much variety. If you know you eat carrots, salad, broccoli, parsley, and chard on a regular basis, it's fine to grow only that.

Many of us have fallen into eating less produce than we ought to. It can take months or years to change eating and cooking habits, so start by growing fruit and salad greens. Eating homegrown fruit is the gateway to increasing your produce intake. Two standard-size fruit trees are all that each person needs poundwise. If you need a bit more variety, you can achieve that by swapping with your neighbors and/or growing multiple superdwarf and/or espaliered fruit trees. There are many easy-to-grow fruiting shrubs and vines. Yours may take a few years to begin producing, so it's a good idea to get them in the ground as soon as possible. Salad is a style of vegetable preparation that does not involve cooking and is generally very easy to grow. Coating lettuce leaves with a delicious sauce called salad dressing makes it irresistible, even to children!

Grow the Most Nutritious Vegetables

A look at calories, protein, and levels of important nutrients, such as calcium, iron, and vitamin C shows that the following crops have high nutritional value.

- Blue and purple berries
- Cooking greens
- Edible pod peas
- Parsley
- Potatoes
- Root vegetables (carrots, beets, turnips)
- Salad greens
- Snap beans
- Winter squash

While we recommend that urban growers avoid wasting space on long-season crops, such as bulb onions, the following alternative onion varieties are urban farm mainstays due to low space requirements, high yields, hardiness, and earliness.

- Bunching onions
- Cipolin onions
- Garlic (takes about nine months to mature but can be planted with other crops to save space)
- Green onions
- Leeks
- Shallots
- Spring onions (bulb onions harvested early)

Vertical trellising (see page 108 for instructions) reduces the growing footprint, turning many low-yielding crops into high yielders. If you put some work into it, the following crops can be tied completely vertical and virtually flat, allowing other crops to be grown below.

- Cucumbers
- Melons
- Pole beans and peas
- Summer squash (especially vining varieties)
- Tomatoes
- Winter squash

Fast-maturing crops increase yields especially when interplanted with slower-maturing ones. The most important of these are:

❉ Arugula and other specialty greens
❉ Cilantro
❉ Lettuce
❉ Mustard greens
❉ Radishes

A few crops and cultivars stand out as very high yielders:

❉ Asian eggplant
❉ Black Seeded Simpson lettuce
❉ Celery
❉ Chayote
❉ Cut-and-come-again lettuce mixes (mesclun)
❉ Cylindra beet
❉ Fordhook Giant chard
❉ Fortex and Cascade Giant snap pole beans
❉ French Breakfast and Icicle radishes
❉ Green sprouting Calabrese and early purple sprouting broccoli
❉ Gypsy pepper
❉ Japanese Giant spinach
❉ Kwintus Romano pole beans
❉ Red giant and Mizuna mustard greens
❉ Lemon cucumbers
❉ Seven Tops turnip (for greens)
❉ Snow peas
❉ Sugar snap peas
❉ Tree collards/Jersey kale
❉ Trombetta/Tromboncino summer squash
❉ Violetto artichoke

Maturing Times

The following table lists the length of time some crops require to mature. For each crop there are many cultivars, some of which mature faster than others. To increase yields it is important for urban farmers to look for quickly maturing cultivars. This is especially true if you choose to grow any of the typically long-maturing crops, such as cabbages.

Crops are listed according to the length of time before first harvest not time in the ground; harvest doesn't always mean removing the plant. With many crops you can harvest leaves or fruits over a long period of time. For example, chard and parsley are in the ground for a long time, yet they can be harvested just a few months after planting. For crops that are planted out as seedlings rather than seeds, the length of time till harvest is given from the time the seedling is planted.

Short Maturing Time (up to two months until harvest)	Medium Maturing Time (two to three months until harvest)	Long Maturing Time (three to six months until harvest)	Very Long Maturing Time (six months or longer until harvest)
Arugula	Beans, fava	Artichokes (perennial in warm climates)	Garlic
Asian greens	Beans, lima	Beans, dry	
Basil	Beans, snap	Brussels sprouts	
Bitter greens	Beets	Cabbage	
Broccoli rabe/ Rapini/Sprouting broccoli	Broccoli	Corn	
Chard	Cardoon (biennial)	Leeks	
Cilantro	Carrots	Melons	
Dill	Cauliflower	Onions	
Green onions	Celery	Squash, winter	
Lettuce and cutting greens	Collards	Sweet potatoes	
Mustard greens	Eggplant		
Okra	Fennel		
Radishes	Kale		
Spinach	Peas		
	Peppers		
	Potatoes		
	Shallots, spring onions, and bunching onions		
	Squash, summer		
	Tomatillos		
	Tomatoes		
	Turnips		

INVASIVE CROPS

The sad truth is, some food crops are highly invasive or get so huge that they shade everything else out. With some, such as tomatoes, their tendency to reseed themselves from fallen fruit can be a happy event, but others, including herbs and edible flowers, can easily become your worst weeds, either through spewing seed or by spreading roots and runners. Be forewarned and ready to do battle.

In the case of seed spewers, take care to cut off seed pods or remove them before they set seed. For the underground spreaders, install an underground barrier or keep them cut back to their area. Be prepared to heavily prune those in the too huge category—they can take it.

Some of the Worst Offenders (This Is Not a Comprehensive List)

Spread by seed

Amaranth ("love lies bleeding"), calendula, cardoon, nasturtium, sunflower, tomatillo, tomato

Calendula, though beautiful, can become invasive.

Spread by roots or runners

Blackberry, chamomile, comfrey, mint, raspberry, strawberry, watercress

Get too huge

Cardoon, castor bean plant, chayote, fig, Jerusalem artichoke, mulberry, rosemary

FRUIT

Like vegetables, growing fruit in the city is all about yield and flavor. As an ideal, we suggest the following plan when it comes to fruit trees.

❖ Grow two to five trees or shrubs per person (if you want to make a dent in your fruit needs).

✥ Choose a wide variety of types of fruits that mature throughout the year.

✥ Select a number of varieties of each type of fruit that mature at different times.

✥ Grow fruit trees that yield lower amounts at a time or yield over a longer period of time.

✥ Plan your garden layout so that trees and shrubs don't shade annual vegetable beds, for example, by planting trees along the northern property line or by pruning trees to a flat, two-dimensional shape.

✥ Grow dwarf trees that can be cared for and harvested without the use of ladders.

✥ Grow fruit with the most delicious flavor (See page 80 for information about taste-test winners).

Small fruiting plants provide ample fruit without shading your growing beds. Something to be considered, however, is the amount of time harvest will take: Know that it will take a long time. Here are some to try.

✥ Currants

✥ Gooseberries

✥ Hardy kiwis

✥ Josta berries

✥ Raspberries

✥ Serviceberry (Juneberry)

✥ Sour cherries (bush form)

✥ Strawberries

✥ See chapter 15 for more detailed information about fruit trees.

LIVESTOCK

Although we do love our vegetables, we realize that the majority of people still tend to eat more meat than vegetables, drink milk or use sweeteners than eat fruit, cook eggs more often than kale. Enter the urban livestock to provide these valuable products to the farm. Rabbits, ducks, and turkeys will provide an ongoing source of hormone- and antibiotic-free meat; dairy goats can provide gallons of milk; chickens and ducks lay eggs; and honeybees create a healthful sweetener on the urban farm. These six animals tend to be productive even in small spaces and are able to be fed and made comfortable with scrounged

Eat locally grown food! This is not by any stretch a new concept, yet in recent years it has gained widespread attention. While some local food activists may be extreme locavores, we suggest steering toward a *more sustainable* food system, not one where food is never shipped long distances.

As a way to define what "local" food means, the term "foodshed" has lately come into fashion. The term describes a region within which food travels from producer to consumer. Food transported within this set region is considered local and more sustainable than food transported between larger regions.

Foodshed mapping tends to ignore the possibility of urban farming contributing to the food supply. We believe that growing more food in urban areas could be the biggest contributor to localizing our food system.

We recommend that urban and rural growers exploit different market niches to create a more sustainable foodshed. This specialization could help cities to be less reliant on using fossil fuels to ship produce from rural areas and could make cities contributors to, and not just consumers of, their foodshed. It could also increase urban and rural farm efficiency. There is absolutely no reason urban people can't grow all the salad greens, cooking greens, and fresh herbs they need, allowing rural farmlands to be used to grow calorie crops such as grains, dry beans, and potatoes and storage vegetables such as cabbage, carrots, and bulb onions. Salad greens, cooking greens, and fresh herbs taste best and have the highest nutrient values when eaten superfresh, so why waste fuel on trucking lettuce?

At the food system level we recommend the following specialization. This list is a generalization—we aren't saying urban growers should never grow cabbage or that rural growers should never grow chard. But an overall move in this direction makes sense.

City Farm Niches

- quickly perishable vegetables, such as greens and herbs
- vegetables that mature quickly

- high-yield vegetables
- small egg, milk, and meat animals
- fruits such as berries that are more perishable and require more labor

Rural/Suburban Farm Niches
- storage vegetables and calorie crops
- vegetables that take a long time to mature
- low-yield vegetables
- large milk and meat animals
- fruits such as those from orchards that are less perishable and require lower labor

city materials. To figure out which animals to raise, keep a journal and mark down how often you buy eggs, milk, honey, or meat. Also keep in mind what is most important to you: If free-range eggs are a priority, hens are a great choice. If you are driven to have access to raw milk, goats might be in your future. If knowing that meat animals have been raised humanely, take the plunge in raising meat birds or rabbits. Use the table below to make some predictions about which and how many animals you should raise to supplant, or at least supplement, store-bought items.

Animal[1]	Standard Output
Chickens (6 hens)	15 dozen eggs per month
Breeding meat ducks (2 females, 1 male)	25 ducks per year for the table
Egg-laying ducks (6 females)	15 dozen eggs per month
Small goat breeds (2 females)	15 gallons milk per month
Large goat breeds (2 females)	30 gallons milk per month
Honeybees (1 hive)	5 gallons of honey per year
Rabbits (3 does, 1 buck)	75 meat rabbits per year
Turkeys (2 females, 1 male)	6 turkeys per year

[1]See part III for more detailed information about each animal.

YEARLY FARM PLANNING

Because many crops can produce for six to twelve months, you will need to plan ahead to have some beds empty, to get your summer, fall, winter, and spring gardens in at the best time. Summer gardens must be started in late winter to midspring to mature at the right time. Fall and winter gardens must be planted by mid- to late summer to produce before the daylight and temperatures drop below the threshold for active growth. An early crop of cool-season spring vegetables must be planted in mid-fall and wintered over to produce in February, March, and April. Plan your garden to have beds come available in early spring, midsummer, and midfall, so you can plant at the right time. A good way to do this is to plant at least a few beds each season entirely with crops that mature quickly, such as salad greens, spinach, and radishes. That way, around March 1, August 1, and November 1, you will be sure to have some space available. In cold climates the beds reserved for your spring and summer gardens will need to be protected by cold frames or temporary low tunnels so you can plant your seeds and seedlings as early as possible.

Like vegetables, livestock will also have seasons and dormant production periods.

Chickens: Highest production in spring and summer; molting with no eggs in the winter

Ducks: Highest egg production in the spring and summer

Goats: Ideal milk production in early spring and summer

Honeybees: Harvest in late spring or summer, and sometimes in fall

Rabbits: Ideal meat production is in the spring and fall; rabbits won't breed in extreme cold or heat.

Turkeys: Ideal meat production in the summer and fall

See part III for more detailed information about each animal.

PLANNING HOW MUCH TO GROW

The number of different vegetables you will actually have space to grow will depend on how much gardening space you have. If you have a small space, you are going to have to prioritize. It makes much more sense to grow enough plants of just a few types of vegetables, so you will have enough to make meals with them, than to grow ten vegetables and harvest tiny bits of each kind. Having a small space may make you reconsider growing lower-yielding crops, such as bell peppers, melons, and corn in favor of the more productive roots, cooking greens, beans, and peas.

SUCCESSION PLANTING

Planning multiple plantings of the same crop during the time of year it can be grown will stretch out the harvest and help avoid gluts. For instance, if you plan to plant a 15-foot trellis with green beans in the spring, you can plant 5 feet every month for 3 months instead of planting them all at once. If you want to have a continuous harvest of head lettuce, plant the number of lettuces each month that you want to eat in the next month. Since some plants yield continuously, or hold in the ground for months, succession planting isn't as simple as just planting a few of each type each month during the growing season. Chard, for instance, can be harvested for up to a year before going to seed; it will probably only need to be planted two times a year to have harvests throughout the growing season. Once mature, many varieties of root crops, such as beets and carrots, can remain in the ground for months and be harvested as needed. Beans and peas yield pods for up to three months from one planting.

Due to this complexity, it's best to create succession-planting and -sowing calendars, so you don't have to break your brain each time you go out to the garden to plant. Begin and stop your succession planting based on each crop's temperature and day-length requirements. Schedule first and last plantings to push the envelope and you will be pleased to have extra early and late harvests.

A few examples may help. Peas require cool weather during the majority of their life cycle, but warmth helps them put on green growth quickly before the cold weather sets in. In our area we plant peas from mid-July through the end of October that we will harvest during the fall and winter. Sometimes, if it's especially hot, our first mid-July planting doesn't survive, but when it does we are the first to harvest fall peas. Tomatoes, on the other hand, like heat. We plant our first tomatoes in February, because occasionally we get a March heat wave. If the early spring turns out to be colder, our next tomato planting will often overtake the first. If the early spring is warm, we are the first to harvest this most sought-after summer crop.

Use the template and examples in appendix 6 and the succession-planting table in appendix 5 to create your own calendar. After one year, adjust. Then track the weather patterns each year, and if it looks like you'll have an especially early or late spring or winter, move the sowing and planting dates forward or back accordingly. If you make a mistake and plant crops too early or late, resulting in crop failure, this calendar ensures you will plant the crop again.

Once you have this calendar, you can create one for succession sowing for the crops you want to start indoors (see the following chapter on starting seeds). Just shift the dates back one to two months, depending on how long it takes the seedlings to mature, and you'll have seedlings ready when your planting calendar says to plant.

We recommend setting a schedule. For a family-size garden you might choose to do your indoor sowing the first week of each month and your outdoor planting and sowing in the third week. For larger gardens or projects you may want to schedule to do both indoor and outdoor sowing and planting each week on different days. We are ignoring the fact that some months have five weeks and organize our propagation and planting calendars into week 1, week 2, week 3, and week 4, beginning again at "week 1" the first week of each month.

The example calendars in appendix 6 will work for most regions of the country, although season extenders will be required in the late winter/early spring, and in the fall, to lengthen the harvest. Southern gardeners may need to shift the schedules to avoid growing in the hottest months and to take advantage of November, December, and January temperatures.

A very wide variety of crops are listed in the example calendar. If you have a limited growing space, you may want to choose only one to two in each of the root, salad greens, cooking greens, and herb categories and limit the number of summer crops. You can begin with these calendars and revise them to add your own crops and varieties or to change planting dates based on your experiences. Adding annual flowers and perennial vegetables, herbs, fruits, and ornamentals can help ensure that they get into the ground at the right time.

Our examples do not list specific varieties—we leave those to you to choose. However, we do list different types of each vegetable. For example, we list hot peppers, green peppers, and colored peppers separately in our succession-planting and -seeding calendars but don't specify "bell" pepper. We also list summer squash and winter squash/pumpkins and types of tomatoes. We want to ensure that we grow tomatoes that are early, yellow and red cherries, for sauces, midseason medium-size, and large (usually interesting heirloom varieties).

Market farmers will need to decide what crops to plant in succession based on their climate, farmland, and customer base.

CROP PLANNING FOR YOUR CLIMATE

Understanding the climate in your region will help you decide when to plant and what to choose for the greatest yields.

While we all swoon over the seasonal joys of a garden-ripe tomato, it's the kale and lettuce we eat every day that make up the greater part of a nutritious diet. We should not only pine for the first tomato of the season but for the first fava bean, the first artichoke, the first sugar-snap pea, and the first frost-kissed collard greens. Luckily, because they don't require as much heat, many of these crops can be grown earlier in the spring and later into the fall and winter than summer, heat-loving veggies.

Most gardening resources suggest planting dates based on the last frost of spring and the first frost of fall. While these dates are helpful when planning your summer garden, they aren't very relevant for a long list of cool and warm season crops. The obsession with frost dates demonstrates a prejudice toward the hot season crops that have tended to be the focus of hobbyist gardeners. We want to help you plan for year-round growing and escape the last-frost-date mania, even if it means growing in cold frames or row covers. In addition, garden planning based on frost dates is not applicable for gardeners in temperate regions, where there may be few or no hard frosts, or in the South, where summer heat makes winter growing a necessity.

There are three environmental factors that effect plant growth: the number of daylight hours; the intensity of the light; and the intensity of the heat, including air and ground temperatures. Annual vegetable crops can grow actively when there are at least ten hours of daylight per day and sufficient heat (65 to 85 degrees during the day) either through natural sunlight or by artificial means, such as the use of cold frames or heated greenhouses. When the number of daylight hours falls below ten in the winter and daytime temperatures drop below 40 degrees, crops don't grow actively, but even so, winter varieties can stay alive in the ground and be harvested throughout the winter months. In the majority of the United States daylight hours are sufficient until November or December, when they drop below ten.

When they rise back up above ten, in January or February, as long as the necessary heat is present, growth can resume. Crops can also be overwintered. This means the crops stay in the ground in a kind of holding pattern, waiting for spring to begin growing again. If the over-wintering crops are large enough, when the daylight drops below ten hours they will have a jump start come spring, and you will have very early harvests.

The period when sunlight drops below ten hours a day varies based on your geographic region (see appendix 8 for information). For an early spring harvest, plan to plant your crops at least a month before the daylight hours drop. In the spring, begin seeding and planting about a month after the last ten-hour day. Season-extending tools, such as cold frames or row covers, can be used to continue your harvest when unprotected crops would be killed by freezes. If you are using these crop-protection techniques, you may be able to plant later in the fall or earlier in the spring.

Some crops, such as onions, won't mature if the day length doesn't reach a certain number of hours. Varieties listed as short-day onions bulb up when the day length is between twelve and fourteen hours. Long-day onions begin to form a bulb when the day length is between fourteen and sixteen hours. Because of this, regardless of the temperature, to form a bulb onions must be planted in time to mature when the daylight hours are longest. Day-neutral varieties will set bulbs during days of any length. If you live in a southern latitude you won't be able to grow short-day varieties; the reverse applies for northern latitudes.

The intensity of the light also varies due to latitude. Crops have different light-intensity needs, which is why it can be difficult to grow tropical varieties in northern areas. Corn, cucumbers, beans, potatoes, and sweet potatoes all require higher light intensity. Onions, carrots, celery, *Brassica*s, lettuce, and spinach are some of those that have lower light intensity requirements. Knowing where crops originated will provide a clue to their needs. For example, potatoes originated in equatorial regions; therefore, they need higher light intensity.

Just as too little heat keeps crops from thriving, too much heat has a negative effect on plant growth. If you live in Los Angeles, a nice level of heat is present all year, so the only factor that would affect planting dates is the day-length rule. Winter tomatoes and pepper plants that turn perennial are common in southern California. If you live in Florida, however, excessive heat will make summer growing difficult at best. In Chicago, you can pretend you're in L.A. if you simply provide the missing heat through hoop houses, high tunnels, floating row covers, or cold-frames. See page 232 for instructions on using these season-extending tools.

Heat over 85 degrees can harm plants and limit growth. Many novice gardeners will think that they are behind on watering when they see leaves wilt in hot weather. Leaves can only absorb so much heat, then they kick into gear to cool down by wilting. Check the soil, and if it seems moist, don't add water. Instead, cover your beds with shade cloths.

Freezing and below-freezing temperatures can affect annuals, perennials, and fruit trees negatively or positively, depending on the type of plant. For example, many deciduous fruit trees need a number of "chill hours" to produce fruit; a variety of apple tree may require four hundred chill hours per year. On the other hand, some subtropical and tropical fruit trees can't tolerate hard freezes well. The flavor of *Brassicas*, such as collards and Brussels sprouts, are improved with cold weather, while chard will be killed by repeated hard freezes. In colder climates, knowing whether a crop is hardy (tolerates freezes), half hardy (tolerates a certain number of hours below 28 degrees), or tender (does not tolerate freezes) is important for extending the harvest. In hot climates, knowing which crops require cool or cold weather is necessary.

Minimum and maximum temperatures vary widely within geographic regions because of microclimates created by geographic features, such as mountains or exposure to sea air. The U.S. Department of Agriculture designates plant hardiness zones with a number from 1 to 11, and knowing yours will be useful for your preparations. Because cities have a lot of concrete and asphalt, they tend to absorb and hold more sun energy. Therefore, their frost dates are often a few weeks later in the fall and frost-free dates a bit earlier in the spring than in surrounding rural areas. You can easily find your specific zone number and recommendations for growing online. Sunset climate zones are a different system developed by the company that publishes *Sunset* magazine and regional gardening guides. These also can be useful and found online. (See appendix 11 for sources.)

While it is useful to know data about your climate, observation is your best friend. Temperatures and conditions vary from year to year (especially with climate change), and while most human beings may be oblivious to such subtleties, other creatures and organisms around us are highly attuned. If you begin to pay attention to when trees start to bud, bulbs begin to poke out of the ground, and seasonal animals and insects appear, you will know when to plant.

TEMPERATURE VARIATIONS WITHIN THE SAME LATITUDE: THE MARITIME EFFECT AND ELEVATION

While many geographic regions sharing the same latitude have similar climates, others have radical differences. The most important causes of these are the proximity to oceans and elevation. People living near the ocean or at high altitude must learn to disregard blanket recommendations for their geographic region and calculate their planting calendar based on an understanding of the effects of these factors.

The Maritime Effect

Large water masses such as oceans and seas do not change temperatures very quickly. In general, this results in more constant temperatures on land next to them. For example, the Gulf Stream in the Atlantic Ocean causes England to be much warmer than areas at similar inland latitudes in North America. The Pacific Ocean has a similar effect on coastal California and the coastal Pacific Northwest.

Urban farmers in coastal cities can expect less variation in temperature than those in areas just a few miles inland. Less variation downward means fewer frosts, if any; and, unfortunately, less variation upward may make growing hot season crops difficult. This decreased variation is true for both daily and yearly temperature ranges.

Gardening on coasts has advantages and disadvantages, and the positive should be exploited rather than the negative fought. Benefits often include a longer growing season, few or no hard frosts, and less extreme summer heat. Disadvantages include difficulties growing heat-loving summer crops and tree fruits requiring chill hours. To succeed, focus on growing temperate, cool, and warm season crops and fruits requiring low chill hours.

The Effects of High Altitude

For every 1,000-foot rise in altitude expect temperatures to be around 3 degrees cooler. As temperatures drop and air becomes thinner, humidity levels decrease. These two factors lead to the cooler, more arid characteristics of alpine climates. The changeable weather at higher elevations leads to the notorious difficulty of determining garden planting dates and the length of the growing season. Be prepared for unseasonable snow dumps and frosts. Constant drastic changes in altitude and slope direction cause microclimates to vary widely in temperature, wind, and moisture levels. In high-altitude conditions, direct observation and noting down temperature readings over an entire growing season are important aids for choosing your garden site. Keys to success are: choosing crops that tolerate both cool and warm weather and are specially bred, high-altitude cultivars; getting a jump on the season by starting seedlings indoors; and using season-extending techniques, such as cold frames or row covers. See page 524 for resources about high-altitude gardening.

Crop Temperature Tolerances

To help simplify seasonal crop timing we have categorized crops according to their temperature tolerances; please see page 85 for a list. Very few thrive in temperatures over 85 degrees, though they may be able to tolerate limited heat spells. For this reason, southern urban farmers will need to become familiar with shade cloth and learn how to time their crops to avoid the hottest months.

Some crops require hot weather to mature and yield. Urban farmers in northern latitudes and coastal regions that lack extended hot weather will need to forgo growing hot season crops or use season-extending techniques to create warmer microclimates. Please see chapter 10, page 232 for season-extending techniques. See appendix 8 for regional guidelines on extending your season.

CROP TEMPERATURE CATEGORIES

Cool season: requires cool weather to be productive

Hot season: requires hot weather to be productive

Overwinter: tolerates below freezing temperatures for extended periods

Temperate: tolerates both cool and warm weather; may or may not tolerate extended hard freezes or heat spells

Warm season: requires warm weather to be productive; usually tolerates hot weather

Crop	Temperature Category	Crop	Temperature Category
Artichokes	Temperate	Fennel, bulbing	Temperate
Arugula	Temperate; overwinter	Garlic	Temperate; overwinter
Asian greens	Temperate; overwinter	Green onions	Temperate; overwinter
Basil	Warm season	Kale	Temperate; overwinter
Beans, bush	Warm season	Leeks	Temperate; overwinter
Beans	Temperate; overwinter	Lettuce	Temperate; overwinter
Beans, pole	Warm season	Melons	Hot season
Beets	Temperate; overwinter	Mustard greens	Temperate; overwinter
Bitter greens	Temperate; overwinter	Okra	Hot season
Broccoli	Temperate; overwinter	Onions	Temperate; overwinter
Broccoli rabe/rapini	Temperate; overwinter	Parsley	Temperate; overwinter
Brussels sprouts	Temperate; overwinter	Peas	Cool season; overwinter
Cabbage	Temperate; overwinter	Peppers	Hot season
Cardoon	Temperate	Potatoes	Warm season
Carrots	Temperate; overwinter	Shallots	Temperate; overwinter
Cauliflower	Temperate; overwinter	Sweet potatoes	Hot season
Celery	Temperate; overwinter	Radishes	Temperate; overwinter
Chard	Temperate	Rutabaga	Temperate: overwinter
Chayote	Warm season	Spinach	Temperate; overwinter
Cilantro	Temperate	Squash, summer	Warm season
Collards	Temperate; overwinter	Squash, winter	Warm season
Corn	Hot season	Tomatillos	Warm season
Cucumber	Warm season	Tomatoes	Hot season
Dill	Warm season	Turnips	Temperate; overwinter
Eggplant	Hot season		

Within these parameters, pay attention to the fact that specialized cultivars are bred to do well in a variety of climates and seasons; for instance, there are sweet peppers that are specially bred to produce high yields in cooler- or higher-altitude climates and lettuce that is bred to tolerate high heat. If you plant that particular lettuce cultivar in the winter it may not do well; to overwinter many vegetables you will need to select a winter variety. Such specialized cultivars can help you extend your season and have a more productive urban farm.

The temperatures referred to in the following table are needed during the main growth period of the crop. Seeds of warm and hot season crops can be germinated in cooler temperatures and seeds of cool season crops can be germinated in much warmer temperatures. Check your local newspaper, cooperative extension service, or a farmers' almanac for the climate conditions in your region.

IDEAL TEMPERATURE RANGES

IDEAL TEMPERATURE RANGES FOR MOST ANNUAL CROPS

Night air	55 to 75 degrees
Day air	65 to 85 degrees
Warm and hot season crop night air	Above 65 degrees
Cool season crop night air	Below 65 degrees
Night range for cool and cold season protected (greenhouse) crops	35 to 75 degrees
Night air temperature range tolerated by outdoor overwintering crops	25 to –20 degrees

DAMAGING TEMPERATURES

Frost damage possible for warm and hot season crops	Below 50 degrees
Killing frost possible for tropical plants	Below 50 degrees
Killing frost possible for warm and hot season crops	32 degrees
Growth-inhibiting high temperatures for most annual crops	Above 85 degrees
Growth-inhibiting high temperatures for cool season crops	Above 70 degrees
Heat damage	Above 105 degrees
Killing heat	115 degrees

GENERAL GUIDELINES FOR SOWING DATES FOR GROWERS IN TEMPERATE REGIONS AND REGIONS WITH COLD WINTERS

Sow early spring through midfall, depending on ground temperatures	Temperate crops
Sow early spring, depending on ground temperatures	Warm season crops
Sow late spring, depending on ground temperatures	Hot season crops
Sow summer through early fall for winter harvest and fall or early spring for spring harvest	Cool season crops
Sow summer through early fall	Overwintering crops

GENERAL GUIDELINES FOR SOWING DATES FOR GROWERS IN HOT AND TROPICAL REGIONS

Late summer through early spring	Temperate crops
Late summer through early spring	Warm season crops
Midsummer through early spring	Hot season crops
Midfall through midwinter	Cool season crops
Not applicable	Overwintering crops

HOW TO EXTEND THE HARVEST IN YOUR REGION

Your climate not only determines the crops you can grow and when you can grow them but those crops that will do best and others you might try as an experiment. Regardless of your region, if you have extended hard frosts you will not be able to grow some of the temperate and cool season crops without protecting them. If you have cool summer weather you will have a hard time growing hot season crops. Overwintering vegetables cannot be grown below –20 degrees. Here we give a general guide, but you will still want to learn about your USDA zone and consult other resources. See page 524 in the appendix for links to climate zone maps.

Cool season and overwintering crops can be grown to make use of cool and cold weather to extend the season. Even in the coldest north, overwintering vegetables enables you to harvest throughout the winter and/or make the early spring months, usually spent preparing for spring and summer harvests, productive (see page 232 for instructions on extending the harvest). Cool season crops—primarily peas—really can't be grown year-round. While it is possible to artificially create heat using passive methods, to cool the air and soil around crops would require a great input of energy, making it impractical. Similarly, in most of the country hot season crops can't be grown year-round as the day length shortens. It could be done with greenhouse heaters and lights, but the energy required makes it impractical.

CLIMATE CHANGE

As we write this book gardeners and farmers throughout the country are dealing with torrential rains, record droughts, new high and low temperatures, and unseasonably warm and cold weather. Recently we have experienced the worst drought on record where we live in Northern California, as well as the rainiest March on record, unseasonable temperature fluctuations in early spring (from near freezing to the eighties), high heat into October, and humid air—this last an extreme rarity in dry, coastal California. Could any of this be caused by global climate change?

Farmers know it's happening—we are outside all the time! We pay attention to the weather (we've never thought of it as a banal conversation topic). Since the current global climate changes are a reality that can't be altered much now, try to see it as exciting. What interesting weather might we have today? But the reality of global climate change affects how we plan for growing crops. Now it's more important than ever to understand how weather and climate affect plants, since frost dates and climate patterns might be changing radically from year to year and season to season.

Vegetables that are tolerant of both cool and warm weather, such as salad greens, cooking greens, root vegetables, *Brassicas*, and herbs should be a main focus of our farming efforts. Crop protection may also become more important as pounding rain, flooding, and high temperatures threaten plants. Building raised planter boxes and using row-cover materials and shade cloth may become more essential.

PART II

RAISING CITY VEGETABLES AND FRUITS

This section is devoted to growing vegetables and fruits. If you have an existing garden, read this section for new ideas about how to build beds, increase soil fertility, best practices for propagation, how to set up irrigation, and suggestions for growing fruit trees. This section also outlines ways to deal with pests and diseases in the vegetable garden and has tips for harvesting the produce you've worked so hard to grow. If you are prioritizing animals, skip to part III, starting on page 359, which goes over how to set up and maintain honeybees, poultry, rabbits, and urban goats on your farm.

Before we jump in, one thing to remember: Perennial optimism is the farmer's friend. To succeed as a farmer you must avoid perfectionism. Making "mistakes" is a big part of learning. Crop failure will probably happen: Your carrot seeds may never sprout because you forgot to adjust your watering system. Or a bug may eat all your cauliflower seedlings because you didn't cover them with floating row cover. Bad weather might ruin your tomato crop just as it's maturing. And yet, you must go on. As you become more experienced you will learn to avoid more and more pitfalls, but if you give up after a few failures you'll never get to the successes waiting just around the corner. As Voltaire said, "the perfect is the enemy of the good." So let's get out there and try!

CHAPTER 6

CREATING FARM BEDS

Once you have chosen your site, tested your soil, and planned what you'd like to grow, it's time to begin building farm beds.

SITE PREPARATION

If you are blessed with a weed-, garbage-, and rubble-free site, skip to page 95 (Creating Garden Beds). Otherwise, read on.

After assessing your land visually and with a dig bar as described in chapter 1, you will have an idea of what needs to be done to clean up your plot. First, remove or move existing perennials according to your layout plan. Next, clear away the top growth of weeds and brush down to the ground.

You don't need to dig your soil now. When you dig your garden beds you will remove remaining weed roots, underground garbage, rocks, bricks, and concrete chunks. If you have pernicious weeds (turn to page 280 for a list if you're not sure), solarizing the roots is your next order of business (see page 281 for instructions). This process can take up to a few months but is well worth it.

What to Do If Your Site Is Covered With a Concrete Slab, Blacktop or Gravel

If your site is covered over you have three options: 1) remove the slab or gravel; 2) build boxed beds (see page 103 for instructions); 3) create unboxed raised beds by piling up organic matter and compost on top of the slab or gravel. In most cases removing the material is the preferable choice. However, even if you can do the work yourself (jackhammer anyone?), the cost of disposing of the waste may be too much for your pocketbook. Research dump fees so you know what you're in for. The shock may stimulate you to find a way to reuse broken up concrete slabs in the garden to build structures.

Boxed beds work well if you don't have the time or money to remove the material. The downside to this method is that even the tallest beds don't allow plant roots to penetrate as deeply as they normally would.

Piling up organic matter is a great option if the slab or gravel area is large, the cost of removing the material or building boxed beds would be prohibitive, or if the soil quality below the slab is poor. This method has the same downside as boxed beds—lack of depth for roots.

Removing material

If you haven't opted to build on top of the existing material, you will need to remove it. If the area is large, an option—especially good for easy-to-cut blacktop—is to cut out and remove the material only in the areas in which you plan to plant, leaving the rest of the slab as pathways. Blacktop or concrete paths will save you a lot of time that would otherwise be spent weeding paths. They will also heat up your garden, which will be good or bad, depending on your climate. Removing gravel or blacktop is fairly simple, if laborious. They can be removed by hand with picks, jackhammers, shovels, and wheelbarrows. Consider hiring a backhoe, dump truck, and operator and the job could be finished in a day or two. Removing a concrete slab is a more technically complicated project.

Materials you may need for concrete removal:

- large and hand-held sledgehammers
- circular saw and diamond masonry blades
- chisels
- jackhammer

- hammer drill and masonry bits
- crowbars
- dig bar for longer prying leverage
- bolt cutter for cutting wire reinforcement mesh
- heavy-duty sacks, garbage bags, or garbage cans for collecting rubble
- flat shovels
- strong wheelbarrows
- 2 inch × 12 inch × 12 inch plank for running the wheelbarrow up into the truck
- truck
- eye protection, ear protection, gloves, and protective clothing and shoes

Steps for slab removal:

As you begin to break the concrete slab into chunks, remember that the maximum carryable weight for most people is about fifty pounds. No one should be a hero on this kind of project. With large pieces it's helpful for two people to lift the chunk and then for one to hug it to their middle to carry it. Once at your destination, either lower yourself by bending your knees and placing the chunk where you want it or drop it (with your feet out of the way). Don't bend over from the waist to lower it. Other safety measures include keeping children away from the site and wearing eye protection to guard against flying bits. Strikes against concrete are very loud—ear protection is a must.

1. Using the circular saw and a diamond masonry blade set as deep as your saw will allow, cut a grid of lines in the slab. Your plan of attack will be to break the material up into manageable chunks, lever them out with a crowbar or dig bar, and carry them to a wheelbarrow or truck.

2. Break up each grid section using sledgehammers or a jackhammer. This is hard work and may take a number of hours or days with rest breaks.

3. The slab will likely be reinforced either with rebar or wire mesh. Use the hand-held sledgehammer and a chisel to break up concrete to expose the wire or rebar.

4. Use the diamond saw blade to cut through rebar or the bolt cutter to cut though wire mesh.

5. Use the crowbar or dig bar to pry out pieces of concrete.

6. Move concrete and metal reinforcement to your truck and keep them separate, since you may be able to recycle the metal. Carry large pieces directly to the truck or place them in a wheelbarrow. If your truck bed isn't too high, you may be able to run your wheelbarrow up on a large wooden plank.

7. Use a flat shovel and bags or buckets to clean up small chunks.

Planting on Top of Concrete, Blacktop, or Gravel

If you are a squat farmer (working an empty lot without express permission), or one with a short-term lease, creating beds on top of concrete slabs, blacktop, or gravel rather than removing it is a good option. We've made amazingly productive farms using this method. The idea is to put down a mulch layer of wood chips, chaff, or straw over the concrete, then create long mounded planting beds of compost and manure over it. The ideal height of the mounds is about three feet. Due to the large volume of organic matter necessary to create your beds, we recommend this method for those who have access to a truck and large amounts of free recycled organic material, such as wood chips, straw, chaff, sawdust, manure, and compost. Having large truckloads of these materials delivered is worth the price if you can swing it. A front-end loader for moving materials to form beds is also extremely useful for larger sites.

1. Map out your beds and paths. It's best to make the beds 3 to 4 feet wide and as long as possible.

2. Plan your work so wheelbarrows or machinery can access the beds as you work. This means not spreading mulch in areas in which you will be using your wheelbarrow or machinery, i.e., creating each bed and path as you go so you have access.

3. Starting with the first bed, spread about a foot of mulch (can be of mixed materials) over the bed area.

4. Pile up layers of compost and composted manure on top of the mulch to about 2 feet to 3 feet high (remember that the material will settle). If you like, you can use raw, uncomposted manure about 6 inches thick for your first layer on the mulch. This will decompose by the time roots reach it. Rake the bed to form edges that gradually slope down to meet the surrounding paths.

5. Spread about a foot of mulch in the paths surrounding the first bed and in the area where you will form your next bed.

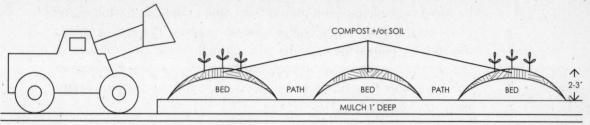

6. Repeat the process for the remaining beds.

7. Water the beds and paths with an overhead sprinkler to wet and help settle the material.

8. Plant your beds with crops, keeping in mind that the depth of your beds may require frequent amending to provide sufficient nutrients to deeper rooting crops.

CREATING GARDEN BEDS

To Box or Not to Box

Contrary to popular belief, the term "raised bed" refers to an unboxed, inground bed that has been cultivated and amended rather than to a wooden box bed. It is considered "raised" because the bed soil becomes higher than the surrounding soil through cultivation. Boxed or edged beds are those with permanent borders made out of building materials.

On land with soil that is safe for gardening, the benefits to boxing or edging beds are simply conceptual and aesthetic. A defined border discourages people from accidentally stepping in it; you may also feel more able to focus your attention when weeding, watering, and planting. These benefits are helpful for novice gardeners and spaces with many community members, children, or pets around. Or you may opt for edged or boxed beds for purely aesthetic reasons.

However, building borders or boxes is time-consuming, expensive, and often uses nonrenewable resources. Constructed beds must be cared for, repaired, and rebuilt from time to time. And while many people think that potting soil or straight compost is a great growing medium for plants, the opposite is actually the case: Garden soil—in the ground—is more complex and nutritive. If you choose to edge or box beds after they have been dug, you can find instructions beginning on page 101.

Creating Raised Beds

After your land is cleared, set up a sprinkler for a few hours the day before you plan to work, or if you have clay soil, two days prior. This will make working your soil much easier, especially if it is compacted. If, however, the weather has been excessively wet and the soil is waterlogged, wait a few weeks, since digging this soil can damage its structure.

Supplies you may need:
- hammer, stakes, and string
- buckets or cans for garbage and pernicious weed roots
- buckets or cans for compostable weed roots
- buckets for rocks or bricks you may want to save
- spades
- digging forks
- bow rakes
- trowels or hand hoes
- gloves and protective shoes
- compost or composted animal manure (amendment)

The day after watering, measure and stake out your beds, running string around the perimeter of each bed four to six inches from the ground. Use a flat shovel and a bow rake to level your paths, lowering bumps and filling in holes. This is really important, since holes and bumps only get more extreme with time and can pose a tripping hazard.

We recommend that you add from three to six inches of compost to the top of the soil and dig it in as you dig your beds. How much to add depends on your soil type and how much organic matter is already present in your soil. See chapter 8, page 147 for more information. To calculate how much compost or composted animal manure to get for 3½-foot-wide beds, measure the length of your beds and use the following table.

DETERMINING AMENDMENT NEED BY BED LENGTH

Desired Amendment Depth	Bed Length	Yards Required (1 yard = 27 cubic feet)†	Bags Required (2 cubic feet per bag)
3 inches	8 feet	½ yard	7 bags
6 inches	8 feet	1 yard	14 bags
3 inches	16 feet	1 yard	14 bags
6 inches	16 feet	2 yards	28 bags

†Note: If purchasing bulk materials, check with your supplier to find out if they sell half-yards.

Now it's time to choose a cultivation method to loosen the soil, remove debris, and incorporate compost. While we generally recommend double digging (see page 98), if your soil is on the sandy side and doesn't seem severely compacted, you may opt to simply turn over the top 8 inches to 12 inches with a spade or mattock, removing debris and mixing in the compost as you go.

Tilling

If your soil is completely clean of debris and pernicious weeds, you can use a rotary tiller if you wish, but in our experience tilling takes just as much work as double digging and is much less effective. Tilling is not recommended for severely compacted or clay soil, since it can damage the soil's structure. If you choose to till, one or more passes will loosen the soil and another will incorporate compost. We recommend you use a lightweight electric tiller, and till only within the confines of beds (you can even till inside boxed beds), not in paths, as you will make a lot of unnecessary path leveling and compacting work for yourself.

If your garden site is large and you plan to use field farming techniques, hiring someone with a tractor to scrape and till your land may be the only time-efficient method. Know that this will stir up dormant weed seeds and chop up pernicious weed roots, multiplying the number of weed plants. Planning to plant a weed-smothering cover crop directly after tilling can help.

Double Digging

If you have pernicious weeds, or if you plan to plant clover paths, you will also need to dig up the soil in your paths when you double-dig beds. After staking your beds, remove any weed tops that might have sprouted up since you initially cleared the land. Spread an even 3 inch to 6 inch layer of compost or composted animal manure over the surface of the bed. Measure the depth with a ruler to make sure you've added the right amount.

Double digging looks very orderly while in schematic form, but in reality dirt doesn't stick to square angles and straight lines. Use your visualizing skills to ensure that every square foot of the bed is turned, loosened, cleaned of weeds and debris, and forked below.

After the bed has been dug, use a bow rake to establish straight sides and ends. To avoid lumpy paths, it's important to bring dirt that may have rolled into them back into the beds. Using the flat side as well as the tined side of the rake, move soil so the bed is level, and carefully shape your bed to the desired shape (to choose an appropriate shape see diagram on page 100). If you plan to box your beds, simply rake the surface flat, keeping dirt away from the edges so the box will sit flat. From this point forward into eternity, never step in the beds you have created, since walking in it compacts the soil you have so carefully cultivated. If you must step in for some reason, use a piece of plywood to distribute your weight over it evenly.

DOUBLE DIGGING DETAIL

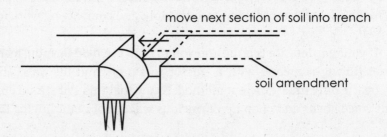

move next section of soil into trench

soil amendment

DOUBLE DIGGING

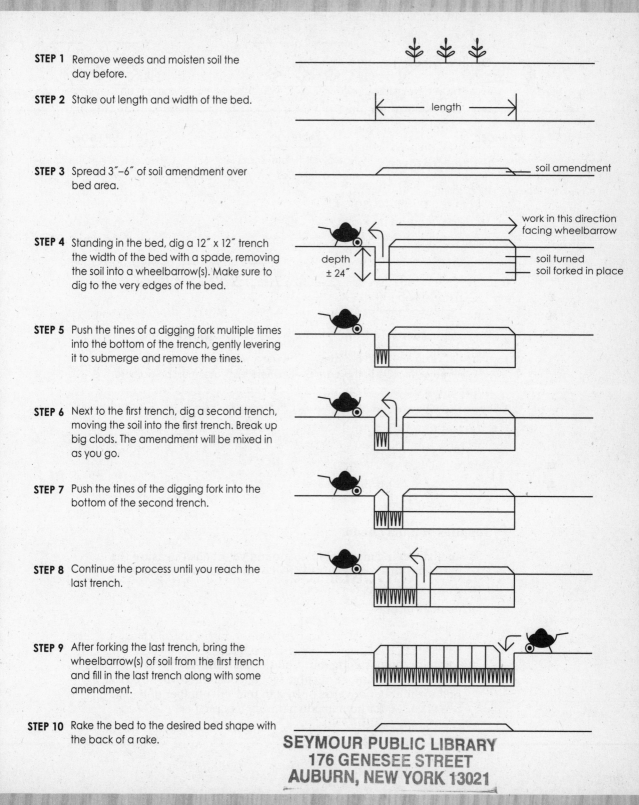

STEP 1 Remove weeds and moisten soil the day before.

STEP 2 Stake out length and width of the bed.

STEP 3 Spread 3″–6″ of soil amendment over bed area.

STEP 4 Standing in the bed, dig a 12″ x 12″ trench the width of the bed with a spade, removing the soil into a wheelbarrow(s). Make sure to dig to the very edges of the bed.

STEP 5 Push the tines of a digging fork multiple times into the bottom of the trench, gently levering it to submerge and remove the tines.

STEP 6 Next to the first trench, dig a second trench, moving the soil into the first trench. Break up big clods. The amendment will be mixed in as you go.

STEP 7 Push the tines of the digging fork into the bottom of the second trench.

STEP 8 Continue the process until you reach the last trench.

STEP 9 After forking the last trench, bring the wheelbarrow(s) of soil from the first trench and fill in the last trench along with some amendment.

STEP 10 Rake the bed to the desired bed shape with the back of a rake.

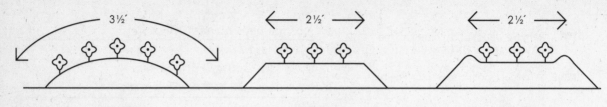

ROUNDED
- works well if water soaks in easily
- most efficient
- largest surface area for planting

TABLE TOP
- works well if water soaks in moderately well

LIP EDGE
- works well if water doesn't soak in well

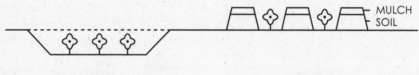

MULCH
SOIL

SUNKEN
- works well in extremely arid climates where rainwater must be gathered

WAFFLE
- works well in extremely arid climates where rainwater must be gathered

Edging Your Beds

Supplies you may need:

- enough edging material to go around your beds (measure the perimeter to know the linear feet of material you will need)
- wood or rebar stakes
- mattock or hoe
- a level
- a hard rake
- compost, composted manure, and/or planting mix to fill the boxes (calculate the number of cubic feet needed and divide by 27 to find the number of yards or by 2 to find the number of bags). See page 130 for homemade planting mix recipes.

Boxing your beds with urbanite, wood logs, bricks, stone, or cinder blocks

First rake some of the soil at the edges of the beds toward the center to ensure a flat surface for the edging material to rest on. Creating a narrow trench with a mattock or hoe for it to rest in will provide stability. After placing the material in the trench, fill the dirt back in around it, and compact it with your feet. Another way to stabilize edging material is by hammering in stakes every few feet on its outside. The soil in the box will keep the edging material from leaning inward, since the direction of the force comes from the weight of the dirt inside the bed. Just make sure your stakes won't be a tripping hazard. Hammer them flush with the edging material or top them with a used tennis ball (simply cut a slit in one side and slip it over the stake).

If dry stacking bricks, urbanite, or stone, put your first layer of material a few inches wider than the final bed width. To create stability you will angle the wall inward just slightly as you stack each layer. Use a string between two stakes at the height of the first layer to ensure that it's flat (it needn't be level per se). Set each subsequent layer just slightly in toward the inside of the bed, say, by ¼ inch or ½ inch. Stagger the material in the rows (courses) so that the joints are covered by a solid piece on the next layer. Only use cleaned bricks, since patches of mortar will make uneven layers (to remove mortar, a cold chisel, hammer, and a comfortable seat are all that's needed). Over time, bricks, rocks, or pieces of urbanite may fall out of place, so be sure to maintain the bed by restacking them.

If you don't need a particularly tall bed, edging with larger pieces of urbanite set in a trench on their thin edge is the easiest and most space-efficient method for this material. Set the pieces vertically in a trench and compress the soil around them firmly (see page 52 for illustrations).

To create edging with recycled or new corrugated steel roofing, cut 26 inch by 10 foot sheets in strips crosswise to make pieces that are 26 inches by 2.5 feet or 26 inches by 3 feet, depending on how high you want your edging. With a mattock or trenching shovel, dig a 2 inch to 18 inch trench and set your pieces in, overlapping by a few corrugations. Line the metal with fabric to delay rusting and rotting. To protect fingers from the sharp top edge, bolt a wiggle board sandwich to the top lip of the metal. A simpler way is to slit an old garden hose

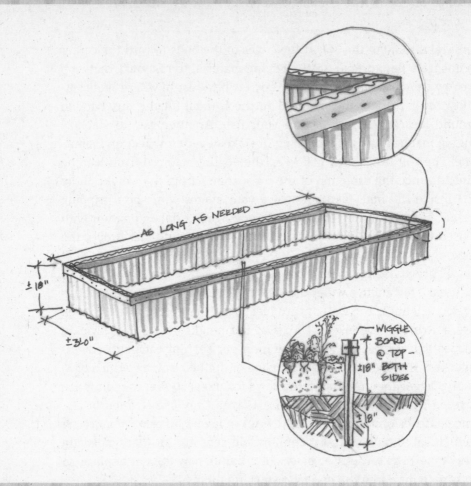

Labels within the illustration: AS LONG AS NEEDED; ±18"; ±3'-0"; WIGGLE BOARD @ TOP — ±18" BOTH SIDES; ±18"

Plan for a corrugated metal bed with the corrugations running vertically

lengthwise and put it over the top, bolting it in place every two feet or so.

Corrugated sheets can also be used with the corrugations running horizontally to build boxes with wooden 2 inch by 4 inch or 4 inch by 4 inch corner supports. To make a (very) tall box, use the full width of the sheets. To make a low box, cut the sheets in half lengthwise. Ends should be cut between 3 feet and 4 feet long. Drill holes in the metal with metal-drilling bits and bolt through the wood corner supports.

CORRUGATED IRON WICKING BED

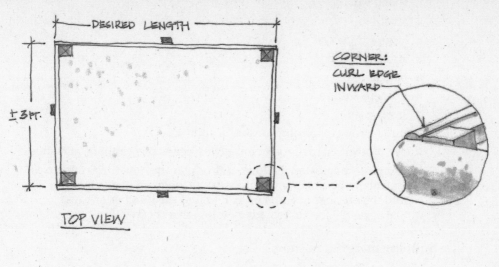

DESIRED LENGTH

± 3 FT.

TOP VIEW

CORNER: CURL EDGE INWARD

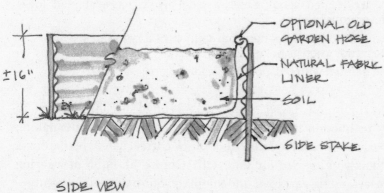

± 16"

OPTIONAL OLD GARDEN HOSE

NATURAL FABRIC LINER

SOIL

SIDE STAKE

SIDE VIEW

Plan for a corrugated metal bed with the corrugations running horizontally

CONSTRUCTING WOODEN BOXED BEDS

Supplies you may need:

- lumber
- drill-and-bit set
- circular saw
- jigsaw (optional)
- chop saw (optional)
- hammer
- cat's claw

- chalk line
- speed square
- level
- large bar clamps
- shims
- pencils
- framers' square
- sawhorses
- three-prong extension cords and splitter
- eye protection, ear protection, gloves, protective clothing, and shoes
- compost, composted manure, and/or planting mix to fill the boxes (calculate the number of cubic feet needed and divide by 27 to find the number of yards or by 2 to find the number of bags). See page 130 for homemade planting mix recipes.

Building material no-nos:

- pressure-treated wood (usually but not always has a "stapled" look and a greenish tinge)
- plywood or other glued-together laminates (if you can see layers that are not wood grain, it's glued)
- wood painted with old, possibly leaded paint

Many community colleges offer classes on basic construction and cabinetry techniques, so if you are a novice, please consider taking one. At the least be sure to know and follow correct safety precautions for any tools you are not familiar with. Get in the habit of wearing safety glasses, using earplugs, and keeping your hair tied up tightly.

The best height for boxed beds is between 4 inches and 21 inches. Unless you have a soil toxicity problem or an ability issue (back problems, etc.) we recommend your boxes be no more than 12 inches high. For boxes 8 feet or longer, cross bracing must be added at any side joins (lumber isn't generally longer than 12 feet; see diagrams on page 106). Stapling gopher mesh to the bottom of the box is wise, whether you think you have gophers or not. Build it and they often come. Place staples close together on the bottom edges of your boards so the critters can't burrow through.

Wood comes in standard dimensions, but the name of the dimension is not actually the size of the wood, since it refers to the boards before they are surfaced in the factory. Following are the nominal and actual sizes of lumber to help you decide what to use for the box height you want. For example three rows (courses) of 2 inch by 8 inch lumber would yield a 21-¾-inch–high bed, not a 24-inch-high bed.

Nominal Size (inches x inches)	Actual Size (inches x inches)
1 x 1	¾ x ¾
1 x 2	¾ x 1 ½
1 x 3	¾ x 2 ½
1 x 4	¾ x 3 ½
1 x 6	¾ x 5 ½
1 x 8	¾ x 7 ¼
1 x 10	¾ x 9 ¼
1 x 12	¾ x 11 ¼
2 x 2	1 ½ x 1 ½
2 x 3	1 ½ x 2 ½
2 x 4	1 ½ x 3 ½
2 x 6	1 ½ x 5 ½
2 x 8	1 ½ x 7 ¼
2 x 10	1 ½ x 9 ¼
2 x 12	1 ½ x 11 ¼
4 x 4	3 ½ x 3 ½

Salvaged wood tends to be quite hard. This is good because it is well seasoned and most likely has a tight grain, resulting in a longer life than fresh wood. If you are used to driving nails into fresh wood, you may end up with a pile of bent nails, split wood, and extremely frazzled nerves. To avoid this problem, always predrill the outside piece of wood of the two you are joining together (don't predrill your corner supports if using nails or screws). Wherever possible, join your projects with bolts, the largest washers you can find, and nuts. Water and soil contact degrade wood faster, making it likely that nails and screws will pull out. To drill for bolts, you will need to use a drill bit just slightly wider in diameter than the bolt. You will need an extra long drill bit to extend through the outside piece and the 4 inch by 4 inch corner support.

Avoid using painted wood. Leaded paint was banned in the United States in 1978, but since we don't date wood, it's nearly impossible to know whether the paint contains lead or not. In addition, contemporary unleaded house paint can still contain toxic substances.

Work on a flat surface such as a driveway or sidewalk. The most efficient way to construct your beds is to first build each end as a unit. Then precut and drill the side boards and any necessary cross braces. The assembled ends are light enough to be carried to the bed location

PLANTER BOX PLAN AND ELEVATION

Wooden planter box plan

- Boxes are placed on dirt (no bottom boards are required)
- Seat frame is attached to the lip of the box
- Additional 2″ x 4″ or 2″ x 6″ seat supports are shown

42″ wide planter box constructed of 2″ x 6″, 8″, 10″, or 12″ lumber (depending on wood available and desired height), with 4″ x 4″ corner supports.

PLAN VIEW: If desired length is greater than length of lumber, use bracing made of 2″ x 6″ ends with a 2″ x 4″ cross brace where side pieces join.

MATERIALS

- 2″ x 4″
- 2″ x 6″
- 4″ x 4″
- Laterals (sides) 2″ x 4″
- Stucko wire mesh
- Bolts, screws, + staples

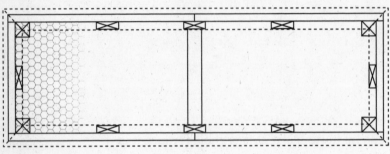

PLAN VIEW

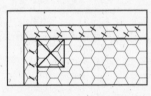

GOPHER WIRE APPLICATION DETAIL
ON BOX BOTTOM

SIDE VIEW: Box is either one or multiple boards high, depending on wood available and desired height.

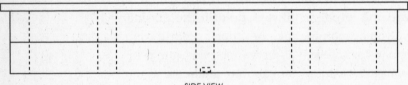

SIDE VIEW

SEAT DETAIL

BRACING SIDE VIEW DETAIL

BRACING SIDE VIEW DETAIL

How to correctly attach bracing boards

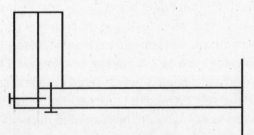

for assembly with precut and drilled side boards, corner supports, and brace pieces.

Construct your ends as follows:

Cut your pieces on sawhorses with a circular saw. For efficiency you can line up three or four boards side by side, mark them with a speed square and chalk line, and cut them all in one go. You may want to use a bar clamp to keep them squarely together, as if they were one piece of wood. To cut your 4 inch by 4 inch corner braces, use the speed square to mark a line, turning your board and "carrying the line over" to each side with the speed square. Even at the deepest setting your circular saw blade won't be wide enough to cut through a 4 inch by 4 inch, so cut on one side and then flip the piece over to cut through.

After cutting out all your pieces, set up a predrilling assembly line, making sure to place scrap wood under your work pieces so the drill doesn't hit the cement or dirt. The location of fasteners on the end boards should be offset from those on the side boards so the fasteners don't hit each other when you fasten your boards to the corner supports. Mark the locations of drill holes with a measuring tape. To be speedy, use a chalk line to snap a line and then use a small board the width of your lumber with drill holes where you need to put them on your pieces as a guide. Just mark through the holes on your guide. Remember when measuring that most lumber is not the full dimension listed.

If your box is longer than the length of your boards—say you have 12-foot-long boards but you want your box to be 18 feet long—you will need to brace it at the joins as shown in the diagram on page 109.

PROTECTING WOOD FROM MOISTURE AND SOIL CONTACT

There are a few safe methods to slow the rotting of wood. With all the ambient moisture outside, water needs to be able to escape from wood so it doesn't rot, so completely sealing it is not wise. Wood sealers you can buy at the hardware store are also probably toxic in the garden—they're bound to peel eventually, and you don't want that stuff in your soil.

Protecting Wood with Fabric

While it is nearly impossible to keep planter box wood dry, you can keep it from making contact with soil and the teeming organisms just waiting to munch on it.

One option is to staple flaxseed oil–soaked burlap, canvas, or second-hand blue jeans to the inside of your container. Linseed oil, a great wood sealant and waterproofing agent, is commonly made of flaxseed oil, toxic solvents, and even heavy metals. While plain flax oil is very thick, it's important to skip the toxins in linseed oil. Fill a container with the oil and submerge the fabric. After the oil has penetrated, wring it out over the container and let it dry as much as possible. This may take a few days. Be sure to spread the fabric out in a breezy, shady area so it won't self-combust.

If this seems too complicated for you, another option is to simply cut up old blue jeans or another durable cloth and staple it to the wood. This provides good protection and lasts quite a while before decomposing.

Another option is to seal the inside of your planter box either with the flaxseed oil itself or with wax. To use flaxseed oil just paint it on the inside surface of the wood, letting it dry as per the fabric instructions above. To use wax, rub some wax, either old candle stubs or bar wax, over the inside surface of the wood.

CONSTRUCTING GARDEN TRELLISES

We have found that constructing vertical trellises down the length of about a third of our beds really increases yields. Portable trellis cages that can be placed over vining plants are also very useful. For the highest yields, plan to install trellises in about one third of your growing beds.

Good crops to trellis include beans, peas, cucumbers, squash, melons, tomatoes—anything that vines.

If you've already constructed wooden planter boxes, building a simple wood trellis frame will be a cinch.

Placing your trellis 1½ feet from the north side of the box, rather than in the center, will make for more efficient use of bed space and an easier reach for the farmer. When building your frame, allow 3 inches to

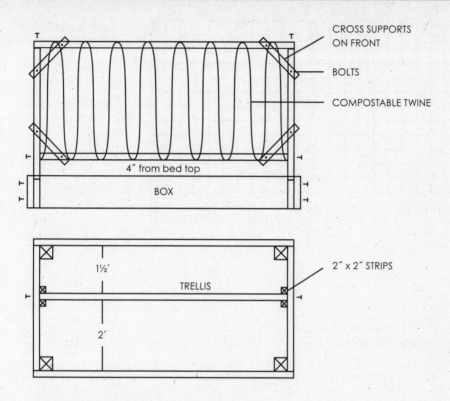

CROSS SUPPORTS
ON FRONT

BOLTS

COMPOSTABLE TWINE

4″ from bed top

BOX

1½′

TRELLIS

2″ x 2″ STRIPS

2′

- 2″ X 2″ strips create a groove for trellis to fit into
- Screw trellis into place after sliding into groove
- Trellis is constructed of 2″ x 2″ boards
- Predrill outside piece holes
- Use screws and bolts

4 inches of space between the bottom trellis board and the top of the bed for your hand and a ball of string to fit through.

When you're ready to trellis your plants, simply run cotton or jute string back and forth from the top to the bottom along the frame. At the end of the season cut down the string along with the plant waste and compost.

Trellises built from stakes and concrete-reinforcing mesh are very economical and easy to construct. Stock panels made of thicker,

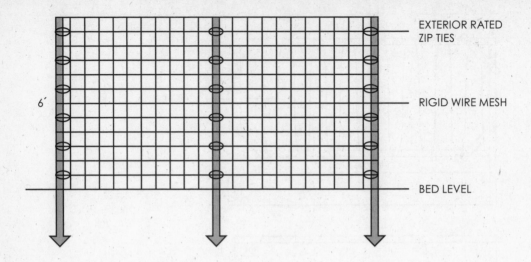

EXTERIOR RATED ZIP TIES

RIGID WIRE MESH

6'

BED LEVEL

MATERIALS

T-stakes or 8'–10' wood peeler core stakes
Zip ties rated for outdoor use
6" square wire concrete reinforcing mesh

plated wire (so it won't rust) make beautiful, durable trellises but are pricier. Both options are great for longer beds. Use either wood peeler-core stakes or T-stakes between 8 feet and 10 feet tall.

- Use a post-hole digger to dig 1 foot to 1½ foot deep holes, 4 feet to 6 feet apart.
- Using a tape measure, set your posts at the right height for your wire mesh. Use a level (post levels are particularly useful) to ensure your posts are straight.
- Tie a string between the first and last posts to set the rest at the correct height easily.
- To create a solid footing for your posts, either fill the bottom of the hole with gravel and then fill in with well-tamped soil or set the posts in concrete.
- Once the posts are set, run the wire mesh along them, securing it firmly with exterior-rated zip ties or plastic coated wire.

TOMATO, MELON, SQUASH, OR CUCUMBER CAGE

MATERIALS
6″ square wire concrete reinforcing mesh
Zip ties rated for outdoor use
Cover with polyethylene sheeting to create a warming season extender

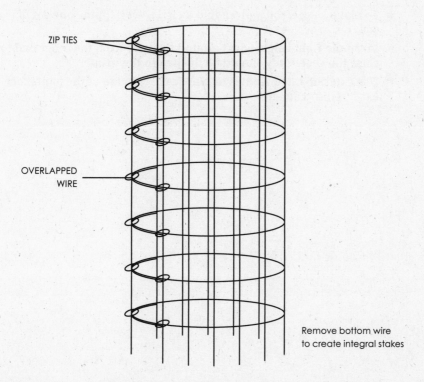

ZIP TIES

OVERLAPPED
WIRE

Remove bottom wire
to create integral stakes

STURDY WIRE CAGES

Purchased tomato cages are notoriously flimsy and often too short. Construct your own and they will last for years. Using concrete-reinforcing mesh, cut lengths 1 foot longer than needed for the desired diameter. An easy way to figure out the length you need is to place string in a circle on the ground, adjusting it until it is the size you want. About 2 feet in diameter is a good size.

❖ Make a knot in your string and use it to measure your lengths, cutting the wire with bolt cutters. Be sure to wear eye protection, gloves, and heavy clothing and shoes, since the wire can be dangerous.

❧ Cut the edge wire off the bottom end of each of your pieces, exposing 6-inch-long cross wires. These will form self-stakes at the bottom of each cage.

❧ Cut the edge wire off one side of each of your strips, exposing 6-inch-long cross wires. Bend these wires inward at a right angle, using a pair of pliers.

❧ To make a cage, roll a piece into a circle, overlapping one set of squares at each end.

❧ Wrap the right-angle wires around the wire from the other end to close the cage. They should be flush with the inside.

❧ Use exterior-rated zip ties or plastic coated wire at the joints for extra reinforcement.

W ← → E

CHAPTER 7

CONTAINER GARDENING

Container gardening may be your only option, or you may choose to use containers to make every bit of space on your farm productive. Either way, to be successful you will need to understand the unique challenges of growing in root-limiting vessels. Read on to learn how to harvest abundant yields using any manner of container, from terra-cotta pots to plastic bags.

CONTAINER GARDEN LAYOUTS

The main reasons for growing in containers is to add square footage to your garden by using spaces you normally couldn't grow in: above the beds, with suspended containers, placing pots in concrete-covered spaces, such as sidewalks, or growing on decks, patios, or fire escapes.

Using Above-Ground Space

Vertical tubes placed in beds or on cement can greatly increase your growing space while themselves having a very small footprint. Cylindrical tube planters with planting holes in the sides can be purchased or easily constructed out of wire mesh. Horizontal structures

Here are some ideas for homemade containers. Pictured from left to right: wire mesh planter, repurposed bathtub, two hanging tube planters, and burlap potato sacks.

suspended above beds, from walls, or on fences are also very useful for expanding square footage. They use the growing space you never knew you had—the air above your beds or fence line (see page 115 for illustrations and instructions).

Useful spaces for containers may also be found on walkways or concrete footings up against the south-facing wall of your house or other buildings. These spaces can be challenging for in-ground growing but are a great spot for containers. Walls can reflect light to plants and store the heat of the sun, creating warmer microclimates.

Using Patios, Decks, Rooftops, Fire Escapes, Window Ledges, or Balconies

Keep in mind that we are not structural engineers and cannot advise you of the weight-bearing capacity of your structure (see appendix 11

IDEAL PATIO OR DECK GARDEN (DECK 20'X18')

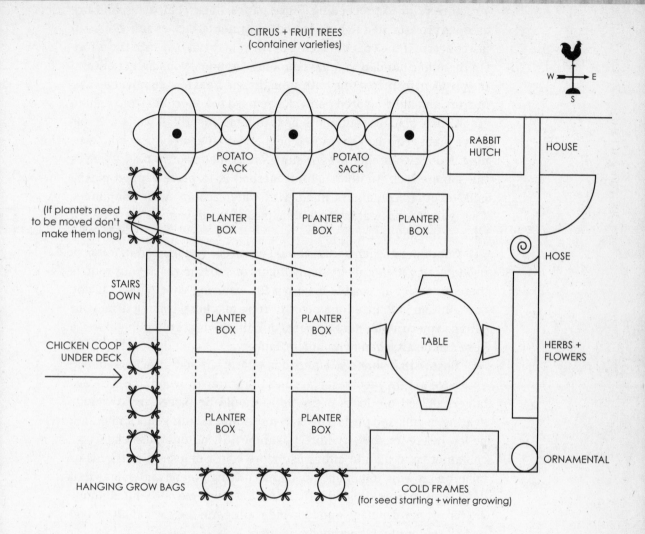

CITRUS + FRUIT TREES
(container varieties)

POTATO SACK

POTATO SACK

RABBIT HUTCH

HOUSE

(If planters need to be moved don't make them long)

PLANTER BOX

PLANTER BOX

PLANTER BOX

HOSE

STAIRS DOWN

PLANTER BOX

PLANTER BOX

TABLE

HERBS + FLOWERS

CHICKEN COOP UNDER DECK

PLANTER BOX

PLANTER BOX

ORNAMENTAL

HANGING GROW BAGS

COLD FRAMES
(for seed starting + winter growing)

POSSIBLE YIELDS

Eggs 1½ dozen per week (4 hens; decreasing in winter)
Rabbit Meat 72 lbs max (bred 3x per year = 24 rabbits @ 3 lbs each)
Tree Fruit 75 lbs per year
Vegetables/Herbs 97½–292½ lbs per year (97½ sq ft including tubes + potatoes)

DIMENSIONS

Planter Boxes 12″–18″ high; 3′ wide
Cold Frames 18″ high with pitched glass top; 2′ wide; 4′ long
Herb Planters 1½′ wide; 8′ long
Tree Containers 2′ x 2′
Potato Sacks burlap coffee sacks
Hanging Tubes 1′ diameter back plastic tubes with side holes for plants;
1′ diameter watering + drain pipe in center

for rooftop-gardening resources). Please evaluate the safety of your structure prior to proceeding. Since patios, decks, balconies, and fire escapes are attached to houses and apartments, lack of sufficient sun is a concern. If the space is facing north, it may not be worth the effort. On the other hand, if you have no other options, you can get around the problem by increasing both sunlight and heat in a number of ways. Mirrors and light-colored paint can be used to reflect heat and light to crops. Also, placing containers next to heat-absorbing walls will improve plant growth as the stored heat is radiated back during the evening. A dark-colored gravel or stone mulch below or in pots achieves the same effect. Another helpful method is to place capped plastic milk or other containers filled with water around your containers. They will absorb heat in the hottest part of the day and then radiate it back as the day cools.

Container garden access should be carefully considered. You may need to carry materials up many flights of stairs to reach your roof. Is there an easily accessible hose bib for watering your plants? If not, one solution is to tie a long piece of rope to a hose coming from your yard so you can pull it up as needed. With the right plumbing parts, a hose can be attached to an indoor faucet.

Ensure that your containers don't harm support structures due to excessive weight, incorrect placement, or water-to-wood contact. Containers placed on decks or rooftops should be placed over weight-bearing beams and posts. The containers and planting mix should not be too heavy for the structure to support. If in doubt, use shallower containers and plan to either use more water-delivered nutrients or focus on growing immature salad and cooking greens and onions that don't require a great root depth. You will need to elevate your containers for proper drainage and to avoid rotting the deck wood. Bricks on end or cinder blocks are good options.

WHAT TO GROW IN YOUR CONTAINER GARDEN

Everything from fruit trees to potatoes can be grown in containers. Finding the right varieties will increase your yields. Choose fast-maturing cultivars or "baby" vegetables. A focus within the commercial seed industry on breeding container-appropriate cultivars has yielded great new options for the microfarmer. Many of the seed resources listed in the back of this book sell container varieties. In addition to seeds specifically bred for containers, look for phrases like "bush habit" or other specific size descriptors.

Do not forget fruit trees in your plan. If cared for they will provide delicious food, beautiful blossoms, and soothing greenery. The key to containerized fruit trees is selecting the right varieties. Citrus trees generally do well, and for deciduous fruits, select genetic dwarf or superdwarf varieties. Fruiting shrubs will also yield well, but most berries are quite low yielding or require special soil conditions.

Because your container garden is likely to be smaller than the average backyard or empty lot farm, in order to grow enough to make meals, choose only a few crops to grow at any one time. Following are some guidelines:

- Many warm and hot season summer fruiting crops, such as tomatoes, squash, and cucumbers, will provide you with a nice harvest from a single plant.

- Fresh herbs aren't needed in large quantity. Many container gardens have too many perennial herb plants that mostly go unharvested. Select only those perennial herbs you use often.

- A number of plants are required to make a nice-size bunch of chard, collards, and kale. If you don't like them mixed, grow only one type at a time.

- Root vegetables require many plants to yield enough to make meals. We suggest choosing to grow only one type of root vegetable per growing cycle. For example, grow radishes in the early spring, carrots in the late spring, and beets in late summer for overwintering.

- Potatoes grow well in pots. Simply fill the pot with six inches to a foot of soil, plant seed potatoes (see page 189), and continue to add soil, leaving 6 inches of the stems and leaves exposed as they grow. Dump out the pot once the tops bloom and die back, and harvest tubers for eating and seeding another pot.

- Beans and peas also require multiple plants (beans less so) to yield much. If you trellis pole-type beans or peas at the back of your containers with other crops in front of them, your garden will be more productive.

- Cut-and-come-again salad mixes do well. Unless it's for fresh eating as part of a salad mix, forget spinach, since you will need large amounts to make a meal.

Container Depth

Annual vegetable plant roots go deeper than we think—3 feet to 10 feet or more. In the ground, plant roots grow deep to forage for nutrients and water. Generally, the majority of nutrients and water are absorbed in the upper sections of the root zone (70 percent of water and nutrients in the top half), but if the soil is poor or water scarce, a deep root system guarantees survival.

Growing vegetables successfully in containers requires much more care than growing them in the ground: Roots can't forage for water and nutrients in shallow containers. When supplies are exhausted, stress sets in, leading to poor growth. You must consistently provide the right nutrients and plentiful water to avoid stunting.

Shallow rooting crops have roots extending between 16 inches and 2 feet deep. Moderately rooting crops have roots that extend between 2 feet and 5 feet deep. Deep rooting crops have roots that extend between 5 feet and 8 feet (or more). The majority of crops fall in the moderately rooting category (roots between 2 and 5 feet deep). Yes, radishes, lettuce, and parsley are all moderately rooting though people often think they are shallow rooting. Tomatoes and corn are deep rooting. The only truly shallow rooting crops are onion family crops and celery.

A pot with one quarter or less of the average rooting depth of the plant will need to be fed weekly with nutrient solution. A pot with one half of the rooting depth of the plant will require little or no extra feeding as long as your planting mix contains nutrients. Regardless, in the second season, you will need to replace your planting mix or fertilize. Most crops have moderately deep root systems, so to reduce the need for fertilizing, most of your containers should be near 21 inches deep. A five-gallon bucket allows for 14 inches of soil, and will work for many crops as long as the planting mix is rich and you attend to replenishing nutrients over time. For tomatoes and trees think recycled garbage can. See page 194 for information on fertilizing your crops.

Container Types

Containers can be found, constructed, or purchased. Our mantra should be clear to you by now: Find it free, invent it, create it, reuse it. Using found materials doesn't mean your containers have to be ugly.

You can find and build aesthetically pleasing containers from free or cheap materials. We list here the most useful containers and materials in each category.

Found containers

When looking for found materials, salvage yards, secondhand stores, big trash pickup days, garbage transfer stations and landfills (but beware: Sometimes employees don't allow you to take garbage from the dump), and garbage areas behind businesses should all be stops along your journey. Other great resources are freecycle.com, craigslist.com, and state reuse network programs (see appendix 11).

If some of these materials seem puzzling, never fear. Construction plans follow.

The best found containers:

- five-gallon buckets
- old municipal recycling bins
- old garbage cans
- old barrels of any kind (you might have to saw them in half)
- old terra-cotta or other types of plant pots (make sure they are sufficiently deep)
- used nursery pots (five-gallon size or greater)
- used wooden nursery-tree pots (reassemble and add a bottom if needed using scrap wood)
- sections of used plastic sewer pipe cut to length
- plastic sacks (find thick, clear plastic sacks used for heavy materials such as gravel; otherwise they'll tear)
- burlap sacks (coffee or cocoa bean sacks are ideal)
- old bathtubs or laundry sinks
- food-grade industrial plastic containers (check with grocers or recycling companies)
- recycled galvanized animal (stock) watering troughs
- hardwood crates such as those used to transport granite countertops

If you can't find them for free, containers can be purchased used or new. Check garage sales, newspaper and online classified listings, secondhand and antique stores, and garden or pottery store sales.

The best purchased containers:

❖ terra-cotta or glazed ceramic plant pots (make sure they have drainage holes)

❖ half wine barrels

❖ galvanized animal (stock) watering troughs

❖ cylindrical cardboard concrete forms (cut to height; add a bottom if necessary)

❖ culvert pipe (cut to height; add a bottom if necessary)

Container and building material no-nos:

❖ lacks drainage and you can't make drainage holes yourself

❖ pressure-treated wood (usually has a "stapled" or green look)

❖ plywood or other glued-together laminates (if you can see layers that are not wood grain, it's glued)

❖ too heavy to move when empty

❖ too big to fit through your door or your gate, or to go around corners on the way to its home

❖ painted with old, possibly leaded paint

Refitting Containers

When you consider a container, examine it to ensure there are no large cracks or gaps. If there are, and you still want to use it, you can line the container with burlap, newspaper, or screen mesh to keep planting mix from falling out.

You may need to cut some plastic or wood containers, such as barrels or industrial ones, in half, or cut off the top to make them usable. This is most easily accomplished with a drill and a jigsaw. Drill a hole wide enough for the jigsaw blade. Take a chalk line and wrap it around your container, lining it up with previously made measure marks. Then snap the line in a few places to mark where to cut. It may take practice to cut straight, especially on round containers. Having a friend hold the container will help if it can't be clamped down.

Make sure to provide sufficient drainage. Most pots intended for plants will have a drainage hole already. It helps drainage in these pots to put a curved shard of a pot in the bottom so compacted dirt won't plug up this one hole.

Use a drill or hammer to create holes in plastic, wood, or metal containers without drainage holes. Instead of making holes on the bottom of the container, the most ideal scenario is to drill holes in the sides up to about one inch from the bottom. The holes should be staggered and be from one to three inches apart. Make sure the holes are not too tiny or too large—⅜ inch to ½ inch is fine. If side holes aren't workable for some reason, drill holes on the bottom. Be sure to put a piece of scrap wood below the container to protect your work surface. For metal containers use a metal bit, or simply hammer in and remove a nail working from the outside with the container upside down.

Old bathtubs, claw-foot or not, and old concrete or enamel laundry sinks make ideal containers. There is only one drain hole in bathtubs. This will provide plenty of drainage as long as it doesn't get plugged with compacted planting mix. To ensure proper drainage, be sure to make a "drain rock sandwich" in the bottom: Place a few inches of gravel in the bathtub. Cover it with a layer of burlap or shade cloth to keep the soil from filtering into the drain rock, and then fill with planting mix. You may want to hook up the drain to tubes flowing into a barrel in order to reuse the water.

CONSTRUCTING CONTAINERS

Building your own containers gives you the freedom to provide the right rooting depth and to custom build for the dimensions of your site, maximizing the use of space. Built or constructed containers can be made of free or cheap materials or of purchased materials.

In constructing containers the main considerations are:

- ease of construction
- sufficient height
- sufficient drainage
- fits through doors and gates
- solidly built and reinforced to support the weight of planting mix
- safe (no splinters or rusted nails protruding when completed)
- ergonomic height and width
- no paint if using salvaged wood
- Bottom or no bottom? If on a concrete patio there's no need, but you may want to sandwich a layer of gravel between the bottom and the concrete to keep soil from washing out of your container

Constructing Wooden Containers

If your space can support the weight, containers built of recycled wood are one of the most attractive and space-efficient options. The following plan allows for drainage, elevates the box off the floor, patio, or roof, and can be built with an optional seat frame (see page 106 for the seat detail). You can choose the exact height, width, and length that fits your space best. For general instructions on building with wood, see page 38.

PLANTER BOX PLAN AND ELEVATION FOR CONTAINER GARDENS

- For boxes placed on decks, patios or roofs (no gopher wire necessary)
- If smaller boxes, you can use 2″ x 4″ corner supports
- For bottom boards, use 1″ x 4″, 6″, 8″
- Fill bottom of box with 1″ of drain rock covered with filter fabric or burlap

42″ wide planter box constructed of 2″ x 6″, 8″, 10″ or 12″ lumber (depending on wood available and desired height), with 4″ x 4″ corner supports.

PLAN VIEW: If desired length is greater than length of lumber, use bracing made of 2″ x 6″ ends with a 2″ x 4″ cross brace where side pieces join. Add a lip of 1″ x 1″ or 2″ x 4″ screwed to the inside bottom of the box for bottom boards to rest on.

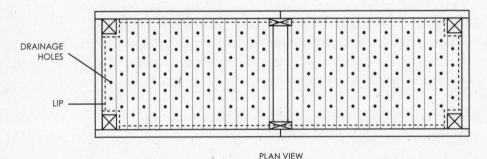

DRAINAGE HOLES

LIP

PLAN VIEW

SIDE VIEW: Box is either one or multiple boards high, depending on wood available and desired height.

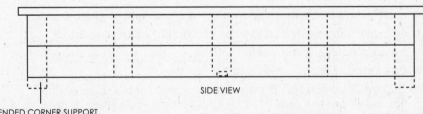

SIDE VIEW

EXTENDED CORNER SUPPORT
(3″–4″ to elevate off deck or roof)

Note that in a container situation you need a bottom with sufficient drainage holes.

Constructing Containers Out of Unconventional Materials

Metal

Salvaged corrugated-steel roofing and sheet metal are wonderful materials for building planter boxes, since they don't eat up growing space with their own bulk like wood and are fairly long lasting (line metal boxes with fabric to extend the life). If you are a metal worker, bending sheets of corrugated roofing to form an oval or round bed is a very simple solution; you can bolt the overlapping edges together. If you're not a metal worker, a simple boxed bed can be constructed using corrugated or sheet metal (cut by the supplier) and 4 inch by 4 inch corner posts. Simply bolt the sheets to them. They are rigid enough to require no more structural support (see page 102 for illustrations).

Aluminum and corrugated-steel culvert and storm drainpipe can be found used or free to make circular beds. It won't degrade in your lifetime and comes in diameters hard to even imagine. You will need to know or learn how to cut the pipe to the height you want. Attach a lip of wood or some other material to protect you from the sharp-cut edge.

Cement column-form tubes

You can easily adapt cardboard cement column-form tubes to make beautiful cylindrical planters. Contractors use these thick cardboard cylinders as forms for poured concrete. While they will degrade over time, they are quite durable, and some brands are waterproof. To increase their life, staple fabric on the inside. They come in diameter sizes from 6 inches to 36 inches, the latter being just about the perfect size for a bed. Since they are sold by the foot by contractor supply companies (for obvious reasons, these forms don't come used), you can cut them to any height with a circular or jigsaw, fill with planting mix, and plant. If you need a bottom, make a square frame covered with planks of wood set ⅜ inch to ½ inch apart for drainage, and set the cylinder on top.

Hanging polyethylene bag planters

The cutting edge of urban farming is in the so-called third world. City dwellers in developing countries have been growing stuff for years using whatever they can get their hands on. Necessity being the mother of invention, many fabulous ideas have resulted. We've been keeping up with the times and have seen and used many of these inventions ourselves. Black polyethylene cylindrical tubing—just like a black plastic garbage bag but formed in a cylinder—is used to create vertical planting tubes that can be hung from railings, windows, or an unused clothes rack. A watering/drainage tube is cut from PVC pipe and drilled all over with holes. This sand- or rock-filled watering/drainage tube is stuck down the center of the bag which is tied at both ends after filling with planting mix. Slits are cut in the sides. Then seedlings are planted in the slits. Nutrient solution and water are poured in the top of the tube, permeate the planting mix, and drip back into a receptacle at the bottom.

Grow bag made out of tubular black plastic. Note watering funnel to access center tube.

Plastic bag planters

Plastic bags, especially heavy-duty bags used to hold stone or other heavy materials, can be used in a similar way. The only differences are that the bags sit on the ground, and a few holes are cut in the bottom and up the sides a few inches for water to drain out. Plastic and burlap sacks can be used as planters for anything but are especially great for growing a sack of potatoes. Since potatoes like to be hilled as they grow, roll the sides of the sack down and fill with 1 foot of soil, plant the seed potatoes, and, as they grow, unroll the sack material and cover the stems and leaves with dirt, leaving the top leaves exposed. After the potatoes have flowered and died back a bit, simply upend your bounty.

These bags can also be used to grow veggies. In this case a vertical

PVC tube with holes drilled down the length and filled with sand is inserted prior to filling with planting mix to help evenly irrigate the lower parts. Then slits are cut in the sides of the bag 6 inches to 12 inches apart in a diamond or staggered row pattern, and seedlings are poked in. Water and nutrient solution are poured into the top of the tube. If using drip irrigation, you can put a length of emitter line (a piece of tubing with tiny holes for water to dribble out) inside the tube before filling it with sand.

"Tabletop" organoponic planters

If your deck or roof can't take much weight, using shallow table beds filled with a lightweight, nonnutritive mix may be your best option. Similar to hydroponic growing, the nutrients are delivered via water solution each time the bed is watered. The height is ergonomic and works especially well for wheelchair gardeners. You can easily construct these tables from deconstructed pallets. If you need to protect the roof or deck from water, line the box tabletop with plastic and put a tube in one corner to create a drain that leads into a bucket. If not, simply line with burlap or used blue jeans.

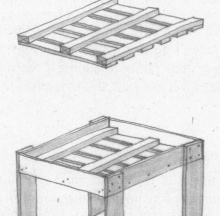

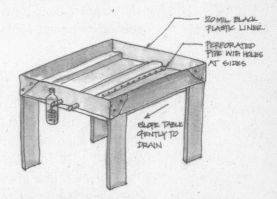

20 MIL BLACK PLASTIC LINER

PERFORATED PIPE WITH HOLES AT SIDES

SLOPE TABLE GENTLY TO DRAIN

Step 1: Use one pallet for the bottom of the box and use another deconstructed pallet for the sides and legs.

Step 2: Line the box with black plastic and construct drain tubes. Note bottle attached to the drainage tube for collecting irrigation water.

PVC planters

PVC pipe makes great horizontal-trough and vertical-tube planters. Unfortunately, it leaches toxic elements into the environment, but if you find it used, the leaching has already taken place, so it will probably be safe for your use. Food-grade PVC pipe is considered safer but harder to find. Horizontal or vertical grow tubes can be constructed easily out of PVC pipe 8 inches or greater in diameter. Simply cut circular holes in the top or sides for the seedlings and drill drainage holes in the bottom or end cap.

Hang horizontal tubes off of a well-constructed existing fence or wall using heavy-duty hooks and chain. Hanging them at the same height on both sides of the fence will balance the weight. If you lack existing structures, construct one by setting metal post-holders and posts or peeler core stakes in concrete about 4 feet to 5 feet apart. Suspend tiers of tubes from both sides of the stakes to balance the weight. This setup can be built right over existing garden beds to expand your growing area, since they don't cast much shade (see illustration on page 44).

Wire-mesh planters

You can also construct vertical cylinder planters out of burlap-lined wire mesh. If you're not using them on a deck or patio, carefully level the ground they sit on well so they don't fall over. Adding a central watering/drainage pipe filled with gravel and sand is advised (see illustration on page 125).

Subirrigated planters (SIPs)

Self-watering or subirrigated planters (SIPs) are all the rage, and rightly so. The concept is that water is wicked up into the plant container through perforations from a water-filled reservoir below. The reservoir is filled via a vertical watering tube inserted down into it through the center of the planting mix. As we mentioned earlier, providing sufficient moisture is a seriously limiting factor in container gardens. With SIPs the water is always there for the plant to use. You can easily make your own SIPs out of found or purchased containers; for instance, five-gallon buckets or plastic storage totes. The easiest design requires two five-gallon buckets and a few other readily obtainable materials.

SUBIRRIGATED PLANTER
(Made From Two Nested Five-gallon Buckets)

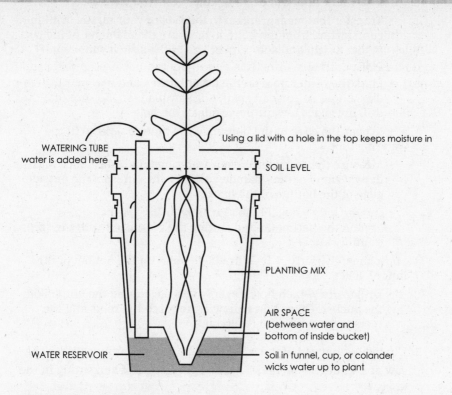

WATERING TUBE
water is added here

Using a lid with a hole in the top keeps moisture in

SOIL LEVEL

PLANTING MIX

AIR SPACE
(between water and
bottom of inside bucket)

WATER RESERVOIR

Soil in funnel, cup, or colander
wicks water up to plant

Materials and tools:

* two five-gallon buckets
* one 1-inch diameter watering food grade PVC or copper tube,
  3 inches taller than the bucket
* one sixteen-ounce or twenty-four-ounce rigid plastic drinking cup
* permanent marker
* drill
* ¼-inch drill bit
* jigsaw
* utility knife
* tin snips—a hand tool that resembles scissors but can cut metal
* hole saw (not required)

1. Drill the following holes:

* a hole in the center of the bottom of the inner bucket to fit the
  plastic water-wicking cup
  * Set one bucket inside the other.
  * In front of a light source, make a mark on the side of the outer
    bucket where you can see the bottom of the inner bucket.

- Measure the height from the bottom of the outer bucket up to your mark (this will be your water reservoir height).
- Transfer that measurement to the side of your plastic drinking cup, starting from the bottom. Measure the diameter of the cup at that height and add ⅛ inch. This is the size of your center hole.
- Mark the center hole on the inner bucket, and use your drill to drill a hole, so you can get the jig saw blade in. Be sure that the drill hole goes toward the inside of the hole.

✣ a hole near the edge of the bottom of the inner bucket to fit the watering tube

- Measure the diameter of your watering tube. Add ⅛ inch and draw a circle of that diameter about 1 inch in from the outside edge of the bottom of the bucket.
- Cut the hole using a hole saw or by drilling ¼-inch holes around the perimeter and cutting the plastic out with the snips or utility knife.

✣ approximately thirty ¼-inch drainage holes on the bottom of the inner bucket

- Drill thirty ¼-inch holes, evenly spaced, around the bottom of the bucket. These holes allow air to reach the roots and the roots to grow into the reservoir.

2. Cut the bottom end of the watering tube with the utility knife or jigsaw at a 45 degree angle, so it won't get clogged when sitting in the bucket.

3. Make four vertical slits with the utility knife, evenly spaced around the sides of the plastic cup, making sure not to cut through the bottom. Once the cup is filled with potting mix, water will be wicked up from the reservoir through these slits.

4. Put the bucket with the holes inside the other bucket. Using the mark you previously made on the outer bucket (marking the reservoir height), mark and drill two overflow holes on either side of the outer bucket about ¼ inch below the bottom of the line. This will ensure that the bottom of the inner bucket will always have air access.

5. With the buckets nested, drop the angled end of the water tube into the hole along the edge of the inner bucket. Fill the plastic cup with wet planting mix and set it inside the large hole in the center (it's okay if it sticks up above the inner bucket).

6. Fill the inner bucket with planting mix, watering it and pressing down every so often so it's fairly compacted.

7. After you plant seeds or seedlings, water them in by saturating the planting mix from above (not by using the watering tube). This is the only time you'll water from overhead. After the first watering, supply water through the tube. Check every few days to make sure the reservoir is full. Nutrient solution can also be added through the tube if your planting mix isn't rich enough.

CONTAINER PLANTING MIXES

Most books and articles on container gardening recommend purchasing bagged soil-less mixes from gardening stores as a growing medium for container plantings. The next step after planting is to begin a fertilizing schedule using organic or nonorganic liquid or powdered fertilizers. This method is certainly one of the options for container gardeners, but not the best. Using manufactured nutrients makes it easy, but we know you're up to the challenge of reducing your carbon footprint by reducing your reliance on industrially produced materials.

Purchased potting mixes

We recommend you buy potting mix in bulk at your local bulk rock and soil supplier or create your own planting mix and fertilizer out of free waste materials. If you do buy potting soil in bulk, ask which is their best veggie mix and get a list of the ingredients, so you will know if and when you will need to add more nutrients. As we've said in previous chapters, soil is a complex geological formation teeming with millions of organisms, minerals, and organic matter. Potting mix made of relatively inert ingredients, such as decomposed wood, peat moss, perlite, vermiculite, and coconut coir, doesn't come close to approximating this diversity. Although some of these products come with added fertilizer or manure, many are intended to be a nutrient-free holding medium for plant roots.

Garden soil

If you have access to good topsoil, by all means use it to fill containers. Rather than using it straight, mix it with a few other ingredients to improve moisture retention and drainage, such as sand and rice hulls.

The idea that containers filled with soil won't drain is simply a misconceived blanket statement. Pure clay soil would be a bad choice for containers, but it is fine to use a balanced garden soil, even if it drains slowly. Soil actually has some benefits for containers, since it retains moisture better than mix and contains mineral nutrients that release slowly and can be lacking in potting mix. With soil, you will still need to periodically replenish nutrients by adding compost, composted manure or cover crops (for nitrogen), and/or nutrient solution.

Homemade potting mixes

There are a variety of homemade mixes to use in containers that have nutrients plants need for a growing season. When you plant a second crop, you will want to observe your plants to see if they fail to thrive. Once you see signs of a lack of nutrients, either replace the mix or fertilize with compost or composted manure or cover crops (for nitrogen). Another option is to use homemade nonnutritive mix and make your own nutrient solution. In this case you won't need to remove the mix and refill your containers, but a periodic loosening of the mix will be helpful.

Compost, either homemade or not, is an important ingredient for potting mixes. If you live in an apartment without a place to make compost it can be difficult to find it for free. Municipal and county waste management companies that make compost from yard waste and kitchen scraps can be persuaded to provide free compost to community gardeners as part of their public relations/community benefit programs. This is already happening in some cities—make it happen in yours, and apartment farmers will be set for compost.

Focus your efforts on finding free materials that are wastes from local industry or agriculture. For example, in California rice is an important agricultural product, so rice straw and hulls can be had for free if you're willing to pick them up. Peanut hulls are a food-processing waste in the South and can be found on barroom floors throughout the nation. Coffee grounds are on offer daily at your local café.

If you make an arrangement to pick them up, stick to your word or you'll be at risk for burned bridges. Major home improvement stores often give away "split" bags of compost, soil mix, or manure if you ask at the beginning or end of the day. If you have to purchase materials, buy them singly and mix them together yourself rather than buying a premade mix. They will be more economical, especially if you buy them in bulk to split with friends.

While the common potting materials peat, vermiculite, and perlite are very effective in improving moisture retention and proper drainage, they are either endangered, as in the case of peat, or produced through high-energy-input mining and manufacturing processes, as in the case of vermiculite and perlite. Coconut coir, made from recycled coconut hulls, sand, and grain or peanut hulls are effective and much more environmentally sustainable.

Homemade compost can introduce problems, such as soil-borne diseases, weeds, and fungal problems (even mushrooms). You can ensure a weed-seed-free mix by making sure the compost and/or manures you use reached high enough temperatures during decomposition to kill seeds. Or, just weed your containers. It's really not so difficult.

Recipes

The general concept is that you want to make a mix that will hold plant roots and retain moisture while allowing for sufficient drainage. Feel free to experiment with proportions of ingredients based on what is readily available in your region. Plants are programmed to grow and reproduce their kind by flowering and fruiting, so they will often still do quite well in less than perfectly ideal conditions. For instance, we've found that light-feeding crops such as onions grow fine in pure composted horse manure, while heavy feeders such as broccoli don't. When horse manure's all we can get, we work with the crops that thrive in it.

Consult pages 174 through 176 for container mix recipes.

NUTRIENTS

To avoid having to change your mix out regularly, you can provide nutrients by mixing them into the top few inches of soil or delivering them in water solution. A number of them can also be added when making planting mix to provide major and minor nutrients that may be otherwise lacking.

While light feeding crops can be grown in the same potting soil successively without fertilizing, heavy feeders (crops with higher nutrient needs) will benefit from fertilizing over the entire season, even if you initially fill your pot with a rich planting mix. Without periodic fertilizing it can be hard to get a cauliflower or broccoli to head up in a pot. The nutrient needs of crops are generally related to the part of the crop that is eaten. The formation of roots, leaves, blossoms, and fruits all require different nutrients at different stages of plant growth. We like to group crops in the following way to get a general sense of these needs:

> *Heavy feeders:* fruiting and leaf crops
>
> *Light feeders:* root vegetables and onion family crops
>
> *Nitrogen-fixing:* legumes

Generally, nitrogen-fixers do not need to be fertilized since they absorb nitrogen out of the air. You can find many nutrients for free in and around cities. Many potential plant nutrients are lost to the landfill, and some are even major polluters of our waterways and water tables. If you have access to the raw ingredients, you can make your own fertilizing "teas" out of compost, manure, worm castings (worm poo), or seaweed. If you've got a fish-processing plant in your city, homemade fish emulsion is an option for the brave of heart.

If you do decide to purchase nutrients, we urge you to focus on those made of waste materials such as fish emulsion, oyster shells, and bone and blood meal. These are all great fertilizers. Especially important times for fertilizing are flower and fruit set. Yellowing, curled, and mottled leaves are a telltale sign of nutrient deficiency (confusingly, these signs can also indicate a problem such as a plant disease). Overfertilizing will result in overly lush green growth that makes plants more attractive to insects and diseases at the expense of flowering and fruiting. If feeding with nutrients in water solution, make sure that the container mix is moist prior to fertilization. If plant roots take in nutrient mix when they are water-deprived it can "burn" them, resulting in stunting.

Nutrient Sources:

- bone and blood meal
- coffee grounds and/or tea leaves top dressing (see page 133 for description)
- compost tea
- compost top dressing
- composted animal manure top dressing
- composted seaweed top dressing or seaweed tea

⚜ cover cropping with nitrogen fixing legumes

⚜ diluted human urine

⚜ fish emulsion

⚜ green manure tea (cover crop greens, green garden waste, lawn clippings, nettles, comfrey, yarrow, or other herbs or wild plants)

⚜ manure tea

⚜ powdered eggshells

⚜ powdered oyster shell

⚜ colloidal or marine phosphate

⚜ worm casting juice (from the tap on your commercially manufactured worm bin)

⚜ worm-casting top dressing

Consult pages 195 through 197 for water solution nutrient recipes.

Using Powdered Nutrients

A variety of powdered fertilizers can be used to add macro- and micronutrients to the soil (see page 138 for recommendations). They must be evenly dug into the soil or mixed into the bottom of a planting hole. Some, such as ground eggshells, you can make yourself, but most are available for purchase at garden stores or on the Internet. For rates of application, follow the instructions given on the package.

Top dressing

Top dressing—adding a layer—of compost to your containers is a great way to add organic nutrients midseason or between crops. The key is to use fully composted animal manures, compost, or worm castings. Coffee grounds needn't be composted and can be applied directly to the planting mix. Make it between 1 inch and 6 inches deep, depending on available space in the pot. More is better. You can either simply layer it over the existing mix or gently dig it into the first few inches or so. Don't worry if you can't dig it in, since the action of watering and of soil organisms will move nutrients down to the root zone.

Cover Cropping

Another aid in fertilizing container plantings is cover cropping with nitrogen-fixing legumes. While this will cut down on your yields of edible plants since your pots will be tied up with nonyielding plants for a number of months, it's a great way to add nitrogen to your soil. See page 149 for instructions on cover cropping.

WATERING CONTAINER PLANTINGS

Containerized soil dries out quickly and can take time to rewet. Either stick to a hand watering schedule or install drip irrigation. When watering containers it helps to cultivate the top inch of soil as you go, eliminating any crust that has formed. This allows the water to soak in more readily. Periodically press the soil around the inside perimeter of the containers firmly down. This will close tiny gaps that form where the soil makes contact with the container as the soil dries and pulls away. If you don't attend to this detail, although the top can look wet, the water will in fact just run down the sides of the pot and out the bottom. To conserve water, once you see water emerging from the bottom, stop. More frequent watering will help your plants thrive, but once the soil is saturated, extra water is just wasted.

AQUAPONIC GROWING

Aquaponics uses water that was used to raise fish to irrigate container plantings. Though we haven't had the opportunity to create aquaponic container gardening systems ourselves, they are a wonderful way to grow food in small spaces. As virtually closed-loop systems, they are an ideal setup for those who lack access to soil and soil amendments. Growing Power in Milwaukee is the expert and provides wonderful training and information: www.growingpower.org.

CHAPTER 8

BUILDING SOIL FERTILITY

One of the biggest mistakes beginner farmers make is to not build soil fertility. Organic farming—which we recommend for urban farmers—begins with the soil. The relationship between soil, compost, organic matter, microorganisms, and plant growth are unbelievably complex. Scientists still can't explain many things about their processes. Industrial farming ignores these interactions and assumes that one can just add nutrients in the form of water-soluble chemical salts, which eventually deadens the soil. In organic urban farming, a basic understanding of the nutrients essential for plant growth and how to provide them is helpful, but not necessary. In most cases, if you have a patch of soil and dig in compost each year, your plants will have what they need, and you will leave the earth, at least your patch of it, better than you found it. The information in the rest of this chapter will allow you to hone your skills so you can grow plants with improved vigor, a greater ability to ward off pests and diseases, and a higher nutrient content.

The light, air, and water surrounding plants supply the building blocks that form 98 percent of the plant's mass. You may remember from grammar school that plants use sun energy to transform carbon dioxide, hydrogen, and oxygen (all from air and water) into starches and sugars to use as food for their growth processes. The parts of photosynthesis that you as a farmer can affect are the amount of light, air, and water plants receive and crops need ample light, water, and good airflow to thrive.

You can also affect how well your crops will do by providing a balanced, nutrient-rich soil, which provides the matter that makes up the remaining 2 percent of the plant. The six primary soil nutrients, in descending order of abundance, are nitrogen, phosphorus, potassium, calcium, magnesium, and sulfur. These six nutrients, along with the previously mentioned carbon, oxygen, and hydrogen, are know as "macronutrients." Other soil nutrients are equally important but are used by plants in smaller quantities. These "micronutrients" include iron, manganese, molybdenum, zinc, copper, chlorine, boron, cobalt, sodium, vanadium, and silicon. Depending on the plant, different proportions of nutrients are needed. Of the 2 percent of the plants' mass made up by soil minerals, about 1.5 percent is nitrogen, which is replenished by the addition of compost as it is absorbed by plants and eaten by people and animals.

Though there can be nutrient deficiencies (this is why we recommend soil testing), often soil has plenty of nutrients. The ability of plants to absorb nutrients depends on the presence of soil microorganisms and mycorrhyzae (organisms related to mushrooms). Soil building consists of enlivening the soil by providing soil microbes with compost to eat. Once the compost is digested and microbes die, the existing minerals are in a form that can be absorbed by plants. Plants, organic material, and microorganisms are interdependent—plants cannot use nutrients in soil solids, only in solution. Soil microorganisms can deliver nutrients in solution to the plants, while plant roots exude carbohydrates that, in addition to organic matter, feed soil organisms—it is truly a miracle.

To create a supportive environment for these amazing interactions in the soil, we suggest three things: First, use organic methods in the garden. Chemical fertilizers and pesticides often throw off the organisms that you are trying to support. Second, you'll need to replenish the soil by applying fully decomposed manure and compost to your beds. Having fully broken-down compost maximizes the nutrients available for your plants. If you add fresh, uncomposted organic matter to your garden, you will need to wait (a few weeks to a few months) before planting as microorganisms do their work of breaking it down (during which time nutrients will not be available for roots to absorb). The third tenet is to disturb the soil as little as possible after building your beds.

Other important factors in plant/nutrient interactions include the "cation exchange capacity" (CEC), the "anion exchange capacity" (AEC), and soil pH (the degree of alkalinity or acidity). The CEC is a measure of the soil's ability to retain cations, or positively charged, plant-available forms of nutrients such as calcium, potassium, and magnesium. The CEC is an indicator of amount of humus, and of the type and amount of clay, present in the soil, both of which retain and make cation minerals available to plant roots. The AEC is an indicator of the absorbability of nitrogen and potassium. Clay is low in AEC, while organic matter is high. CEC and AEC can be improved with the addition of decomposed organic matter.

Much as we love organic matter, it shouldn't be used alone, particularly with container gardening, because it can be deficient in nutrients and might lack qualities that make them available to roots. This is why actual soil, formed from mineral-rich rock through geological and organic processes over eons (even, yes, in your urban backyard) is necessary for growing crops. It also brings up the question: Can your soil have too many nutrients? It *is* possible. For instance, an overabundance of some can cause too much green growth as opposed to fruits and seeds, and can also cause an increase in pests that feed on leaves. If you can't find your tomatoes amid a jungle of green growth, it's a tip-off that there's too much nitrogen, potassium, or phosphate in your soil.

SOIL TESTING

If you followed our advice and tested your soil for heavy-metal contaminants, you will also have received nutrient, pH, AEC, and CEC test results as well. Many urban soils have never been farmed and have rich soils. On the other hand, many cities have sprawled out to cover land that was once farmland and may be depleted. Testing soil nutrient levels will let you know if anything is missing or if your soil lacks the necessary CEC to hold bioavailable nutrients. The results from your soil tests should include recommendations for how to address any problems with CEC and pH.

In addition to the most conventional soil test (strong acid method), you might consider getting a weak extraction, or Morgan, test. This test, developed by Dr. Morgan at the University of Connecticut in the 1940s, measures the bioavailability of nutrients to plants more accurately, and it is usually only a wee bit more expensive than the strong acid test.

A word of caution: We are advocating the practice of finding and making your own nutrients in the form of composted organic matter and homemade nutrient teas. Soil-testing companies may offer various services and products to improve your soil fertility. If you need to add or balance nutrients, you can purchase amendments directly from farm supply stores and/or make your own, saving money and avoiding overdoing the fix with the industrially produced fertilizers usually advocated by the testing companies. Improving and maintaining soil fertility in an environmentally sustainable way is an ongoing, long-term project, not a quick fix.

CORRECTING FOR LOW NUTRIENT LEVELS

Low nutrient level results on your soil test are most often simply an indicator of low organic matter and microorganisms rather than an actual nutrient deficiency. We highly recommend adding organic matter after your first soil test, waiting six months, and then retesting prior to adding individual nutrients. If indeed your soil is deficient in nutrients, you can correct the deficiency according to the recommendations in this section. You may want to test and correct every few years.

Many of the organic and mineral amendments we recommend will increase micronutrient levels as well. How much of each to add will be included in your soil test or on the package, but since most of these are natural, balanced substances, exactitude isn't especially important.

The following table lists the telltale signs of the most common specific nutrient deficiencies. Though diseases and pests can also cause copycat symptoms, nutrient deficiencies tend to show up more uniformly in all the plants of the same type and age in your garden. They can also appear due to the pH balance of the soil being too high or too low, or to a lack of sufficient water.

Nutrient	Sign of Deficiency
Nitrogen	poor plant growth and yellowing of leaves, especially in older leaves
Phosphorus	poor plant growth and purplish leaf and/or stem color, often showing on the underside of leaves; stiff, brittle plant tissue
Potassium	leaves showing yellow between the veins; brown scorching and curling of leaf tips and edges; can be accompanied by purple spots on the undersides of leaves
Calcium	stunting, death, or hooked appearance of new growing tips (never mature leaves); blossom-end rot in fruiting vegetables; tip burn in cabbage family plants; interior browning in celery and other rosettelike vegetables
Magnesium	leaves showing green veins with yellow spaces in between; young leaves yellow first, followed by older leaves
Sulfur	newer leaves are yellow while older leaves remain green

Nitrogen

Aside from carbon, oxygen, and hydrogen, nitrogen is the plant nutrient in greatest demand relative to other nutrients; its levels must be maintained continuously over time for healthy plant growth. As the major constituent of chlorophyll, nitrogen is what makes plants green (a lack of green in the leaves of your plants can indicate a nitrogen deficiency). Your soil may show an initial problem because nitrogen is removed from the garden when plants are harvested for eating or green garden waste is removed. Each growing season removes some nitrogen from the farm, and it simply must be replaced for yields to be maintained. In traditional rural farming, leaving the soil fallow allowed microbes time to take nitrogen out of the air and concentrate it in the soil. Because urban farmers lack space, adding nitrogen to the soil is accomplished by breaking it out of green waste and animal manures through composting. If your soil is nitrogen deficient, add good quality compost.

Phosphorus

Phosphorus is the second most important plant nutrient derived from the soil. Organic matter and mineral particles both release it when the soil has enough organic matter and is warm, moist, and well aerated, which encourages microbial activity. If yours is deficient, add bone meal or soft rock phosphate.

Potassium

Potassium is the third most important plant nutrient and usually occurs naturally in the soil. Potash is the name given to the potassium salts necessary for plant-cell functioning. Adding compost will meet potassium needs unless there is a serious deficiency. If you need to add potassium, wood ashes or bone meal are good options. Ideally, the amendment should be added to the compost pile. If using wood ashes, make sure the wood used to make the ash was glue-free—in other words, solid wood. The often recommended green-sand can contain asbestos and other toxins.

Calcium

In addition to being a nutrient, calcium buffers the soil pH (makes it more alkaline) and helps rootlets absorb soil nutrients. Even with sufficient quantity, calcium absorption can be low in poorly draining soils, so ensure good drainage. Add finely ground eggshells, limestone, or gypsum to increase levels.

Magnesium

If the CEC is good, even relatively low magnesium levels will be sufficient to grow healthy plants. If you also need to add limestone, magnesium-rich dolomitic lime should be added to provide the correct balance. If you don't plan to add limestone, Epsom salts will provide the necessary magnesium as well as increasing sulfur. Adding Epsom salts is also rumored to improve the flavor of strawberries and tomatoes.

Sulfur

Deep, sandy soils may lack sulfur through the leaching action of rain. Otherwise, sulfur is often applied to correct highly alkaline soils. Add elemental sulfur or gypsum at the recommended rates on the package and take care, since too much can make the soil too acidic.

MICRONUTRIENTS

Your soil test will likely yield normal levels of micronutrients. If you want to make sure your produce has a high nutrient density, you may want to dig in calcium rock powders, such as limestone, gypsum, or soft rock phosphate. These will ensure sufficient levels of micronutrients.

SOIL FERTILITY IN BOXED BEDS

There's nothing like real soil, and while many think bringing in potting mix will be better than using native soil, it's almost always worse. Soil is a geological and organic formation with a truly staggering liveliness and complexity. If you plan to use boxed beds, maximize the use of your own soil, which is rich in mineral nutrients, by building boxes no more than a foot high and filling them with your compost of choice. The plant roots will reach the soil below. See chapter 7 on container gardening, if you plan to build boxed beds with bottom boards.

BUILDING MORE CATION EXCHANGE CAPACITY

Most soils actually have ample levels of nutrients, but these are simply unavailable to plant rootlets. Increasing the CEC is key to releasing and maintaining the supply of nutrients stored in your soil. The primary method for doing so is to increase the organic matter. Since both the CEC and the soil structure are improved by adding it, unless you have a very good soil texture and CEC, you will probably need to add between 3 inches and 6 inches of finished compost at the initial dig.

Your soil test from the lab may also include the percentage base saturation level (cations are also known as base cations). This will let you know if some of the major nutrients that should be made available through the cation exchange process are in the right proportions to each other. The goals for the base saturations are 70 percent calcium, 12 percent magnesium, and 4 percent potassium. Adding organic matter may also help regulate the percentage base saturation level.

CEC below 15: incorporate up to 6 inches of organic matter when double digging.

CEC above 15: incorporate 3 inches of organic matter when double digging.

Retest six months after adding organic matter to see if the situation has improved.

CORRECTING YOUR SOIL PH

Most vegetables like to grow within a pH of 6 to 6.5. Garden plants can't grow in severely acidic soils (below 3.5) or highly alkaline soils (above 8.5). If your soil is slightly acidic or alkaline, simply adding compost will buffer the effects of the pH imbalance. If your pH is highly acidic add ground limestone to help buffer the pH. If it is highly alkaline, add elemental sulfur or gypsum. Your soil test should include information on what amounts to use. Retest after adjusting.

To understand better how to make the major and minor nutrients available to your garden plants, you will first need to know what type of soil you have.

TESTING THE TEXTURE AND DEPTH OF YOUR TOPSOIL

Soil Texture

Soils fall into four basic categories: **sand, silt, loam,** and **clay.** Soils contain three mineral particles: sand, silt, and clay, as well as decomposed organic matter (humus) in varying proportions. Because loam soil contains sand and silt in roughly equal proportions and about 25 percent clay, it is the ideal gardening soil.

The three mineral particles started out as rock and were broken down over time. Sand particles are the largest, silt particles are in the middle, and clay particles are the smallest by a considerable order of magnitude. Sand particles don't hold together well, silt particles hold some shape, and clay particles stick together tightly. Loam is ideal because each of these particles lends good qualities: Sand and silt allow for good infiltration and drainage; clay plays a pivotal role in making minerals available to plant rootlets, helps soil retain moisture, and binds well with other particles. There are two easy ways to find out about your soil texture.

The rope test

Make a 1-inch ball of moist soil in your hand, and then rub it between your fingers. The feel will help tell you the texture. Sandy soil feels gritty and won't hold a ball well. Loamy soil feels slick, smooth, sticky, and slightly gritty, and forms a crumbly ball. Clay feels sticky and smooth, and you may have the urge to make a pot out of it. You will be able to make a ball out of clay soil easily.

Try to press the ball into a ribbon. If you can't form a ribbon, you have sandy soil. If your ribbon measures less than an inch long before breaking, you have silt or loam. If your ribbon measures one to two inches long before breaking, you have clay-loam. If your ribbon measures more than two inches long before breaking, you have clay.

The jar test

You will need a clean quart jar and lid, a marker, clean water, and some of your soil (don't use the topmost layer of leaves, grass, or weeds, just mineral dirt) to do the jar test. Fill the jar about two-thirds with water. Put enough soil in to fill it, leaving about an inch at the top. Add a few tablespoons of dishwashing detergent. Screw on the lid and shake it really hard for a while. Really shake, or the particles won't separate properly. Let the soil settle for about a minute, and make a mark on the outside of the jar marking the top of the layer that has settled—this is the sand layer. Set it down carefully and go do chores for an hour. Make a mark on the outside of the jar at the top of the next layer that has settled—this layer is silt. Set it down carefully again and go about your business for five days (long enough to see fairly clear water at the top). Make a mark on the outside of the jar at the top of the last layer—the clay layer. Estimating the percentages of each layer will tell you what type of soil you have.

> **Sandy soil:** approximately 80 percent to 100 percent sand, 0 percent to 10 percent silt, and 0 percent to 10 percent clay
>
> **Silt soil:** approximately 25 percent to 50 percent sand, 30 percent to 50 percent silt, and 10 percent to 30 percent clay
>
> **Loam soil:** approximately 40 percent to 60 percent sand, 40 percent to 60 percent silt, and 40 percent to 60 percent clay
>
> **Clay soil:** approximately 0 percent to 45 percent sand, 0 percent to 45 percent silt, and 50 percent to 100 percent clay

Knowing your soil texture will help you know how to garden your land. What weeds can you expect? How much and how often should you water? How much compost should you add? When should you work the soil?

IMPROVING THE STRUCTURE OF SPECIFIC SOILS

You can't really change your soil texture, because to do so would require bringing in epic volumes of material. Who would be willing to give you 20 yards of sharp sand, and how would you deal with transporting and incorporating the material into your clay soil? You can, however, improve the *structure* of your soil.

Soil structure refers to the tendency of your soil to aggregate into larger pieces and the degree to which it is interspersed with pockets of air. For instance, compacted soils have poor structure because the aggregates have been smashed together and the soil pores destroyed by the pressure of foot or car traffic, a very common problem in the urban setting. This can happen even with the ideal loam, and can make water infiltration and root growth very difficult. Extremely sandy soils, and those that have an extreme amount of clay, will require the addition of compost to improve the structure and/or make soil nutrients available to plant roots. A good soil structure will have plenty of larger soil aggregates interspersed with pores that allow easy air and root penetration and increased water-holding capacity.

Sandy soil can contain the necessary minerals for plant growth, but it lacks the organic matter to store nutrients so they are available to plant rootlets. Because sand particles are larger, nutrients are easily leached out by the rain. On the up side, it is very easy to work, even without damage when wet, and drains well. It does not retain moisture well. To improve sandy soil, double digging may not be necessary, since even compacted sandy soils are easy to dig. Incorporate as much organic matter as you can manage, up to 6 inches when you first dig your beds. Continue to add compost to the top of your beds each time you replant.

Loam soil is the ideal soil for most vegetables. It usually has plenty of nutrients, is easy to dig, and retains moisture in pores while draining well. Loam retains nutrients well. Up to 3 inches of compost can be added, but don't add more than is needed to replace depleted nutrients each season.

Clay soil is very high in vitality and available nutrients, but it is so heavy that plant roots cannot grow down easily. It's very hard to work unless almost dry, but equally hard to work when baked or compacted. If worked wet the structure of the soil can be damaged. It drains poorly but retains moisture well. When working in clay it's easy to contract a strange foot disorder, "clay platform boot syndrome," when the wet soil sticks to your shoes. To improve clay soil, incorporate up to 6 inches of organic matter by double digging when it is not waterlogged, and add organic matter to the top of beds often.

TOPSOIL DEPTH

Soil is formed of distinct layers called "horizons." The first, topsoil, is a darker color and is composed of weathered rock, organic matter, and living organisms. The next layer, subsoil—is a lighter color and is composed of only partially weathered rock lacking organic matter and biological life. Topsoil is needed for crops to grow. You can usually determine its depth by digging down until you see a clear color change in the layers. However, in urban soils, because of construction and the use of subsoil as fill–dirt around construction sites, it is sometimes difficult to determine where the topsoil ends and the subsoil begins. You need a minimum of 6 inches of topsoil to grow crops, but a few feet or more is best. In deep topsoil plant roots can forage for nutrients and water, resulting in less work for you, the farmer, in providing nutrients and frequent water.

Around recently constructed houses topsoil will most likely be shallow or absent because it was removed and never replaced. In this case you will have to layer in a lot of compost or use boxed beds. Another potential problem is called "hardpan," a layer of compacted subsoil that doesn't allow any water to drain through. If this is the case you will have to dig down and break it up, being careful to mix as little subsoil in with your topsoil as possible. In all these cases, if you keep adding organic matter, you will have pretty good soil in a few years.

MAINTAINING SOIL FERTILITY OVER TIME

The organic farmer's tools for maintaining soil fertility over time are *continuously adding the right amount of organic matter, cover cropping,* and *crop rotation.* Adding organic matter is just what it sounds like— amending the existing soil with composted organic materials, including animal manures. Crop rotation means alternating among crops that have different nutrient needs in order to avoid excessive depletion of them and to ensure that the cycle of plant diseases and insect pests is disrupted. Cover cropping involves planting soil-building crops and incorporating their roots, leaves, and stems into the soil.

Fava beans are a beautiful cover crop that adds nitrogen to the soil.

Note that these methods are contrary to industrial agriculture's focus on adding chemical forms of isolated nutrients. Industrial farming's answer for home gardeners, Miracle-Gro, offers an "insta-boost" of water-soluble mineral salts. This is analogous to human nutrition. Consider the difference between an intravenous drip of nutrients versus a good meal and exercise. In addition, this form of plant nutrition is derived from petroleum and requires a lot of energy to obtain, as opposed to using sun energy to fertilize, like with cover cropping. Chemical fertilizers are fast acting but peter out quickly, unlike organic methods, which release nutrients slowly and last a long time. Our three recommended soil-building methods are great at preventing disease and pest problems, eliminating the need for chemicals to deal with these problems. Note that homemade organic fertilizers can be delivered in water solutions; see pages 174 through 176 for these recipes and methods.

COMPOST

Compost is the stock and trade of the sustainable farmer. It is made by aging a mixture of natural materials containing large proportions of carbon and/or nitrogen as well as smaller amounts of most of the necessary plant nutrients. As the materials age, micro- and macroorganisms break the materials down, producing compost that smells much like a handful of soil from the forest floor.

High-carbon materials are called "brown" waste and high nitrogen materials are called "green" waste. Your compost pile or bin should contain roughly two-thirds browns and one-third greens. Typically materials are laid down in alternating layers of browns and greens. The mass of the compost pile should be at least 3 feet square but can be much greater. A compost windrow is a pile at least 3 feet wide and high, of whatever length you want.

Compost organisms need air and water to proliferate. Even if enclosed in a bin, the pile should receive airflow and be about as damp as a wrung-out sponge. Depending on the weather, compost piles should be watered, so they don't dry out, or covered, to keep out excessive rain. There are two basic compost pile styles. After being created, slow piles can be left to age until finished (the material will smell like soil), taking three to eight months, depending on conditions. Fast piles can be turned over with a garden fork one or many times to speed up the decomposition process, and take one to four months, depending on conditions. Slow piles retain more nutrients but take longer to mature. Fast piles result in a larger amount of compost from a smaller space. Take your pick, depending on your needs and available space—but we highly recommend at least making some "slow" compost. If you are short on space, go up instead of out. In the fall, make a four foot by four foot pile that goes up as high as you can manage and leave it to decompose until the next spring or summer, watering or tarping as needed. Please see appendix 11 for composting resource materials.

Compost ingredients themselves may not actually be brown or green. Kitchen scraps are a green but actually come in many colors, and straw is a brown but can be yellow in color. A good way to judge is to ask yourself whether the material became dry or was alive and fresh when it was harvested. Animal manures are often brown in color, but because animal waste contains large amounts of nitrogen, they are considered green. Bread is a "green" because, by the time the grains used to make it have ripened, soil nitrogen has been transferred to the grains.

Since animal manures contain large amounts of nutrients, especially the nitrogen that is in such high demand by plants, they should be a part of your compost pile. Traditionally animals were an integral part of farming, providing work (plowing), food (meat and eggs), and fibers for clothing and fertilizer. Farms with animals were self-sufficient; farmers didn't need to exploit many mineral resources to provide nutrients. In exchange, the animals received their share of farm-grown food (grain, hay, and scraps) and hopefully the love and appreciation of the farmer. Whether you are a meat eater or vegetarian, animals can have a place in the nutrient cycle of your farm. They can provide meat and eggs or simply serve to recycle garden scraps into high-nitrogen fertilizers.

In cities the manures that are most readily available are avian (chicken, duck, etc.), horse, rabbit, and those generated at zoos. Manure from cows and goats, though great for your garden, may require a trip to the country. Urine-soaked bedding (straw, sawdust, or shredded paper) often accompanies it. Bedding is high in carbon and relatively low in nutrients, though the urine provides quite a bit of nitrogen. Bedding-free manure is ideal, though rare.

Since human food scraps attract rodents, we recommend the following system: Use worm composting for food waste and add all other materials to your compost piles. Be careful about animal feed falling into bedding. Rodents will be attracted to the food when you put the bedding in the compost. Some feeders are designed to reduce feed spillage. It's hard to find fully rodent-proof manufactured compost bins. Sealable plastic barrels suspended on a frame for easy flipping (to turn the material) are pretty effective. For those of you who are builder types, feel free to take up the challenge—it can be done with fine, strong wire mesh and wood.

The "castings," or poo, from red wiggler worms is one of the most effective and balanced organic soil amendments. Worm composting, or "vermicomposting," relies on red wiggler worms (*Eisenia foetida*), rather than the earthworms you may be more familiar with, to create a top-quality fertilizer from rotting food scraps, manure, or other green wastes. Worm bins can be built or purchased, and can be small or large depending on your needs. They eat the waste, and the resulting poop is full of easily absorbable plant nutrients. Worms can most easily digest materials that have already been broken down by microorganisms. After collecting food scraps, put a lid on the container and let it rot for a few days before adding it to your worm bin. It may be stinky to you, but it will be delectable to your worms. Please see appendix 11 for books about making vermicompost.

COVER CROPPING

Cover cropping is done for the sole purpose of feeding the soil. In practice this means seeding and caring for a crop, and cutting it down once mature but while it is still green, whereupon it is worked into the soil or added to the compost pile. Cover cropping conserves existing soil nutrients; the upper part of the plant protects the soil and the plant's root system prevents the leaching of nutrients by the rain. This process transfers nutrients from the soil to the plant's own stems and leaves, which are then returned to the soil either immediately or after being composted. After this process the nutrients are in a more balanced and easily absorbable form. Deep root systems tap nutrients from lower soil strata that then feed your crops once the cover crop is composted and returned to the soil. Finally, legume cover crops can actually take nitrogen from the air and transfer it into the soil.

All legumes (beans, peas, clover, vetch, alfalfa, etc.) "fix" their own nitrogen by absorbing it out of the air and concentrating it in root nodules to be used by the plant at flowering and fruiting time, when it's most needed. Nitrogen-fixing legumes are amazing. As food crops they are stellar, because they produce protein-rich foods without the farmer having to add nitrogen to the soil. As cover crops they are used to draw nitrogen out of the air and into the soil for use by successive crops. An example would be growing vetch prior to growing cauliflower (a heavy feeder) in the same bed.

For urban farming we recommend either exclusive use of nitrogen-fixing legumes or a mixture of legumes and deep-rooted winter grasses, such as oats or cereal rye (not ryegrass). A mixture is great, because the legumes add nitrogen while the grasses pull up nutrients from deep in the soil and provide a lot of green matter for composting. Fava, or bell beans, are ideal nitrogen fixers, since they also produce a large amount of green matter; the leaves also provide a delicious cooking green. Harvest some of the leaves to eat after they have grown at least a foot high (they must be cooked, and some people of Mediterranean descent are allergic). Vetches provide the maximum amount of nitrogen fixation.

To add nitrogen to the soil for the next crops you plant, allow legume cover crops to grow until they just begin to set flower buds. Cut them down before they form beans or peas. Once they form, the nitrogen in the root nodules is transferred to the fruit and is in a much less available form for subsequent crops. Cut the stalks off an inch *below* the soil (this keeps them from resprouting), leaving the roots where the nitrogen nodules have formed in the ground. To save time, rather than mixing the green growth into the soil to decompose there, as is often recommended (this can take weeks), chop it up and add it to your compost. Your bed is ready for immediate replanting.

Cover crops can also be planted alongside heavy feeders to improve nutrient availability and save time. For example, plant Dutch white clover in the late fall in beds intended for spring *Brassica*s. In the spring, mow the clover and plant directly into the lawn. The clover will continue to grow and fix nitrogen for your crops. Another option is to sow nitrogen-fixing cover crops into beds a few weeks to months prior to removing the crops growing there. This way your cover crops will get a jump start on growth and save you time.

CROP ROTATION

To rotate crops with different nutrient needs we group them into four categories—heavy feeders, light feeders, nitrogen-fixers you plan to eat, and nitrogen-fixers you plan to grow as a cover crop. Simply rotating these categories is quite easy and effective: All you need to do is pay attention to what type of crop you're removing (postharvest) and follow it with a crop from the appropriate category. Taking garden notes, which is a great practice but can fall by the wayside in our busy lives, is unnecessary with this system.

The nutrient needs of crops are generally related to the part of the crop that is eaten. The formation of roots, leaves, blossoms, and fruits all require different nutrients at different stages of plant growth. We like to group crops in the following way to get a general sense of these needs:

Heavy feeders: fruiting and leaf crops

Light feeders: root vegetables and onion family crops

Nitrogen-fixing: legumes

If you have the time or are having persistent pest or growth problems, do take comprehensive notes on what you plant, when, and where. Use these notes to avoid planting the same crop in the same bed in a three-year period. The *most* effective crop rotation is to avoid planting crops from the same plant family in the same bed within a three-year period (see appendix 2 for a table of crops and their families). Why do we recommend a simple crop rotation rather than the most effective crop rotation? Busy lives require workable solutions, and rotating crops in a simple way is preferable to not rotating at all.

Simple Crop Rotation

1. Follow a heavy feeder with a light feeder.

2. Follow a light feeder with a nitrogen fixer you plan to eat.

3. Follow a nitrogen-fixer you plan to eat with a nitrogen-fixing cover crop.

4. Return to the beginning and plant a heavy feeder.

- In the field, either label each bed according to the above categories or mentally note the category of the crop in the bed.
- Plant a crop from the next category in the sequence when you replant the bed.
- When you reach the category "nitrogen-fixing cover crop" for a bed, plant a nitrogen-fixing cover crop.
- For greater effectiveness, treat the cover crop seed with legume inoculant (a rhizobium with a symbiotic relationship to legumes) prior to planting (see appendix 11 for a list of farm supply stores).

Just as you begin to see blossom buds forming on the cover crop, cut the green part of the plant off 1 inch below the soil surface and compost the tops in your compost bin.

FINDING NUTRIENTS

Cities are chock full of vast amounts of organic materials that can be composted and added to urban farms to feed plants and people. As an urban grower, you are living in a cornucopia of fertilizer that's actually much richer than the suburbs or rural countryside. The waste of urban-produced nutrients sent to the landfill is tragic—food scraps make up the majority of the garbage there! Many counties and municipalities are finally implementing large-scale food and yard waste composting programs, but since participation in these programs isn't universal, a huge amount of nutrients still end up in the landfill. Rather than using fossil fuels to truck these wastes long distances to large composting operations, urban farmers should take advantage of these resources and make their own compost right in the city.

Urban Compost Ingredients

Please research the practices of businesses or factories you plan to use material from. For example, mushroom compost can contain fumigants and factory materials can contain dangerous chemicals. Only use plant or animal wastes—nothing synthetic.

Material	Green, Brown, or Other
Kitchen scraps	Green
Yard prunings	Green
Dry leaves	Brown
Restaurant scraps	Green
Used coffee and tea grounds and their filters from cafés and restaurants	Green
Animal manure and bedding (stables; zoos; pet stores)	Green
Straw (often used at outdoor events)	Brown
Wood ash	Supplies potassium, phosphorus, and magnesium
Cotton, wool, silk, or felt waste	Brown
Grocery store produce-section scraps	Green
Food-processing-factory waste	Depends
Used mushroom compost	Already composted
Used indoor growing operation compost	Already composted
Solid wood sawdust (furniture makers)	Brown
Seafood waste or seaweed (a marina or processors)	Green
Olive oil pomace	Green
Used brewery mash or grape pomace	Green

If you can help it, don't spend your hard-earned money buying bagged compost from your local garden store. Our planet needs you to recycle the nutrients that go unused every day and often even pollute ecosystems, such as waterways and oceans. Of course we know it's a question of time or money. Sometimes, when you just don't have the time, you may opt to spend the money to purchase what you need. Also keep in mind that bagged products will likely be of lower quality than what you make at home—they simply lack the biological activity of your compost pile or worm bin and can actually be toxic. In one study done by the U.S. Commerce Department of sixty commercial bagged products, half of them failed to even grow radishes successfully they were so toxic.

COLLECTING WASTES

Without systems and habits, using your own household waste can be difficult to keep up. Get a tightly sealing container to keep indoors for collecting scraps. Decide whose chore it is to empty the container into the compost or worm bin. If your household collection system becomes gross, you'll probably stop, so the chore should include scrubbing out the container.

Collecting Animal Manure

Zoos, tot zoos in city parks, the circus, horse-racing tracks, police departments that use horses, city park riding stables, private riding stables, and nearby agricultural operations, such as chicken, goat, or cow farms can all be great sources of animal manure. Ideally, locate a source that provides aged manure in an open pickup area, so you don't have to coordinate pickup times or age the manure yourself.

Having a truck or station wagon is a must for pickup. Bring what you will need, including wheelbarrows, shovels, rakes, and a 2 inch by 12 inch plank to run your wheelbarrow up into the truck. If using a car or station wagon, bring buckets or bags.

Create Animal Manures

Instead of going to the zoo, you can start raising animals on your farm. The least noisy and most productive manure makers are rabbits. Poultry manures need to be composted but are very nutrient rich. Depending on the size of your garden, you may be able to gather all the nutrients you need from your animals. See part III for more information.

When transporting materials by truck, the law requires you to tarp and tie down your load correctly. Regardless of the law, the horror of possibly injuring someone or causing an accident should be enough incentive to use safe practices each and every time you transport building and gardening materials. Here are a few tricks of the trade for tying down your load:

CORRECTLY TARPING YOUR LOAD

The first rule of thumb is, your tarp should be the size of your truck bed, not larger or smaller. Fold it to the correct size if you need to. Ropes and tarps don't always make good contact. We've all seen people driving along with a tarp hanging precariously from the end of their truck. All it takes is a good gust of wind and this tarp could end up on someone else's windshield. To ensure that your tarp stays put, carry four wood or metal stakes the length of your truck bed and six rocks or chunks of concrete with you at all times. When tarping a full load, first lay your tarp over the material, then lay the stakes down lengthwise over the tarp, then rope your load. The pressure caused by the ropes pushing down the stakes will keep the tarp in place. When tarping a partial load, lay your tarp over the load, then weigh it down in the corners and center with your rocks or concrete chunks. Last, tie your rope over the top of the truck bed as insurance.

TRANSPORTING LUMBER

Lumber doesn't necessarily need to be tarped, but it does need to be securely roped and tied. To keep lumber from shifting, tie your rope to a hold-down (the metal brackets on the truck bed). Then loop the rope tightly around the entire stack of lumber a few times, effectively making it one unit. Then, using a truckers hitch, tie your rope to another hold-down. If the lumber sticks out past the tail gage, tack a red flag to the longest piece—it's the law.

TRUCKERS HITCH

It's very disturbing to arrive at your destination and find that the rope securing your tarp is flapping in the wind. Take the time to learn the truckers hitch knot, and you will be able to transport tarped loads without worry.

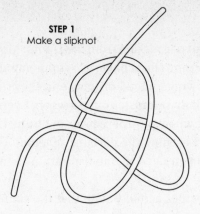

STEP 1
Make a slipknot

STEP 2
Loop the loose end around your
hold-down, then loop it through
the slipknot and cinch it tight

HOLD–DOWN

1. Ensure that nothing in your load could shift and create slack in your rope.
2. Make a slipknot and slip it over the first hold-down on the truck, pulling it tight.
3. Run your rope over your tarp from hold-down to hold-down, removing the slack and pulling it tight as you go.
4. Make a 6-inch slipknot in the rope about 2 feet from the last hold-down. The slipknot must be made the right way round so it will hold its shape.
5. Take the loose end of the rope and loop it through the last hold-down loosely. Then loop it through the slipknot.
6. Pull the loose end of the rope to tighten it as it passes through the slipknot. Check the rope at each hold-down to make sure it's slipping freely, and then pull the loose end again to make your rope as tight as possible.
7. Hold the loose end and the rope between the slipknot and hold-down together in your hand. Wrap the loose end around these two widths of rope till you have a manageable amount left.
8. Make another slipknot with the loose end to hold the rope in place.
9. Fold any extra rope and shove it under a tight spot between the rope and tarp to keep it from flapping around.
10. When you arrive at your destination, simply pull the loose end to release the first slipknot. Remove the second slipknot as you wind up your rope.

Dumpster and Green-bin Diving

This can be done either with or without permission from the Dumpster/bin owner. We use whole grain bread from a local bakery to supplement our chicken feed. We called up the owners ahead of time, and they're happy to have us rooting around in their Dumpster once a week. Even just entering the business and asking whoever's there prior to hitting it can often yield cheerful permission. Learning the garbage schedule is a real help in getting all the material you need in one trip rather than arriving at an empty Dumpster/bin.

Most contracts with waste companies actually state that the community owns the garbage of their community. The waste company can argue that Dumpster diving and the like violates their contract. However, many of these contracts also have a caveat that anyone in the community can use garbage for community benefit. If challenged, read the fine print. After all, it is your garbage until it hits the curb.

Make Arrangements with Urban and Suburban Businesses to Pick Up Materials

If you can be reliable, it's often quite easy to set up a routine pickup of material from any number of businesses. Think like the business owner. They are overworked and have a list of tasks longer than the day is long. Your phone message isn't an important one. Go in person and try to make contact. Leave more messages. Go in person again. Repeat. The best thing is to appeal to their altruism and make their pickup system so easy for them that they can't refuse. Many business owners believe in recycling but don't have the time to do it. They will feel great to know that some of their waste is helping to grow food to feed people.

For years, City Slicker Farms employees picked up hundreds of pounds of coffee and tea grounds each week from a local café. After piquing their interest, we discussed the best way to make it easy for their employees to separate the grounds and filters from the rest of the trash. We helped them set up a shelf behind the café that could hold fifteen five-gallon buckets and left empties for them to fill. Each week we entered the front of the shop, dropped off empty buckets, one of which was kept under the counter, and proceeded to the back to gather all the full buckets. We drove them away in a bicycle cart rated to carry six hundred pounds. We were faithful to our pickup day and time and *never* bothered them to change it, for any reason. We let them know when we were closed for the holidays, and that they should just throw away the grounds for that period.

Infrequent pickups can often be arranged on an as needed basis if you make a good contact within a business. Again, try to think about what their life is like and don't be disgruntled if they can't take the time to do your bidding. You get more flies with honey, and with persistence you will find what you need.

Gather Materials in Urban Nature Spots

Dry leaves are a wonderful compost ingredient in the city, where it can be hard to find browns. Instead of buying straw as your source of compost "browns," offer to rake up leaves on your street. Find a church, apartment complex, school, or university with deciduous trees and ask if you can rake some up. You will most likely be met with bemused gratitude. Ask the gardener at your local city park if you can gather leaves. If you live near the sea, seaweed is a fabulous compost ingredient. Simply gather it at the beach and layer it into your compost pile. The iodine and minerals will increase the nutritional value of your produce.

Arrange with Your Waste Authority to Receive Donated Compost

Most urban garbage companies are now picking up not only green waste but also kitchen scraps. These materials are trucked to rural areas and turned into finished compost using large machines and windrow systems. The resulting compost is then sold back to consumers. Although city garbage companies are privately owned, they are usually governed by the county through a contract, since they usually have a monopoly. As large corporations benefiting from exclusive contracts, they are often encouraged or required to engage in community benefit programs. If a program for residents and community gardeners to receive donated compost doesn't already exist in your area, work to create one, or make an individual arrangement with the waste company to receive donated compost. Promising positive publicity in exchange for their benevolence can really help the negotiation along: "I scratch your back, you scratch mine." If you have trouble, go to your waste authority board and enlist their help in guilting the company into donating compost to the community.

IF YOU OPT TO PURCHASE MATERIALS

Shop at Your Locally Owned Garden Center Rather than a Corporate Garden Center

This will keep money in your community.

Purchasing in Bulk

Bulk suppliers of compost often use at least some locally sourced ingredients, reducing the carbon footprint of their compost mix. Ask the company for their recommendation on the best mix for growing vegetables. If you decide to use purchased mix for raising seedlings, they will most likely also have a greenhouse seedling mix. Buying bagged materials is the most expensive and inefficient method of getting nutrients. For the environmentally concerned, who knows how far they've been trucked. Even if you lack a truck, you can still purchase bulk material from your local rock and soil company by having them dump material into buckets or other liftable containers. Borrowing or renting a truck is another good option. Or you can pay them to deliver.

Bulk materials are purchased by the cubic yard. One cubic yard is 27 cubic feet. One yard of compost will cover a 10-foot-by-10-foot area with 3 inches of compost. Let's face it: You're probably going to need at least a number of yards, and it will cost you double to buy it in bags. Most trucks will hold 1 to 3 yards of compost. If you still feel that you don't need that much yourself, split it with another gardener.

COLLECTING YOUR WASTES

This is the extreme on the spectrum of techniques you may or may not want to use on your urban farm. But many people, including us, started to question the wisdom of voiding our bladders and bowels into perfectly clean drinking water during the third year of a drought in California. Water conservation is a worldwide issue now. Not just a problem for drought-prone California.

The simplest solution is to pee into a container, dilute with water, and then water the garden with this nitrogen-rich liquid. Simply place a watering can or bucket in the bathroom, fill half of it with water. As the day goes on, pee into the container. At the end of the day, empty the urine-water into the compost pile or use it to water fruit trees. Urine is actually sterile, and when diluted with water, is virtually odorless. But if you let the urine sit for more than twenty-four hours before putting it into the garden, it will begin to ferment, smell very, very bad, and will lose much of its nitrogen.

For those who don't want to empty the urine container every day, you can pee in a sawdust-filled bucket. Set up an empty bucket with a few handfuls of sawdust at the bottom. Place a toilet seat on top of it. Next to that, have a container filled with sawdust with a scoop. After you pee, simply sprinkle a scoop of sawdust into the bucket. Depending on how much you pee, this can take up to a week to fill. Once it's full, take the urine-soaked sawdust out to the compost area and empty it. To prevent odors, mix the uriney sawdust with other kitchen compost, straw, or green materials. Wash out the bucket, place a little sawdust at the bottom, and start again. Though you are mixing the nitrogen-rich urine with a carbon-rich material like sawdust, you will still get a net gain of nitrogen in your compost, and you won't be wasting any water to dispose of your urine!

Humanure, yes, humanure! This is definitely an advanced topic, as human fecal matter is a disease vector. We urge you to read the *Humanure Handbook* and Sim Van der Rin's *The Toilet Papers* for a complete portrait of composting human solid waste. As an example, we built an outhouse in our backyard and void into a thirty-five-gallon container that is seeded with red wiggler worms and sawdust. Use this setup for poo, never urine. A scoop of sawdust goes into the container after each deposit, and the lid to the toilet is shut. After the container is almost full, we add finished compost or potting soil, drill some holes in the bottom of the container, plant a nitrogen-loving plant such as a banana or citrus tree in it, and put it in the garden. According to the World Health Organization, humanure is safe for handling after two years, so after two years the plant can be removed and planted in the garden, and the soil from the container can be placed in the garden.

Whatever materials you choose to boost fertility in your soil, you'll notice a marked improvement in the quality and quantity of your garden produce. In the end, building your soil over time will become your legacy—enjoy it!

CHAPTER 9

PROPAGATING SEEDLINGS

Once you've built up soil suitable for growing vegetables, fruit, and herbs, it's time to move on to the dazzling world of seeds and propagation. Starting seeds, either indoors or outside, is the most magical experience. Dried genetic material in a tiny food-filled time capsule comes to life in warm conditions with the simple addition of water. We can't tell you where that animating life spark comes from, but we *can* tell you how wonderful it is to be in the presence of such a miraculous process. Mother Nature and generations of plant-breeding farmers have created the seeds. All we have to do is provide the right environment for new life to begin. Starting your own seedlings and plants in a controlled environment is perhaps *the* easiest way you can save money on your urban farm.

You can start your own seedlings in two ways: in a controlled environment, such as an indoor windowsill, greenhouse, cold frame, or hoop house; or outside, directly in your garden beds. In the first scenario, seeds are often started in trays of potting mix and then potted up (transplanted into larger containers) to give them an extra boost of nutrients and extra room as they grow. The controlled environment protects the seedlings from pests, wind, and cold. See page 169 for outdoor seed starting.

The seedling propagation method we recommend was famously named "breakfast, lunch, and dinner" by the originators of the Biointensive™ method, Alan Chadwick and John Jeavons. Just as a seedling begins to exhaust the available soil nutrients, it is transplanted into fresh, nutrient-rich soil in a new container or garden bed. The "breakfast" is given when seeds are put into an open tray full of rich planting mix, the "lunch" is given when the seedlings are transplanted into a pot, and the "dinner" is given when the seedlings are finally transplanted into rich garden soil.

Organic farmers take a holistic approach to plant health. They connect plant pathologies with early deficiencies in nutrients, light, or water. Healthy, stress-free seedlings result in healthy, high-yielding plants. Growing the best seedlings will help you avoid later insect and disease infestations, low yields, and early bolting (premature flowering).

In the second scenario, called "direct seeding," which is covered in the next chapter, the seeds are planted directly into the garden beds and closer together than they ultimately should be, which ensures even germination and survival. After the seeds germinate they are thinned to their correct spacing. Given the benefits of sowing in trays, the direct seeding method is usually reserved for crops that suffer from transplanting, often those with large taproots, such as carrots, or those that grow extremely rapidly, such as beans.

In fact, there is a third way to start seeds. Sometimes seeds fall into garden beds from plants that were grown to maturity. They "reseed" themselves. As long as they were from an open-pollinated variety (see page 163), these happy accidents can either be grown in place or transplanted into other areas of your farm.

Other plants, like thyme, lavender, and rosemary, are difficult to start from seed and are thus best propagated by vegetative cuttings grown in a controlled environment. If you've ever been given a sprig of a houseplant and rooted it in a glass on your windowsill, you've done vegetative propagation.

SELECTING, PURCHASING, AND STORING SEEDS

Farmers, gardeners, plant breeders, and people who save seeds have been breeding new crop cultivars and conserving valuable varieties for thousands of years. In the last few hundred years, especially productive, adaptable, tasty, disease- and pest-resistant varieties have been bred. Lately, many vegetables have been bred in surprising shapes and colors. While commercial farmers may sacrifice flavor for shipping tolerance, the urban farmer's sole goal is flavor—we won't be sending our produce to distant shores, and so we can experiment with growing rare and delicate fruits and vegetables not often seen at the grocery store. One of the most entertaining aspects of gardening is trying out different cultivars and finding ones that have great flavors and thrive where you live. Once you find a well-adapted cultivar for your climate and soil, you can even develop your own specially adapted variety by saving seed (see chapter 14 to learn how to save seed).

Throughout the country there are specialty seed companies that grow varieties well adapted to unique climactic conditions (see page 545 for a full list of our favorite seed companies). If you can, take the time to find one that specializes in seed grown in and adapted to your region. You'll also want to look for varieties that do well in your soil and microclimate, because these other factors also affect the flavor and productivity of veggie plants. As previously mentioned, urban farmers should focus on varieties that mature quickly. If you live in the extreme north, at high altitude, or in a cool coastal region, find varieties adapted to your conditions. Similarly, if you have clay soil you will want to look for root vegetable cultivars that can tolerate heavy soils.

While many seed companies geared toward the home gardener are great, it's also worthwhile to peruse the pages of commercial seed suppliers. Many companies actually have separate catalogs or Web pages for home gardeners and commercial farmers. Guess what? You're not going to find the deals on the home gardener pages. When you buy fava bean seed, believe us, you need pounds, not packets. If you're buying seed for staple crops that require quite a bit of seed, such as beans, beets, carrots, corn, green onions, peas, salad greens, and turnips, purchase in bulk quantities, not packets. There are often a number of quantity choices. One recent look at prices for salad mix in a catalog for commercial growers yielded the following: 1 gram for $0.90, 2 grams for $1.70, and 4 grams for $2.80. That comes out to saving $0.05 per gram on 2 grams and $0.20 per gram on 4 grams. All of these are much, much cheaper than a typical off-the-rack seed packet of salad mix—1 gram for $2.79.

Another innovation on offer from commercial suppliers is pelleted seed. Quite a bit is wasted when seeding varieties with tiny seeds, since it's difficult to place them exactly where you want them. Small ones, such as those for lettuce and carrots, now come coated with clay to make a larger, more visible seed that can be spaced the right distance apart. The clay also retains moisture, aiding germination. The extra cost is made up by the much more economical use of seed.

Heirloom, hybrid, organic, genetically modified, open-pollinated, biodynamic—seed can be labeled with any of these monikers, but what do these terms mean? To choose which seeds to buy, it's important to understand how they are produced, especially if you're hoping to save some.

For millennia, **Open-pollinated** seeds were the only kind available. For thousands of years farmers all over the world saved seeds from plants with larger, more delicious edible parts; this resulted in varieties that were adapted to specific climates and conditions and were disease resistant. **"Heirloom"** is a term used to describe cultivars that have been saved for a number of generations or more. People grow **Heirloom** seeds to preserve traditional crops and maintain an extremely wide variety of cultivars adapted to many conditions, creating a safety net in the unforeseen event of climactic, pest, and disease problems. **Heirloom** seeds are by definition **Open-pollinated**. **Heirloom** and other **Open-pollinated** seeds can be saved every year, and each generation they will be more adapted to your particular climate.

Hybrid seeds, on the other hand, are man-made crosses within the same plant species that create a distinct variety. Sungold tomatoes are an example of a hybrid seed. Hybrids are usually denoted in catalogs as "F1". If you saved a seed from an F1 hybrid, the resulting offspring would not necessarily look, grow, or taste like its parent plant. Thus, you have to buy seed every year. Other than the financial ramifications of not being able to save your own seed from plants grown from **Hybrid** seeds, hybridization is an environmentally safe process that leaves the plants' DNA intact.

Transgenic or genetically modified (GMO) seeds were developed by petrochemical corporations so that crops can survive the application of pesticides and herbicides applied to fields. Rather than using techniques to affect the natural process of sexual reproduction, as in selective and hybrid breeding (pollen from one plant's flower fertilizes another's ova), genetic modification techniques rely on various chemical and mechanical methods of breaking apart and recombining DNA in a lab to create new cultivars. Currently, up to 45 percent of U.S. corn and 85 percent of U.S. soybeans are genetically modified.

Because, as with hybrid seeds, genetically modified seeds are patented, farmers must purchase seed each year rather than simply save aside a percentage of the harvest for next year's seed. The corporations selling transgenic seeds should be required to prove they are safe for human and animal consumption as well as for the environment before releasing them. Woefully, this has not been the case.

Biodynamic seeds are grown using biodynamic methods—a standard even higher than organic—developed by the first modern scientist to introduce natural farming methods—Germany's Rudolf Steiner.

When buying seed, remember that the difference is in the method of growing the parent plant. Was the plant grown using conventional, organic, or biodynamic methods? Organic seeds are grown using strictly organic methods and are most often open pollinated, but don't assume that just because the seed is organic it can be saved from year to year. Some hybrids are now organically grown.

STORING, PROTECTING, AND CULLING SEEDS

Seeds should be kept dry, both when in storage and when you're outside planting them in the garden. It's really useful to have a protected, well-organized seed-storing system, so you can find what you need when you need it and keep your seeds fresh. They should be kept in the house, away from the steamy kitchen and out of the sun. A dark hall or laundry closet is your best bet. If you don't live near your garden, store them in a dry, dark shed. You can use shallow cardboard boxes or plastic totes with your own homemade cardboard dividers set at the width of seed packets; cut-out card-stock alphabet dividers that stick up above the packets and cardboard dividers will help you stay organized. Winter is a good time to reorganize your seed boxes.

If you have a lot of seed, make separate boxes for the following categories:

- Greenhouse-seeded vegetables
- Direct-seeded vegetables
- Beans and peas

- Culinary herbs
- Medicinal herbs
- Ornamentals
- Bulk seed that won't fit in the other boxes

Your seed containers should be labeled on all sides and alphabetized based on the vegetable (not the cultivar; for example by "Kale", not "Lacinato Kale"). Keep open packets closed with paper clips. If you plan to save seeds from garden plants, you can purchase small manila packets at office supply stores or store them in jars.

Seed packets should be thrown in the compost if they are over three years old. Although seeds can be viable much longer, it really sucks to spend a lot of time preparing and planting a bed only to have a 25 percent germination rate. If you just hate throwing things away, you can take a small number of seeds and see how many germinate by setting them on a wet paper towel.

A really easy way to ruin your seeds is to get packets wet when in the garden or greenhouse and then returning them to the box. Not only will they sprout or rot, but their neighbors will too. Develop the habit of thoroughly drying your hands after you've prepared the bed or pots and before you drop in the seeds. Also get in the habit of keeping your seed box in a safe place while working and tossing packets you're using onto the top of the box, rather than in the middle of a wet bed or path, until you need to pick them up again.

PROPAGATION SETUP

Seedlings and cuttings (plant stems that are rooted to form a new plant) require the right amounts of heat, soil moisture, humidity, light, nutrients, and airflow to thrive. While one can get very technical about it, we recommend a simple setup. Except in the case of a grow-light setup, you will want to pay attention to orientation, making sure your propagation area is facing south. If your setup is outside, trees or shrubs shouldn't block the southern exposure. An area that gets full sun for the seed-starting setup with a partially shady area next to it for hardening off seedlings is best. One worthwhile technological innovation are under-plant heating mats. These warm your trays by a number of degrees and can cut germination times in half. Purchase them through farm supply companies or indoor growing stores.

Windowsill, Window Box, or Sheltered Porch

Though requiring the least work, this is probably the least ideal propagation setup. While your house may provide enough heat for seeds to germinate, once the seedlings have poked through their seed coats they will forever be searching for and leaning toward the light. If there isn't enough light, they will be pale and spindly. If they are near a window with ample light, they may be hardier but will be what's called "leggy." Their stems will be longer than normal, leading to weakness throughout their life cycle. When you plant leggy plants out in the garden, they often bend or break, or their narrow stem constricts water and nutrient flow to the plant. Given the above challenges, if you plan to use a window, make sure it's facing south and that your trays are as close to the light as possible. Turn them regularly to strengthen the stems as they lean toward the light. Put watertight trays under your seedlings to avoid damage to the windowsill, or use a table you don't care about.

Grow Lights

Another option, and one that is so easy it makes windowsill growing seem like a waste of time, is using special lights to provide the right amount of "sunshine" to your indoor seedlings. Since the lights don't provide all the necessary warmth, you must still situate such a setup in a warm area, such as a corner of your kitchen, dining room, laundry room, or sheltered porch. A table you don't much care about is a must. The easiest to install are special lightbulbs that fit regular light-fixture sockets. These can be ordered online, purchased at hardware stores, or at indoor growing stores. Set a number of lamps or utility lights above your table and keep them on for eighteen hours per day. You can purchase a simple timer if you don't want to bother with remembering to turn them on and off.

Another, perhaps cheaper, setup is to suspend salvaged standard-tube fluorescent light fixtures above a table. The seedlings should be about three inches from the lights. It's best to set a piece of plywood up on bricks or blocks of wood on top of a table so you can lower the seedlings to the right height as they grow.

Greenhouse Box or Minigreenhouse

These are easy to build and much cheaper to buy than a full-size greenhouse. The simplest design: Make a box out of four or more

SALVAGED WINDOW GREENHOUSE BOX

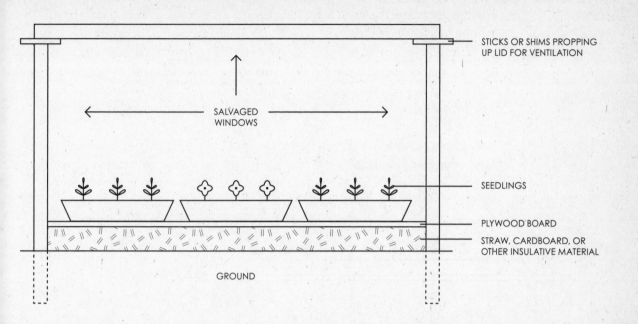

STICKS OR SHIMS PROPPING
UP LID FOR VENTILATION

SALVAGED
WINDOWS

SEEDLINGS

PLYWOOD BOARD

STRAW, CARDBOARD, OR
OTHER INSULATIVE MATERIAL

GROUND

Salvaged window greenhouse box cross-section

windows of the same height by burying the bottom quarter of each window in the ground. A fifth window serves as a removable lid. Care must be taken not to break the glass and to make sure possibly leaded paint doesn't flake into the ground. Also, with this setup you should monitor temperatures with a thermometer, because glass can concentrate too much heat and burn plants. Place a piece of shade cloth over the structure on hot days to prevent this. Vent the greenhouse to let out extra heat and provide airflow by resting the lid on 2 foot by foot boards or shims laid across the top.

Another great small design is an A-frame made by hinging two same-size windows together at the top and attaching polyethylene sheeting to the triangular side openings. An imaginative person could come up with many more cold-frame and greenhouse window designs made from recycled material.

A more permanent solution is to build a box of 2 foot by 4 foot lumber and attach pieces of twin-wall polycarbonate sheeting, or the

A-FRAME SALVAGED WINDOW GREENHOUSE BOX

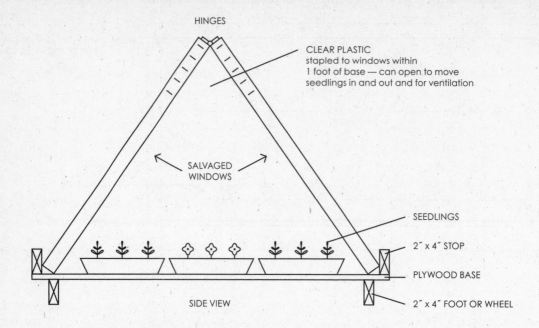

HINGES

CLEAR PLASTIC
stapled to windows within
1 foot of base — can open to move
seedlings in and out and for ventilation

SALVAGED
WINDOWS

SEEDLINGS

2" x 4" STOP

PLYWOOD BASE

SIDE VIEW

2" x 4" FOOT OR WHEEL

Salvaged window A-frame greenhouse box cross-section

less expensive polyethylene sheeting. The semirigid polycarbonate material, sold in 4 foot by 8 foot sheets, is specially formulated for greenhouse glazing. Its twin-wall design lets in the right amount of light for plants, so they won't be burned, and insulates them from the cold. Although the sheets aren't cheap, they're cheaper than buying a ready-made greenhouse.

To build a really useful greenhouse box, find a set of old shelves, preferably made of open-work metal, so light can reach the lower shelves. They should be deep enough to hold seedling trays, at least sixteen inches. Build a frame of wood that fits around the shelves. Then attach polycarbonate sheeting to the frame on the roof and three sides. Use roundhead screws, with a washer between the screw and the sheet to keep it from pulling away. Make a hinged door for the front (along the length of the shelves) and attach polycarbonate sheeting to the door. Add a latch to close the door, and you're done. To vent, prop the door open. To use the less expensive polyethylene sheeting, simply staple thick mill plastic to your frame and nail thin strips of wood over the staples to keep the material from pulling away.

Cold Frames, Seedbeds, and Nursery Beds

Cold frames, seedbeds, and nursery beds are all variations on a theme: outdoor beds to start and hold seedlings. Seedlings are started in the cold frame, seedbed, or nursery bed and transplanted directly into garden beds whenever you deem them big enough. Although these beds are outdoors, they are protected from the elements. Cold frames, seedbeds, and nursery beds should have a well-prepared, nutrient-rich soil. You will need to amend often to provide the necessary fertility to generations of seedlings. It's easiest to protect seedlings from pests if your beds are boxed (cold frames are boxed by definition), so you can cover them with protective material or shade cloth if needed.

Cold frames can be used to start seedlings for later planting, or they can be used to grow plants through to harvest if the climate is too cold for outdoor growing. It is typically a simple box with sloped side-boards and a taller backboard that allows water to run off the pitched cover. The sides are usually solid wood (1x or 2x), the bottom is usually open to the ground, and the top is usually made of a window or glazed frame hinged to the back of the box. Vent by setting shims along the edges of the boards to elevate the lid or by propping it up with stakes (see page 235 to learn how to use cold frames to extend the season).

Seedbeds are often used in more temperate or tropical areas, where the outdoor temperature is warm enough for germination. They can also be used for starting seedlings in cooler climates later in the season, when outdoor temperatures warm. They can be boxed or not and of any size, and should have a south-facing or full-sun orientation, unless you're in a tropical region, in which case partial shade is helpful. Seedbeds can be warmed up by covering them with polyethylene sheeting, twin-wall polycarbonate sheets, or salvaged windows. (If you cover them, make sure the level of the soil is low enough to give the seedlings growing room.) They can be protected from leaf-scorching rays by fixing shade cloth over the bed. Cover the seedbeds with a well-staked floating row cover to protect them from pests.

Greenhouse beds, like seedbeds, can be of any size and boxed or not. They are typically used to grow larger perennials before planting them in the garden. Fruit tree root stocks, newly grafted fruit trees, shrubs, and perennials started from seeds or cuttings may take a long time to grow big enough to plant in their permanent location. Greenhouse beds allow you to keep track of your plants, make sure they get enough water, are properly shaded, and are protected from pests such as gophers, squirrels, deer, and moles. For these reasons, greenhouse beds should be protected with underground wire mesh. Since perennials can put down deep roots, the wire should be set a few feet below the surface. Ideally, the beds should have partial shade provided by trees, a lattice, or a shade cloth structure and be near a water source. If you plan to save seeds, greenhouse beds can be a great holding bed to transplant seed crop into that would otherwise take up growing space.

Greenhouses and Hoop Houses

A purchased greenhouse is a thing of beauty but can be quite expensive and take building know-how to erect. Many newer designs are more durable, since they use twin-wall polycarbonate panels in place of glass for glazing. A good greenhouse will have a manual or electronic method of opening and closing vents, built-in tables, and an optional heater, often gas-powered. For most small-scale farmers in temperate areas the heater isn't necessary. Heat in northern or mountainous regions can also be supplied by layering uncomposted manure covered with straw below the tables in the winter. The manure will give off heat as it decomposes. Heating mats can also be used.

You can build a greenhouse yourself as a simple pitched roof structure from 2x wood, and cover it with twin-wall polycarbonate sheets or thick polyethylene sheeting (be sure to nail strips of wood over the sheeting to keep it from pulling away in the wind). The bottom can be open to the ground. Cover the floor with layers of cardboard or weed-blocker fabric and gravel.

A PVC hoop house can be made any size, from a small one to cover a garden bed to a walk-in. PVC is not an environmentally friendly building material. However, since it's used in abundance in the construction trade it can often be found salvaged. The brilliance of the PVC hoop design is that it's strong, durable, lightweight, and easy to construct with three- and four-way PVC connectors and PVC glue. Special clips can be purchased from farm suppliers to hold polyethylene sheeting to the PVC tubes. The most difficult part is building a door, and this is usually best accomplished with wood. In northern climes where snow drifts pile up, PVC just doesn't hold up under the weight. Metal electrical conduit is a good alternative for constructing inexpensive greenhouses in this case.

PVC HOOP HOUSE

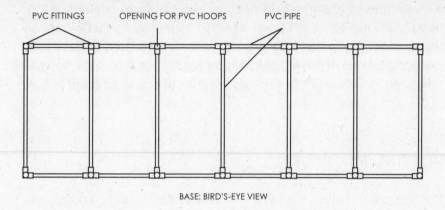

PVC FITTINGS   OPENING FOR PVC HOOPS   PVC PIPE

BASE: BIRD'S-EYE VIEW

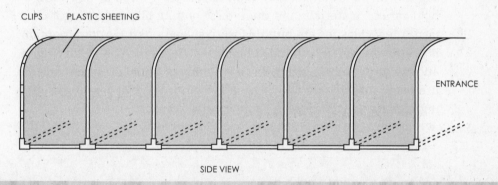

CLIPS   PLASTIC SHEETING

ENTRANCE

SIDE VIEW

Step 1: Construct the floor of the hoop house on the ground.

Step 2: Attach PVC pieces to both sides of the floor to form arches.

Step 3: Clip on plastic sheeting.

Step 4 (not shown): Construct a door.

PROPAGATION TABLES

Tables are an important part of your propagation area. Because they provide an ergonomic work space and elevate your seedlings from the cold ground and protect them from pests. You will probably need them for indoor growing, outdoor hardening-off, and transplanting space. You can use found and salvaged tables or construct them yourself. Be sure to snatch up anything made of unrottable metal. The tables should allow for drainage. The best are made with strong metal mesh tops, so the moisture doesn't rot the table when the trays are watered.

If you're going to build tables, they should be sized to hold the materials you are using: soil blocks (pure soil hardened into a block with a special press), plug trays (solid Styrofoam trays with multiple cells), or flats. Plastic flats are of two sizes: 10 inches by 16 inches and 16 inches square, respectively. The width of the table should be sized to fit two or three rows of trays, depending on your space. If you plan to build tables, be sure they will fit through the door, or construct them inside.

POTTING MIX STORAGE

For larger operations, the most economical way to make potting mix is to purchase various bulk ingredients and mix them together yourself. For this reason, you may want to construct three or four bins, to hold various materials, and then make potting mix by mixing it on a tarp. A few ideas for bin building were given in part I. Another option is to lay a tarp on the ground, dump premade mix on it, and cover with another tarp that you weigh down with bricks. Whichever method you use, make sure that it's easy to back up your truck to deliver these bulk ingredients.

MATERIALS STORAGE

Most of your propagation materials can be stored in a toolshed. Because some effort is involved in acquiring flats and pots, it's a good idea to clean and store them when not in use. Seeds, which can be easily damaged by moisture, should be stored indoors or in a truly waterproof container in the shed.

PROPAGATION MATERIALS

Potting Mixes

These can be bought in bags, in bulk from soil suppliers, or mixed yourself out of a variety of found or purchased ingredients. If you're going to use a premade mix, buy it by the yard, in bulk, to save money.

Purchasing single materials in bulk and mixing them yourself will save you even more. While the common materials—peat, vermiculite, and perlite—are very effective in improving moisture retention and drainage, they are either endangered, like peat, or produced using high-energy input mining and manufacturing processes, as with vermiculite and perlite. Coconut coir and sand are just as effective and much more environmentally sustainable.

Mixing Potting Mix from Purchased Ingredients

Nutritive mix 1:

10 gallons coconut coir

5 gallons sharp sand

10 gallons compost (make sure it's basic and not a mixture itself)

1¼ cup blood or fish meal

1¼ cup bone meal

1¼ cup kelp meal or ashes

Nutritive mix 2:

10 gallons coconut coir

5 gallons sharp sand

10 gallons compost (make sure it's basic and not a mixture itself)

5 gallons fully composted animal manure

Nonnutritive mix (if you prefer to fertilize your seedlings with liquid fertilizer):

20 gallons coconut coir

10 gallons sharp sand

Making Your Own Potting Mix from Found and Free Ingredients

Some feel that purchased mixes are better because they are sterile and weed-seed free. Homemade mixes can introduce soil-borne diseases, weeds, and fungal problems (even mushrooms). To sterilize your own mix, first lay a tarp on the ground and pour the mix out. Moisten it and cover with clear plastic sheeting, weighing it down at the edges with bricks or branches. Leave for one to two months in the summer, moistening every two weeks. If you don't want to go through this rigmarole, you can ensure a weed-seed free mix by making sure the compost and/or manures you use reached high enough temperatures in their composting process to kill weed seeds. Since most diseases arrive with purchased plants and used pots, a much easier way to ensure disease-free seedlings is to sterilize your pots prior to using them.

The general concept when making your own mix is that you want it to hold plant roots and retain moisture, allow for sufficient drainage, and, if desired, provide nutrients (these can also be supplied in water solution). You can experiment with proportions of ingredients based on what is readily available in your region. Following is a list of ingredients grouped for:

- Group 1: moisture retention
- Group 2: drainage
- Group 3: nutrients

Recipes

Compost, either homemade or not, is an important ingredient. If you live in an apartment without a place to make it, or simply can't make enough for your needs, it can be difficult to come by for free. Your garbage company may offer it free to community gardeners (if not, ask them to start a program).

Ingredients

Group 1: For moisture retention

grain hulls

ground peanut hulls

composted ground wood chips

composted solid-wood sawdust (make sure no pressure-treated or glued woods are included)

mushroom compost (free or cheap from mushroom growers, but make sure they use organic methods)

Group 2: For drainage

bagged sand (many cities give out free sandbags to residents to prepare for storms)

beach sand

river sand

sandy soil

Group 3: For nutrients (may also help retain moisture)

compost

clean topsoil or garden soil

coffee and tea grounds

composted animal manures (horse manure is usually most readily available near cities)

composted beer brewery mash, wine grape pomace, or other agricultural waste

composted seaweed

composted fish guts

leaf mold (composted fallen tree leaves)

worm castings

ground eggshells, to supply calcium (use in lesser quantity)

ashes, to supply potassium (use in lesser quantity)

Nutritive mix

10 gallons group 1 ingredients

5 gallons group 2 ingredients

15 gallons group 3 ingredients

Nonnutritive Mix

20 gallons group 1 ingredients

5 gallons group 2 ingredients

The greater variety of ingredients in your mixes, especially of those providing nutrients, the better. However, mixes can be made from a single ingredient from each listed group, depending on availability. Just be sure to follow the proportions. For instance, to make the above nutritive mix from ground peanut hulls, mushroom compost, beach sand, coffee grounds, and clean topsoil would require the following proportions:

5 gallons ground peanut hulls

5 gallons mushroom compost

5 gallons beach sand

7½ gallons coffee grounds

7½ gallons clean topsoil

Use the nonnutritive mix for vegetative propagation of cuttings and fertilize with nutrient solution.

Containers

To get free containers, go to nurseries in your area and ask if they have any old pots, flats, and trays they're getting rid of. You might need to find out when garbage day is and come back then. We recommend using open seeding flats, six packs, 4-inch pots, or plug flats. You will also need rectangular trays for holding seeding flats and six packs and square trays for holding 4-inch pots. It's easier to use containers that are of the same size and height rather than a mishmash of types and sizes. Plug flats are a great way to start seedlings that are okay planted out small, such as salad greens, basil, and herbs, or light feeders, such as onions. These plug trays can hold 32–350 seedlings per tray, depending on the size of the plugs. They are made of Styrofoam or solid plastic. Occasionally you can find free beat-up plug flats from professional farmers. Otherwise, buy them online from a farm-supply company. If you plan to propagate ornamentals, shrubs, or trees you will also need some gallon or bigger pots, but be careful not to stockpile too many—it's easy for your garden to become an eyesore this way.

Seeding Tools

There are many wonderfully designed, manually operated seeding tools that help you place seeds where you want them. From five dollars and up, there's no excuse not to have one. These are also available from farm suppliers.

Labels

Plastic labels are gross garbage, so whatever you do, resist the temptation to buy them new. Wooden plant labels or large popsicle sticks work fine, and can be thrown in the compost or reused. Larger quantities of popsicle sticks can be found at craft supply stores, or you can buy wooden plant labels in bulk from farm suppliers. Old venetian blinds—a plentiful urban waste product—are another option. Just cut them into 4-inch lengths, and they can be used like plastic labels.

Almost all pen ink, even permanent marker, eventually washes off or fades in the sun. Use regular old pencil or grease pencil, or risk producing a lot of mystery plants. Print clearly. You will also need to have a pencil sharpener on hand.

Pest and Fungus Management Materials

See chapter 13 to learn how to use the following materials:

sticky bug paint

yellow card stock

fungicidal baking soda spray

insecticidal soap spray

insecticidal pepper spray

WHICH CROPS TO PROPAGATE

Many crops can be started successfully both indoors and by direct seeding outside. Some experts suggest that seedlings started outside in garden beds have better vigor. However, since tiny sprouts in the garden are susceptible to many a demise, we suggest that you only direct seed the crops that hate being transplanted. The rest should be propagated ahead of planting time. If you understand the pros and cons to greenhouse and direct seeding, you can make your own choices. See chapter 10 to learn how to plant seeds directly into garden beds.

Reasons to start seedlings in a controlled environment:

✣ It provides a uniformly warm environment to support and speed up seed germination.

❉ It's easier to maintain the right amount of water during germination and youth.

❉ Plants are protected from insects and extreme temperatures.

❉ Setting out mature seedlings rather than seeds gives you a lead on the growing season and earlier harvests.

❉ The process of transplanting provides more concentrated nutrition, resulting in healthier plants and higher yields.

❉ Larger starts planted in garden beds are more likely to survive pests than direct-seeded sprouts.

❉ When you finally plant your seedlings in your garden beds, they will be exactly where you want them and there won't be gaps where direct-seeded seeds have failed to sprout, or died.

❉ No time-sensitive and time-consuming thinning is required.

Reasons to direct seed into garden beds:

❉ It's less costly, because it doesn't require special materials and equipment.

❉ There's no danger of stunting root growth when roots hit the bottom of the pot. (This is especially important with larger seeds that can reach the bottom of a pot in a few days.)

❉ There's no shock and stunting due to disturbed roots or leaves during transplanting.

❉ There's no shock or stunting due to moving plants from the shelter of the greenhouse to the outdoor elements.

Rules of thumb:

❉ Root vegetables hate to be transplanted and must *always* be direct seeded.

❉ If you can protect them from insects, it *is* a good idea to direct-seed large-seeded vegetables.

❉ Most slow-growing perennials and ornamentals are best greenhouse seeded or propagated from cuttings.

WE SAY, "BETTER TO START THEM IN THE GREENHOUSE"

We have definitely developed a preference for starting seedlings in a controlled environment over our years of farming. While the pros of direct seeding are good to consider, our experiences with the human factor lead us to weigh in on the other side. Direct seeding is hard for most people to do well. Many people think it takes less time and work, but both methods take a lot of time to do correctly. If you broadcast seeds (aka sprinkling them liberally) into garden beds, you will spend a lot of time thinning after the sprouts come up. On the other hand, placing seeds exactly where you want them also takes a lot of time. And then there's the watering, which you must do every day, sometimes twice. Thinning direct-seeded beds often damages seedlings and is hard to get around to at exactly the right time. Spending the time to start your plants in a controlled environment probably takes about the same amount of time as direct seeding and is usually more successful, since the plants you will plant out will already be big enough to withstand pests and less than daily watering.

Another pro is that many plants gardeners are in the habit of purchasing can be started from seed easily, such as artichokes, cardoons, celery, leeks, onions, parsley, potatoes, and strawberries. Perennial herbs and ornamentals can easily be started from seed or cuttings (an exception is tarragon, which is always started from cuttings). In addition, many plants that gardeners consider to be difficult to start from seed are actually quite easy if you have a good propagation setup. We suspect that difficult usually means that people are simply impatient with seeds that take longer to germinate. Celery is a good example. We've found that while celery takes a while to germinate, it is so hardy you almost can't kill it.

Recommended crops for greenhouse seeding:

- artichokes
- arugula
- asian greens
- basil
- bitter greens (endive, escarole, radicchio, frisee)
- broccoli (heading and sprouting)
- brussels sprouts
- cabbage
- cardoon
- cauliflower
- celery
- chard
- chives
- cilantro
- collards
- cucumber
- dill
- eggplant
- green onions
- kale
- leeks
- lettuce and salad mix
- melons

- mustard greens
- okra
- onions (if starting from seeds)
- parsley
- peppers
- potatoes (if starting from seeds)
- shallots
- squash (summer and winter)
- sweet potatoes (slips)
- tomatillos
- tomatoes

We also start the following categories of plants in a controlled environment:

- edible flowers
- ornamentals
- perennial culinary herbs (from seed or cuttings)
- perennial fruiting vines and shrubs
- perennial medicinal herbs (from seed or cuttings)

Either start the following crops directly in the ground *or* in the greenhouse, depending on outdoor moisture and pest conditions:

- beans
- broccoli rabe/rapini
- corn
- peas
- spinach

PLANNING HOW MANY STARTS TO PROPAGATE

In the previous chapter you will have decided what you want to grow. If you're planning on doing succession planting, you may have created a seeding calendar. See page 201 to calculate the number of plants you will need to fill your beds. Plan to start 10 percent to 25 percent more seedlings than you need to allow for low germination rates. This will also give you extra seedlings to use for filling in if plants die in your garden beds.

The following table will help you to calculate the number of seedlings to propagate and to know how many flats to transplant to meet your goals. For example, to fill a 3½ foot by 8 foot bed with collards you will need approximately twenty-seven seedlings. That works out to 1.68 flats, so plan to transplant 2 flats of collards to ensure you have enough, in case some don't make it. See page 201 for a table listing the number of seedlings you will need for various bed sizes and plant spacings.

Type of Container	Number of Seedlings in a Tray of Containers
Open flat (varies)	200 to 1,500
Speedling plug flats (varies)	32 to 338
Six packs (varies)	48 to 240+
4-inch pots	16

SEED STARTING

You will follow the same basic propagation techniques whether you have a greenhouse, a sunny window, a sheltered porch, indoor grow lights, or a greenhouse box. Different crops are started in different ways as outlined below. See appendix 7 for a table of crops with the recommended propagation method. Basically, the heavy feeders (crops requiring more nutrients) should be given more space and more new soil through transplanting prior to planting out. Avoid transplanting crops that don't like to be transplanted or are light feeders. (See page 506 for a table of crops and their nutrient needs.)

Propagation options for heavy feeders:

- ✜ Seed in an open flat, then transplant up to six packs, and then plant out in the garden.
- ✜ Seed in open flats, then transplant up to 4-inch pots, and then plant out in the garden.

Propagation options for crops that don't like to be transplanted or are light feeders:

- ✜ Seed in an open flat and then plant out in the garden.
- ✜ Seed in six packs and then plant out in the garden.
- ✜ Seed in 4-inch pots and then plant out in the garden.
- ✜ Seed in plug flats and then plant out in the garden.

Propagation options for most perennial plants:

- Seed in open flats, then transplant up to 4-inch pots, and then plant out in the garden.
- Start cuttings in open flats, transplant up to 4-inch pots (possibly gallon or bigger pots), and then plant out in the garden.

SOAKING SEEDS

Constant and even moisture are so important to seed germination that it can help to get a jump-start by soaking seeds for up to forty-eight hours. Air is also a key ingredient in germination, so the soaking must be of a finite duration, or the seeds will rot. Soaked seeds must be planted within a few hours of draining, or else the germ can begin to grow but then will die in the absence of water.

To soak seeds, place the number you plan to start in a shallow dish and cover with water. The next day, strain them. If you like you can dry them for an hour on a paper towel, so they don't stick to your hands.

Seeds to soak: beans, cilantro, corn, cucumber, melons, okra, peas, squash

These crops should be seeded directly into six packs, 4-inch pots, or plug trays.

Once you've decided what method you will use, it's time to gather your materials. First, sterilize your flats, pots, and trays (see text box on page 183). You will need to fill both some open-flats and trays of 4 inch pots and six packs. A number of flats and/or trays of pots can be filled with potting mix at a time and stacked for use when you need them. Preparing a bunch at once saves time. To save your back, once the flats and trays are filled with potting mix, seed them dry. Only water after moving the trays to their permanent spot. If the mix won't hold holes or furrows in this dry state, you may water them before seeding. Water an hour or two ahead, and they will have drained a bit of that heavy water by the time you have to carry them into the greenhouse.

STERILIZING PLANTING CONTAINERS

Sterilize your trays and pots prior to each use to avoid disease and fungal problems. To sterilize terra-cotta or metal containers, simply heat them in the oven for ten minutes in 180 degree heat. To sterilize plastic containers, wash in a bleach or iodine solution. First clean the pots of any soil particles, then let soak for fifteen minutes in a wash basin with five gallons of water to ½ teaspoon of chlorine bleach (sodium hypoclorite 5.5 percent) or five gallons of water to 1 tablespoon plus 2 teaspoons of 10 percent iodine solution. Rinse them off and let air dry. The solution can be labeled and saved for reuse.

Seed-planting depth

Seeds should be planted three to four times as deep as their width. For larger seeds, poke holes with your finger and drop seeds in. For smaller seeds, you have two options: make rows of furrow depressions with your finger and sprinkle them in the rows; or sprinkle the seeds in rows on the surface and cover the flat with a fine dusting of planting mix.

How to sprinkle seeds 1:

* Dry your hands.
* Open the seed packet.
* Tip a small pile of seeds into the opposite hand from your dominant hand.
* Cupping the hand with the seeds, pinch out single seeds or a few seeds with the pointer finger and thumb of your dominant hand.
* Drop the pinched seeds into the hole or furrow, aiming for about ¼ inch apart.

How to sprinkle seeds 2:

* (Use the seed packet itself as a seeder.)
* Make a very hard crease down the length of one side.
* Hold the packet in your hand, tapping the back of your hand gently.
* The seeds will line up single file down the crease.
* Tap them into the hole or furrow.

How to sprinkle seeds 3:

❖ We also highly recommend using a mechanical or manually operated seeding instrument, which is available from seed-supply companies. If you are using one of these, just keep everything dry.

PATTERNS OF GROWTH AND SEED LEAVES

Once a seed germinates, a root begins to grow downward, and the first leaf or leaves, called "cotyledon(s)," begin to reach up toward the light. These "seed leaves" were present in the seed in a nascent form. Most seedlings have two seed leaves, while grasses such as corn and *Allium*s like onions have one. With those that have two seed leaves, those of many species look very similar to each other and very different from the crops' "true leaves." The seed leaves often die back as the true leaves begin to form, and on seedlings with a distinct stem, it's okay to bury the seed leaves at transplanting or planting time.

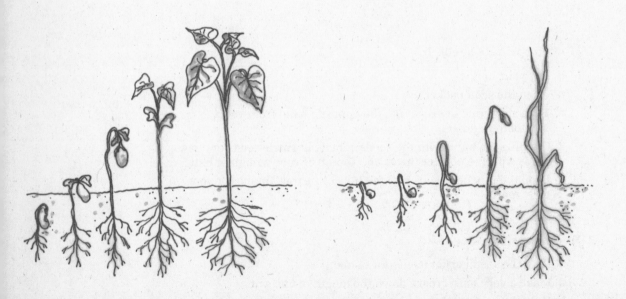

Left: A bean is an example of a dicotyledon. Right: An onion is an example of a monocotyledon.

Among the seedlings with two seed leaves, broadly speaking there are two patterns of growth: an up-reaching stem, with leaves growing off it at intervals; and a rosette, with new leaves forming close to the soil level in the center of the seedling. Seedlings with a stem-growth pattern should be transplanted or planted quite deep, covering the seed leaves and a few layers of true leaves—as long as the growing tip is exposed. Seedlings with a rosette pattern, and those with one cotyledon, must be transplanted and planted carefully at just the right height, ensuring that the rosette, where new leaves are forming, is not buried, but also that the part of the root below the rosette is not exposed. This usually means that the dirt level, once planted out, should be the same as the dirt level was when the plant was in the pot.

Top: Rosette-type seedling transplanting depth.
Bottom: Stem-type seedling transplanting depth.

Preparing and seeding open flats:

✢ Put a solid flat (the kind with drainage holes) inside an openwork tray of the same size, for stability.

✢ Fill the flat with potting soil to the top and tap the bottom on a hard surface a few times, adding more if needed.

✢ Most flats have a ridge down the center. It's best to make four rows of seeds in each flat, two on either side of the ridge.

✢ You can fill the flat with one type of seed or make different sections for different veggies.

✢ The ideal seed spacing is ⅛ inch to ½ inch apart in the row, depending on the seed size.

✢ Scatter potting soil into the furrows and press down firmly over the seeds.

✢ Label with one label per flat or per section.

✢ The seedlings can be transplanted from the time that the cotyledon leaves emerge, up to when the first set of true leaves emerges.

✢ Seedlings that will be planted into the ground directly from the flat are ready when they have their first or second set of true leaves and are 1 inch to 3 inches tall.

SEEDING AN OPEN TRAY

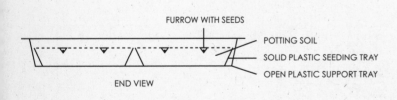

FURROW WITH SEEDS

POTTING SOIL
SOLID PLASTIC SEEDING TRAY
OPEN PLASTIC SUPPORT TRAY

END VIEW

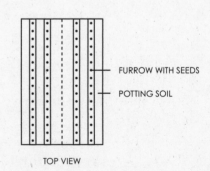

FURROW WITH SEEDS
POTTING SOIL

TOP VIEW

Preparing and seeding trays of six packs and 4-inch pots:

❖ Fill the trays with six packs or 4-inch pots. The pots should be snug and the six packs should be oriented in the same direction (4-inch pots fit in square trays, six packs in rectangular trays).

❖ Set the first tray of pots on the ground and mound potting soil over them with a shovel or trowel until they are overfilled.

❖ Tamp down by tapping the tray on the ground a few times, and brush off the excess with your hands.

❖ Stack the filled trays near your potting table.

❖ Poke holes into each pot or cell to the correct depth and drop in the seeds.

❖ Label with one label per pot or one label per tray, depending on your needs.

❖ The plants are ready to plant in garden beds once they have a few layers of true leaves. You can also check for readiness by temporarily popping a seedling out of the pot or cell to see if it has a well-developed root system.

How to Take a Seedling Out of a Pot:

- Dig a hole for the seedling.
- Make sure the potting soil in the pot is moist.
- Gently squeeze the sides of the pot.
- Make a V sign with your dominant hand.
- Place the V on either side of the plant stem, moving your other fingers next to the V to cover the pot opening.
- Turn the pot over and tap or squeeze the pot until the seedling is released onto your palm.
- Turn the plant over into your prepared planting hole.

How to Take a Seedling Out of a Cell:

- Dig a hole for the seedling.
- Make sure the potting soil in the cell is moist.
- Take a butter knife or other flat implement and slide it along one side of the cell.
- Lift the soil upward to pop the cell out.
- Cradle by the root ball and place into your prepared planting hole.

Rooting hormone is a powdered or liquid substance that encourages latent bud/root nodes on plant cuttings to form roots. You should dip or soak cuttings in rooting hormone prior to laying them in planting mix. Although you can buy rooting hormone, it's also easy to make your own; all varieties of the willow tree can be used. You may have noticed that willow trees often choke creeks and rivers. That's because they contain a lot of natural rooting hormone, and root easily.

HOW TO MAKE YOUR OWN SOLUTION

- Find any variety of willow tree (Salix) and cut about fifteen or twenty thin twigs.
- Cut the twigs into small pieces and put them in a five-gallon bucket; cover it with about a gallon of water.
- Soak the twig pieces for twenty-four hours.
- Strain the water into another container, and put the twig pieces in the compost.
- Soak the cuttings you want to root in the solution for a day before potting.
- Either water newly planted plants with the remaining solution or store it in the fridge for no longer than a month.

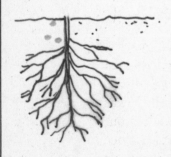

If you're having a hard time finding a willow tree, you can grow one on your property, to have a constant supply of sticks. Just go to a creek in a city park or out in the countryside and cut a few branches. They root so easily you can stick them in moist ground, especially near a water tap, to establish your own plants. Not only will you have a supply for rooting hormone, but the branches also make great kindling and firewood for your fire circle or barbecue when dried. You'll need to cut your tree back, since they can become invasive or huge.

Preparing trays of cuttings:

- Cuttings are taken from an established parent plant and should be at least 3 inches long and contain a number of nodes (spots where leaves are growing).
- Remove the lower leaves, making sure not to damage the bark.
- Dip the cuttings first in water, and then in powdered rooting hormone or let them soak in liquid rooting hormone solution overnight.
- Fill an open flat half full with potting mix.
- Lay the cuttings in rows with the stems at an angle, starting with the back row and moving forward. The cuttings should be about 1½ inches apart.
- Cover 1 to 2 inches of the cutting stem with potting mix as you go.
- You can fill the flat with one type of cutting or use different sections for different types.
- Label with one label per flat or section.
- After a few weeks, gently uncover the lower portion of a cutting to see if roots have formed, and if so, transplant up to 4-inch pots.

TUBERS AND BULBS

Some crops are grown from tubers, bulbs, or slips (sprouts started from tubers) rather than seeds, for example potatoes, sweet potatoes, yams, bulb onions, and garlic. Purchasing these seed or slip potatoes and onion or garlic sets can be quite expensive. They can also be hard to find, since they are usually available for purchase only at specific times of year. It makes sense to make an initial investment on guaranteed disease-free seed potatoes. If you work it right, you won't have to buy potatoes again, since you can save tubers or bulbs from the crop you harvest for the next year (see chapter 14 for information on seed saving). Why not just use potatoes and garlic from the store to start your plants? There is a danger: If the farm they were grown on was infected with a viral or fungal disease, you could import it to your farm. If you decide to take the risk, use organic tubers or bulbs, since organic farmers tend to be more careful about soil-borne fungal and viral diseases. And if you have time and want to save money, all the *Allium*s except for garlic (onions, leeks, shallots, green onions, and spring onions) can be started easily from seed; they will just take longer to mature than those started from onion sets.

Making Your Own Disease-Free Potato or Sweet Potato Slips

Potatoes can be grown for much of the year in temperate areas. It can be difficult to find seed potatoes throughout the year, since they are a seasonal crop for non-temperate areas. Grocery store potatoes can carry diseases. However, if you start your potatoes from slips, you can safely start with grocery store stock.

1. Purchase potatoes at the store.

2. Organize your work so you can keep track of the variety names.

3. Cut the potatoes into pieces, each having at least two "eyes."

4. Dip the cut ends in ashes and allow to "heal over" (dry) for about three days.

5. Plant the pieces of potato for each variety in one or more gallon pots and label them.

6. Keep moist, but allow to dry out between waterings so the roots don't rot.

7. Allow the vines to grow to 6 inches to 8 inches.

8. Cut off the shoots at least 2 inches above the top of the pot, being careful not to touch the soil with your hands or tools (this could spread a disease).

9. Plant each shoot in a 4-inch pot, with at least one set of leaf nodes below the soil surface. It's okay to bury leaves, or you can carefully remove them.

10. Throw the potatoes and soil from the mother pot into the garbage.

11. Sterilize the mother pot.

12. After the shoots in the 4-inch pot establish roots, plant them in the garden bed. Pat yourself on the back: You avoided disease and saved some serious cash!

Note: The same method can be used to start sweet potato slips, but unlike regular potatoes, sweet potatoes and yams need hot weather to mature.

TRANSPLANTING

Once seedlings in open flats have grown one set of true leaves, they can be transplanted into trays of six packs, 4-inch pots, or a garden bed:

- With a knife or your fingers, loosen a 2-inch-square section of seedlings from the flat and gently remove it.
- Try not to touch the roots, as this will damage root hairs. Hold seedlings by their leaves—not the stem!—to separate. *Separating seedlings by the stems can damage them and lead to damping off disease. Seedlings can grow more leaves but only have one stem.*
- Gently loosen a seedling from the group by tugging the leaves, leaving roots to hang free or cradling the root ball with your hand if the soil is heavy.
- If you damage a seedling, discard it. You will most likely have more seedlings than you need.
- Make a hole with one hand while holding the seedling. Drop the seedling in to a depth deep enough to cover the seed leaves, but leaving the true leaves and growing tip exposed. *To fit all the roots this often means making a hole down to the bottom of the pot or cell.*
- Label, using one label per pot or one label per tray, depending on your needs.
- Your transplanted seedlings are ready to plant out when they have two to five sets of true leaves and are 2 inches to 6 inches high.

LABELING SEEDLINGS

Depending on who gets plants from your garden, you may be able to get away with one plant label per tray of six packs or 4-inch pots. If many people use the seedlings, you will need to do the tedious work of labeling each pot or six pack. It pays off, though, because unlabeled seedlings often end up in the compost. If you are a seed saver, unlabeled plants make it impossible. When you plant the seedlings out in the garden, you will label the bed with a larger stake or paint stick with the same information, so you can track which cultivars you like best and which to save seed from.

It is especially important with tomatoes, peppers, and squash to include the type of tomato, pepper, or squash the plant will produce.

The cultivar name often provides no information. "Chianti Rose" doesn't tell you if the tomato is a beefsteak or cherry, red, or yellow, or a determinate or indeterminate type (Chianti Rose is an indeterminate beefsteak type). "Zephyr" squash could be summer or winter squash, and if you're mistaken, you will miss the right harvest time (Zephyr is a hybrid summer squash). One thing we have often done is made a list of all the tomatoes, squash, and peppers we started in the greenhouse, with descriptions of each, and laminated and posted it for reference. Alternatively, you can keep copies of the seed catalogs you ordered from in your toolshed to look up names you may have forgotten.

What to include on the plant label:

- seeding date
- crop
- cultivar
- type (can include as many aspects as you want to include)
- open pollinated or hybrid
- seed supplier

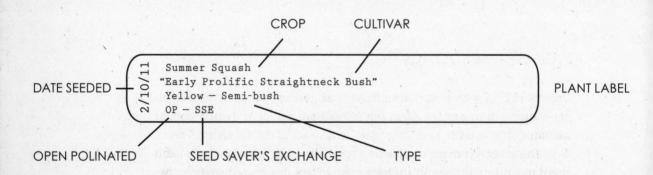

CROP CULTIVAR

DATE SEEDED — 2/10/11
Summer Squash
"Early Prolific Straightneck Bush"
Yellow — Semi-bush
OP — SSE
PLANT LABEL

OPEN POLINATED SEED SAVER'S EXCHANGE TYPE

An ideal, if not always possible, way to label plants

For example:

Seeding Date	Crop	Cultivar	Type	Pollination	Seed Company	
7/12/10	Broccoli	Diplomat	Heading: fall crop	OP	FS	Fedco Seeds
2/20/10	Pepper	Serrano del Sol	Hot	H	JSS	Johnny's Selected Seeds
1/18/11	Tomato	Stupice	Medium red; early indeterminate	OP	BG	Bountiful Gardens
2/16/11	Squash	Ronde de Nice	Summer	OP	RGS	Renee's Garden Seeds

WATERING YOUR SEEDLINGS

Supplying the correct amount of water is essential to their survival and will make your plants more robust throughout their life cycle. You often get an inaccurate picture of how well seedlings have been taken care of at garden centers or hardware stores. Store-bought seedlings are often old and root-bound, and have been allowed to dry out one time too many. They look good on the shelf because employees water them first thing in the morning and often pump them full of artificial fertilizers to perk them up. Plant vigor and productivity is severely reduced when seedlings are repeatedly allowed to wilt from lack of water or don't have enough root space. In addition, serious fungal and soil-borne diseases that can't be eradicated once in your soil, such as verticillium wilt, can easily be brought to your farm in a cheap tomato seedling from Home Shmeepo. Now we're not saying all garden stores are alike. Your locally owned nursery probably has much better practices than your chain megastore, but no one can keep quality as high as you can at home.

Depending on air temperature and humidity, you will probably need to water your seedlings between twice a day and once every two days. They should be kept moist but not sopping wet. Proper drainage is key, so make sure they are in trays with holes, and that you use or make a well-draining potting mix. Trays and seedlings should be allowed to dry to the point of slight dampness in between waterings, as soil oxygen is essential to root function. However, if allowed to dry out fully for too long, seeds won't germinate, so take extra care until the sprouts are up. If you see the growth of mossy green algae, or an explosion of whiteflies or gnats, it's probably an indication that there's too much soil moisture. You can reduce watering, and perhaps run a fan. If your trays and pots are constantly drying out due to intense heat, increase the ventilation and water more often. A good way to gauge how well your watering system is working is to periodically pick up a tray. You will come to know the difference in weight between a dry and a moist tray.

If hand watering, use a watering can or a hose watering wand that has a gentle spray. The mist setting, however, almost never gives enough water. Water everything once. Then make another pass to make sure enough water has soaked in. The first time you water do a finger poke test to see if the soil is moist to the bottom of the trays and pots, and adjust as needed.

If using drip irrigation, be sure to select high-quality misters designed for greenhouse applications. Farm suppliers can be much more helpful in this regard than irrigation suppliers geared toward the ornamental landscaping market. When choosing a timer, be sure to select one that has a water-resistant box, or keep it outside. To increase the amount of water your seedlings get, increase the frequency of watering rather than the duration. For instance, in hot weather set two watering times per day of five minutes each rather than one of ten minutes. Most of the water in a long-duration schedule leaches nutrients and goes into the ground, watering greenhouse weeds.

If you really stressed your seedlings by forgetting to water them repeatedly, you might consider starting over rather than trying to save them. The connection between how well the plant's been treated early and its future ability to ward off insects and diseases and produce high yields is very significant.

FERTILIZING

If you are using a nutritive potting mix and transplanting seedlings into new pots, you will most likely not need to fertilize. For slow-growing perennials, trees, and seedlings that end up sitting around a while before you can get them into the ground, weekly or monthly fertilizing is a good idea. Be sure that they aren't dry when you apply any fertilizer or the roots may be shocked and burned by the nutrients.

Purchased Fertilizers

Since it is made of fish-factory waste, fish emulsion is the most environmentally friendly manufactured fertilizers. It's also one of the most economical. To use it in the greenhouse, mix at half strength for what's recommended on the bottle for vegetables in a watering can or with a siphon mixer, and water your seedlings with it.

Homemade Fertilizer Recipes

One aspect of human waste that is completely safe is your own urine. Dilute it with two parts water and irrigate with it. It is especially high in nitrogen and will burn plants if used full strength. This is a great argument for having a secluded nook, possibly with a curtain, outdoors, so you can deliver the goods without sloshing through your house.

Nutrient "teas" can be made from worm castings, manure, fish guts, compost and seaweed, spent brewery grain, or just about any other organic matter. There are two methods for making them. The first is potentially smelly but very easy; the second requires daily hand aeration or aeration equipment but smells like roses (well, not *quite* like roses, but certainly better than the other kind). Concentrated fish emulsion can also be made from fish guts if you fish or have a nearby fish-processing plant to get some entrails. Some will also think making this concoction is unbelievably disgusting, but hey, if you can bear it, you'll have the most amazing veggies around.

Anything with sugar in it can be added to homemade fertilizers to speed up decomposition; i.e., fruit juices and molasses. These also cut down on the bad smells.

The easy tea recipe uses anaerobic (airless) decomposition and produces a liquid fertilizer that is rich in nutrients but smells bad. Relations with the neighbors may become strained. However, you may be won over by how easy it is to prepare and the low cost.

To use nutrient teas, water your seedlings with a watering can or a siphon mixer. You can also use a spray bottle or siphon mixer, with the mist setting on your watering wand to use as a foliar fertilizer.

Anaerobic nutrient tea

- ❖ Find a five-gallon plastic container with an airtight lid.
- ❖ Fill the container one-third to one-half full of compost, animal manure, fish guts, or worm castings, or to the top with fresh, tender

green plant clippings or seaweed. You can make straining easier by making a burlap or floating row cover "tea bag" to hold your materials.

❖ Fill the container with water.

❖ Cover tightly.

❖ Steep for five to ten days (less in warmer weather).

❖ Strain and dilute one part tea to two parts water.

Aerobic nutrient tea

Method 1:

❖ Find a five-gallon plastic container with an airtight lid.

❖ Fill the container one-third to one-half full of compost, animal manure, fish guts, or worm castings, or to the top with fresh, tender green plant clippings or seaweed. You can make straining easier by making a burlap or floating row cover "tea bag" to hold your materials.

❖ Fill the container with water.

❖ Cover tightly to keep out pests.

❖ Stir once a day for about two weeks.

❖ Strain and dilute one part tea to two parts water.

Due to the disgusting nature of anaerobic teas, and the somewhat disgusting nature of aerobic teas made by method number 1, some ingenious gardener came up with a new method, called actively aerated compost tea. This type doesn't smell as bad because it uses aerobic decomposition by pumping air continuously through the water. To make your own rig isn't too difficult, and you'll be pleased with the results.

Actively aerated compost tea (AACT)

❖ Find a five-gallon plastic container with an airtight lid.

❖ At a pet store, purchase an aquarium aerator, a three-gang valve, and 3 feet to 5 feet of flexible hose tubing to fit the valve.

❖ Cut three lengths of tubing and attach to the valves.

❖ Plug the aerator pump into the valve.

❖ Hang the valve on the edge of the container and put the tubes into it so that air will bubble from the bottom.

❖ Using burlap or floating row cover, make a "teabag" of compost, animal manure, worm castings, fish guts, fresh green manure, or seaweed.

❖ Fill the container with the teabag and water.

- Put the container in a dark, preferably warm place and aerate for twenty-four to forty-eight hours.
- Lift out and drain the teabag.
- Dilute one part tea to two parts water.

Concentrated aerobically decomposed fish emulsion

While you can make fish tea by the above methods, you can also make homemade fish emulsion paste. Homemade concentrated fish emulsion is of better quality than what's made in the factory, because the latter don't usually have the beneficial fish oils or bones after their manufacture.

Method 1:

- Find a five-gallon plastic container with an airtight lid.
- Fill the container half full with brown compost materials, such as sawdust, leaves, or straw. Top off the bucket with fish guts, heads, etc.
- Cover tightly to keep out pests.
- Stir once a day for about two weeks.
- Apply by diluting with equal parts water, up to five parts water, depending on how much fertilization is needed.

Method 2:

- Find a five-gallon plastic container with an airtight lid.
- Fill with fish guts, heads, etc.
- Cover with water.
- Cover with an airtight lid to keep out pests and to allow it to decompose anaerobically.
- Store in a cool, dark place for three to four months (it shouldn't smell too bad by that time).
- Strain and, if needed, dilute with water.
- Store in airtight containers and apply by diluting with equal parts water, up to five parts water, depending on how much fertilization is needed.

"HARDENING OFF" SEEDLINGS BEFORE PLANTING

So you've successfully germinated your seed trays and fertigated (fertilizer plus irrigation, get it?) them with some homemade nutrient teas, and we bet the seedlings are looking great. Now's the time to plant them out into the big bad world. Seedlings are ready to plant out when they have a number of sets of true leaves and a well-developed root system. This perfect window of development can really vary between crops. The most important factor is invisible below the soil—the root development. If the root ball is small and hasn't filled out the pot or cell, the seedling should remain in the shelter of the greenhouse. However, if the root ball or taproot is reaching out the bottom hole or circling around the bottom of the pot or cell, stunting may have already set in—you missed your window. Until you are practiced at knowing how the leaves will look when the plant is in the perfect stage for transplanting, get in the habit of popping one seedling out of its pot or cell to check root growth.

When a set of seedlings is ready to plant out it's best to harden them off. They are used to a cushy life of tropical ease and are unprepared for the real world of temperature fluctuations, wind, rain, and direct sun. Planting these seedlings without a few days of adjusting to the real world might result in the seedlings turning yellow, wilting, dying, and generally acting shocked. When colder temperatures still reign, set seedlings you want to plant on a shady table each day and return them to the greenhouse at night for a few days. In warmer weather, you needn't return them to the greenhouse at night. Simply keeping them shaded and elevated will be sufficient. After a few days of adjustment, the seedlings are ready to be planted in the ground. The next chapter will go over the best practices for planting these seedlings.

W   E

CHAPTER 10

PLANTING AND MAINTAINING CROPS

This is where it all comes together. You've created your farm beds, amended the soil, started seedlings in the greenhouse or on the windowsill, and now it's time to finally plant out your garden. This is an amazingly satisfying time—tucking plants and seeds into newly prepped beds—but it's also a time to make some informed decisions: How far apart should you space your crops? Which plants can be grown together? Which plants should be direct seeded? It can be a time of anxiety for the first-time farmer, or when working with a new crop, because errors in placement can reduce yields or, in the worst case scenario, cause crop failure. Don't panic, though; this chapter will cover all of these issues, with the idea of maximizing your yield in smaller spaces.

INTENSIVE PLANTING FOR HIGH YIELDS IN SMALL SPACES

While rural farmers with land can afford to have lower yields per square foot of growing space, urban farmers certainly cannot. To receive worthwhile yields from small gardens (anything under an acre), intensive planting is a must. This means growing as many plants as possible for as much of the year as possible. This chapter will give you the tools and knowledge you need to plant crops close together while ensuring that they are healthy enough to bring you an abundant harvest. Intensive planting goes hand in hand with the intensive soil-building practices described in chapter 8. To sustain continuous yields you must replenish some key nutrients periodically.

The three most important intensive-planting techniques:

1. Intensive plant spacing, where they are set as closely together as possible so that when they reach maturity the leaves just barely touch. This technique is also known as a "living mulch," because the cover of green growth doesn't allow the sunlight to reach the soil, which makes for fewer weeds and less evaporation.

2. Interplanting two or more crops that fill different areas of the vertical space in the same growing bed; for instance, combining low-growing radishes with taller collard greens and vining beans on a trellis. This allows you to take advantage of all areas of the aboveground space.

3. Interplanting two or more crops in the same growing bed that grow at different rates and mature at different times. While slow-growing plants aren't yet taking up much room, fast-growing crops can be grown to maturity.

Intensive Plant Spacing

Intensively planted garden beds are incredibly productive. If you use these techniques you will be pleased to find that one bed will yield enough carrots for half the year or lettuce for the entire neighborhood. Plants should be spaced so that their leaves will just touch at full maturity. An easy way to achieve this is to use the suggested *plant* spacing and ignore the *row* spacing given on the seed packet. Space all plants equally. See appendix 4 for correct intensive plant spacing.

Unlike with field farming, intensively planted beds don't have rows of crops. It is more space efficient to plant seedlings or seeds in what are called "staggered rows" in 3 foot to 4 foot wide beds. If you look at one of these beds, you will see a triangular or hexagonal pattern. This ensures that all seedlings are equidistant. If, on the other hand, you use a square planting pattern, the distance between plants diagonally will be greater. This wastes an inch or two of space every foot or so. While this loss may not seem great at first, over the length of a garden bed, it adds up.

The closer together plants are supposed to be spaced, the more you will save with staggered-row planting. With plants spaced 12 inches apart you will only gain about 8 percent more plants; with plants spaced 4 inches apart you will gain about 20 percent. For example, staggered-row plantings of cauliflower will fit in only 8 percent more plants; staggered-row plantings of spinach will allow you to plant 20 percent more. Obviously, hexagonal spacing of crops such as lettuce, radishes, carrots, and beets are a must for increasing yields.

The number of plants or seeds needed to fill an intensively planted bed can be surprisingly high. The following table will help you plan to start the right amount of seedlings or have the right number of seeds on hand.

For beginning and experienced gardeners alike, eyeballing the correct spacing can be quite difficult. You've got this tiny seedling that will one day have a wingspan of 18 inches. It's just very hard to imagine how much space it will one day need. To make your plant spacing accurate, use a set of measuring sticks, a ruler, or a spacing jig (see diagram, page 204).

APPROXIMATE NUMBER OF PLANTS OR SEEDS NEEDED FOR A HEXAGONAL PLANTING PATTERN

Bed Size	Spacing							
	2 inches	3 inches	4 inches	6 inches	8 inches	12 inches	18 inches	24 inches
3.5 feet wide by 4 feet long	554	243	140	59	35	12	4	3
3.5 feet wide by 8 feet long	1,128	500	270	117	70	27	10	6
3.5 feet wide by 16 feet long	2,276	1,000	550	234	140	54	20	12

Type of container	Number of Seedlings in a Tray of Containers
Open flat (varies)	200 to 1,500
Seedling plug flats (varies)	32 to 338
Six packs (varies)	48 to 240
Four inch pots	16

TRADITIONAL ROW SPACING

Plants spaced 8 inches apart

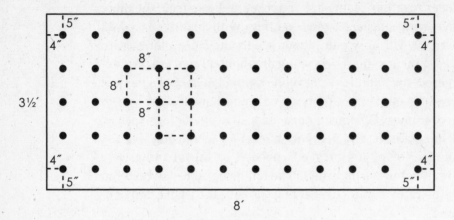

60 PLANTS

Each dot represents a plant, planted in rows 8 inches apart. Sixty plants are possible with this method.

INTENSIVE PLANTING PATTERN WITH HEXAGONAL SPACING

Plants spaced 8 inches apart

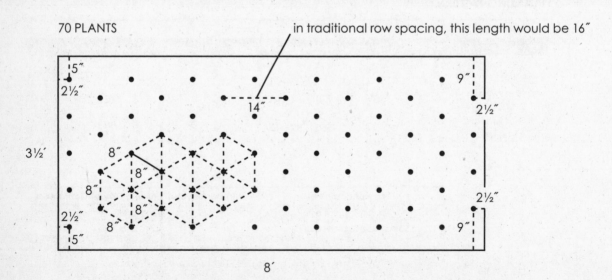

70 PLANTS

in traditional row spacing, this length would be 16"

Each dot represents a plant, planted in offset rows 8 inches apart. Seventy plants are possible with this method.

To lay out a hexagonal pattern, set out seedlings or poke holes in a row across the width at the correct spacing distance. Now measure from the first seedling in the first row to the first seedling in your next planned row. Since it's a staggered row, the first seedling or hole should be halfway between the first and second seedlings in your first row. In other words, the first two seedlings in your first row and the first seedling in your second row should form a triangle. Complete the second row by measuring from seedling to seedling in the row. At the end of each row check that it's straight by measuring your triangles, and adjust as needed.

To make a set of measuring sticks, find some shims, paint-stirring sticks, lath, or other strips of wood and cut one of each plant-spacing length you may need: 2, 3, 4, 6, 8, 12, 18, and 24 inches. Make a bundle of the sticks with one end lined up. Clamp them to a sawhorse or piece of scrap wood and drill a hole near the top. Find or buy a large-size key ring or a piece of twine and attach your sticks together like a set of measuring spoons. If running a larger garden, why not make three or four sets?

PLANT SPACING STICKS

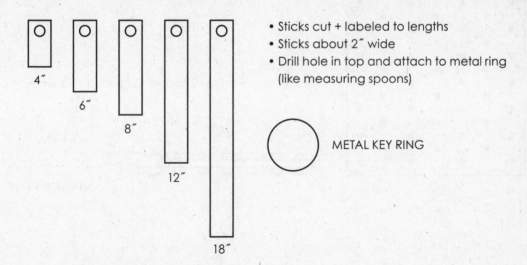

4"

6"

8"

12"

18"

- Sticks cut + labeled to lengths
- Sticks about 2" wide
- Drill hole in top and attach to metal ring (like measuring spoons)

METAL KEY RING

Simple to make plant spacing sticks are used in the garden to make sure you're correctly spacing plants

To make plant spacing jigs, cut plywood triangles to the same dimensions as listed above for the spacing sticks. Cut them out with a circular or jigsaw. Use a trowel or stick to mark the corners of the triangle in your bed as you go.

For speed planting you might want to construct a set of planting/seeding jigs that actually poke the holes for you. Planting jigs are for transplanting seedlings and seeding jigs are for the direct sowing of seeds. Creating these jigs will require a bit more work, but your workload will be reduced for years to come. Seeding and planting jigs are boards the width of your beds with pegs at the correct spacing. Simply press them into the soil to create planting or seeding holes. Make separate planting jigs for 4, 6, 8, and 12 inch spacings. Seeding jigs are most important for closely spaced crops, so make separate ones for 2, 3, 4, and 6 inch spacings.

To make jigs, cut pieces of ½-inch plywood to the width of your

SEEDING JIG FOR 2-INCH, 3-INCH, 4-INCH, AND 6-INCH SPACING

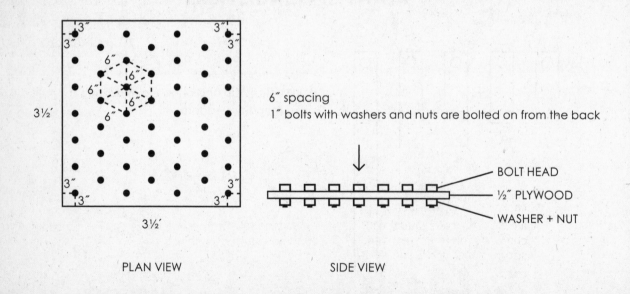

6″ spacing
1″ bolts with washers and nuts are bolted on from the back

BOLT HEAD
½″ PLYWOOD
WASHER + NUT

3½′

3½′

PLAN VIEW

SIDE VIEW

Each dot represents a seed. This jig is great for planting root vegetables, like carrots and cut-and-come-again salad mix. This plan shows 6-inch spacing.

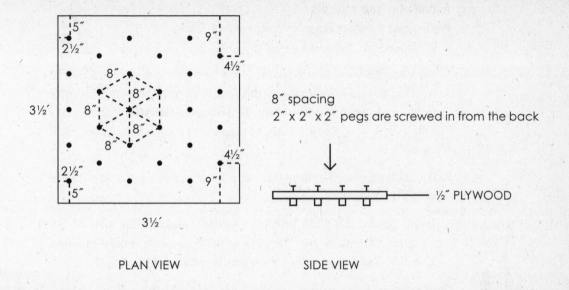

8″ spacing

2″ x 2″ x 2″ pegs are screwed in from the back

½″ PLYWOOD

PLAN VIEW   SIDE VIEW

This jig uses wooden pegs to create a hole for your seedling. This plan shows 8-inch spacing.

bed (if it's 4 feet wide no cut is needed, since that's the width of plywood). Then cut it in half to make a 4 foot long jig. Let's take 6-inch spacing as an example. If these instructions seem overwhelming, read each next step when you are ready to do it. It will make sense as you go.

Tools and materials you may need:

- graphite pencil
- cardboard or cardstock
- a chalk snap line (a tool with a spool of string that sits in powdered chalk. Unroll it on your wood, pull it tight and snap it to make a straight line) or a T square
- circular saw
- chop saw (optional)
- drill
- ½-inch plywood
- 2 inch by 2 inch boards for pegs (for planting jigs)

‡ 4¾-inch bolts to fasten pegs (for planting jigs)

‡ 1-inch bolts to make seeding holes (for seeding jigs)

‡ bolt nuts and washers

‡ Two metal D handles per jig for ease of lifting

1. For each jig you plan to make, cut a hexagon out of cardboard or cardstock where each side is the length of your planting distance (6 inches long in our example). To accommodate your pencil, poke a hole in the center of each hexagon and snip a tiny bit of each corner off.

2. For each jig you plan to make, cut a 3½ foot square board out of ½ inch plywood.

3. Using your chalk snap line, or T square, make a line half of the planting dimension from the edge of the board, in our case 3 inches. This first line can go either the length or width of the board.

CARDBOARD MARKING JIG

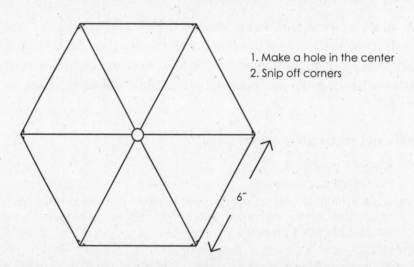

1. Make a hole in the center
2. Snip off corners

6˝

A hexagonal cardboard marking jig with 6-inch spacing. Note the center hole and corner snips to accommodate your pencil mark.

4. After you have made the line, line the center hole of your cardboard hexagon along the line, hanging the cardboard 3 inches off the plywood at the top and side (see the diagram below).

5. Mark dots in the center hole and at the corners of the cardboard hexagon that are resting on the board.

6. Move the cardboard down and line up the top with your lower dot, making sure to keep the hexagon square with the board. Mark dots in the center hole and at the corners that are resting on the board.

7. Continue until you have filled your board with marks. Check your marks with a ruler once in a while to make sure you are keeping your jig square to your board.

MARKING A JIG

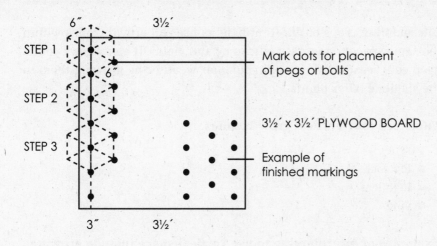

Make a line 3″ from the edge of the board.
Line up the center of the marking jig with this line.
Continue to fill board with markings by moving the marking jig.

How to mark drill holes for a planting or seeding jig with 6-inch spacing

8. To make a planting jig, count your dots and cut the same number of 3-inch lengths from a 2 inch by 2 inch board. Once attached to your dots, these will poke holes for your seedlings.

9. Predrill holes in the center of each 3-inch piece of wood and at each dot on your board, using a bit one size larger than the bolt size.

10. Bolt a 3-inch piece of wood to each dot, attaching the bolt with a washer and nut on each side.

11. To make a seeding jig, predrill holes one size larger than your bolt size at each mark on your board.

12. Thread 1 inch long bolts through the board at the dots and attach with washers and nuts. These bolts will poke seeding holes.

13. In order to be able to carry and place your board, attach two D handles to the back of the board at a comfortable distance apart for your arms.

INTERPLANTING TO USE MORE VERTICAL SPACE

Interplanting crops of different heights uses all growth zones within the same space to increase efficiency and yields. It also allows you to keep cool season crops cool in summer weather by growing them in the shade of taller plants.

There are four vertical growth zones:

- root
- low leaf
- high leaf
- vine

Examples of interplanting to use vertical space include growing:

- low-leaf lettuce with high-leaf collards;
- high-leaf kale with root carrots;
- vine tomatoes with high-leaf peppers and root beets; and
- vine snap beans with high-leaf eggplant and low-leaf mâche.

Either cages or trellises are needed with vining crops.
To interplant for different zones increase the spacing of tall-leaf

INTENSIVE PLANTING PATTERN
WITH HEXAGONAL SPACING
INTERPLANTING KALE (8 INCHES APART)
AND CARROTS (3 INCHES APART)

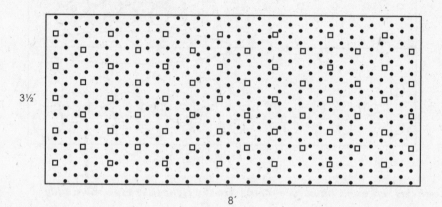

• = CARROT (3″) +/– 300 PLANTS
□ = KALE (8″) 63 PLANTS

3½′

8′

Interplanting high leaf kale with root carrots. The squares represent kale planted 8 inches apart and the dots represent carrots, planted 3 inches apart by seed.

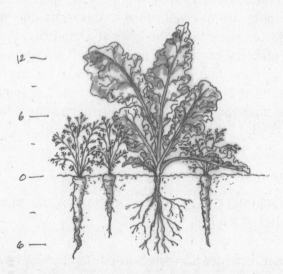

INTENSIVE PLANTING PATTERN
WITH A TRELLIS INTERPLANTING

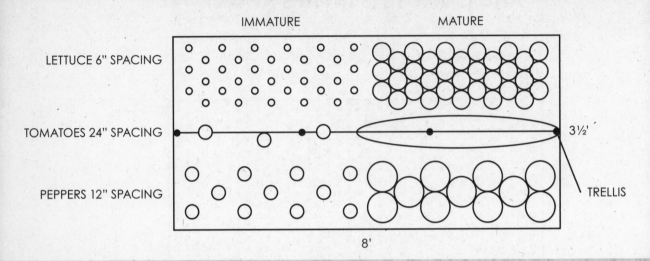

On the left the circles represent immature plants. On the right they are shown fully mature with leaves just touching. The tomatoes climb vertically up the trellis.

crops, eggplant, and peppers by half of their normal spacing so they don't completely shade lower crops and there are enough nutrients for all your crops. Plants that are bushy and take up a lot of space, such as squash or okra, are not well suited to this technique; while it can increase yields, in some cases the layer of complexity it adds to harvesting or to planning can make it unworkable. For instance, if you are planning to grow cut-and-come-again salad mix, having other crops in the bed will just get in the way of your scissors. If you're growing head lettuce, however, it won't slow your harvesting down to have chard in that bed.

INTERPLANTING CROPS THAT GROW AND MATURE AT DIFFERENT RATES

Plants that mature or are harvested over different lengths of time can be interplanted to increase efficiency and yields. By the time slow-growing crops, or crops that have a long harvest period, are maturing and beginning to fill in their spaces, quick-maturing crops have already grown and been harvested. It is important to the technique to growing quick-maturing plants that are harvested during a relatively short period, such as radishes, spinach, or arugula. For example, summer squash wouldn't be appropriate, because it wouldn't give up its space to a slow-maturing crop at the right time: While summer squash has a short period to the first harvest, it has a wide bushy habit and is harvested over a long period of time.

Both kale and lettuce are planted in the bed at the same time, the kale 12 to 18 inches apart and the lettuce 4 to 6 inches apart. By the time the lettuce is harvested, the kale will have more room to grow.

To use this method, plant a crop from column A with a crop from column B at the same time in the same bed on a hexagonal spacing plan. Plant the crop with the larger spacing need first, and fill in the rest with the closer-spaced crop. At first the newly planted or sprouted seedlings will be about the same size. Just as the column B crop begins to block sunlight to the column A crop, the latter should be ready for harvest.

A	B
Fast-Growing Crops	Slow-Growing Crops or Crops With Long Harvesting Periods
Asian greens	Artichokes
Bitter greens	Basil
Green onions	Broccoli
Mustard greens	Brussels sprouts
Radishes	Cabbage
Salad greens	Cardoon (biennial)
Spinach	Cauliflower
	Celery
	Chard
	Cilantro
	Collards
	Dill
	Eggplant
	Garlic
	Kale
	Leeks
	Onion
	Shallots
	Tomatoes

PLANTING SEEDLINGS

Now that you know how to space your plants, it's time to actually plant them! Make sure they have been hardened off outside the controlled propagation environment for a few days, so they can acclimatize to the temperature and weather. Give them a good watering a few hours before and then immediately prior to planting. If it is hot outside, plant during the cool morning or evening. If it's hot or especially windy, you may want to temporarily cover tender seedlings with shade cloth as you plant to protect the leaves from drying out. If it is cold, plant in the first half of the day, so the seedlings will have a chance to warm up.

First, place the seedlings where you want them in the bed on the soil surface, and then come back and plant them. Dig the first hole with your hand, a trowel, or using your jig. In the case of rosette-type seedlings it must be deep enough so that the soil level in the pot or cell is at the same level as the soil in the bed. The spot where new leaves emerge shouldn't be covered. For seedlings with a central stem and leaf nodes, the hole should be deep enough to cover the seed leaves and the stem up to the top two sets of true leaves. Tomato seedlings should be buried as deeply as their top set of leaves. Don't worry—it's okay to bury leaves.

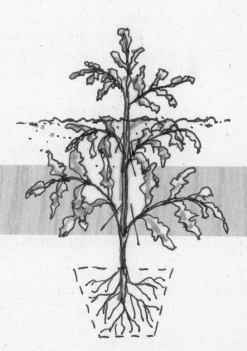

To encourage root growth, tomato seedlings should be planted deep. Snip the leaves or branches that would be buried prior to planting.

Don't pull on the stems to remove seedlings from pots. Instead, make a V sign with your dominant hand. Place the V on either side of the plant stem, moving your other fingers next to the V to cover the pot opening. Turn the pot over and tap or squeeze the pot until the seedling is released onto your palm. If the plant isn't root-bound, turn it over gently into the hole. Push the dirt from the hole back around the plant, and press down firmly. Roots like good soil contact. To take plants out of six packs or Styrofoam trays, use an old table knife to gently pry them out.

Roots: to loosen or not to loosen. It's best not to disturb plant roots at all unless the plant is seriously root-bound. Roots have microscopic rootlets that absorb nutrients; if damaged they can take weeks to regrow. Root-bound plants form a thick mat of roots at the bottom of the pot or cell, or have roots running around the pot sides and bottom, forming several circles. In these cases it will help the plant to loosen or cut off the bound-up roots. A great way to loosen them with less shock and damage is to dip them in a bucket of water and gently swish them around. After doing this you may find your seedling has very long roots indeed. In that case, dig a deeper hole, since the roots should be pointing down.

If the bound-up roots are fibrous, you may need to unwind them with your fingers or even cut some of them off to expose the inner roots. This is preferable to leaving them bound, and the plant will probably survive.

Depending on the weather, water your seedlings either as you go or after you have planted out the entire bed—see chapter 11 for complete instructions for best watering practices. Use a stick and pencil to label what's been planted—the crop, variety, date, and seed company, if you like. Be sure to give your babies water every day or every other day for the first week. You may also want to plan your plantings for a time when you can monitor pests. Small seedlings can be razed to the ground by caterpillars, slugs, and snails. The first few evenings are good times to hunt snails. Keep extra seedlings on hand to fill in any spaces left by failed or eaten seedlings. Another highly recommended option is to cover susceptible young seedlings with floating row cover to keep out pests until the plants are established.

DIRECT SEEDING

Starting seeds outdoors—direct seeding—is integral to farming in small spaces. Using direct-seeded crops to fill in unused space (interplanting) efficiently uses bed space and can increase your yields exponentially. Direct seeding follows a pattern: bed preparation, seeding, daily watering until germination, and then thinning and/or filling in badly germinated areas. See page 183 to learn how to get your seeds out of the packet and into the ground.

How do you know whether to start something in the greenhouse or direct seed it? The gardener's rule of thumb is to direct seed root vegetables, large-seeded vegetables, and vegetables that are sensitive to "early bolting" (going to seed early). In our world we've got a lot of snails and slugs to battle and we've found that most of the large-seeded vegetables can be started successfully in the greenhouse. We often try to direct seed them anyway, but after trying with a row of peas three times we inevitably remember, Hey, didn't we have this problem last year? What are we doing? Back to the greenhouse.

In most cases, we like to direct seed the following crops in garden beds:

- beans
- beets
- broccoli rabe/rapini
- carrots
- chayote
- corn
- garlic
- onion sets
- peas
- potatoes (seed)
- radishes
- turnips

Tip: Sow a six pack or two in the greenhouse of the very same seeds you direct seed outside in the garden. That way, if some of your seeds don't germinate or are munched, you can fill in the gaps later.

In general, direct seeding requires concentration, since the seeds are often too small or darkly hued to see well once dropped in the soil. Seeding is a tranquil activity we prefer to do without distractions. Remember: You want to use the spacing between plants descibed on your seed packet, and ignore the spacing between rows, in order to make more efficient use of your space. Although direct seeding will seem like a lot of work at first, be comforted that with the methods you will learn here, thinning, the most time-consuming part of direct seeding, will be minimal or eliminated entirely.

Beds for direct seeding should be cultivated carefully so that the surface is free of rocks or large clods of soil. Any crust should be broken up so that the bed will accept water readily. Any compacted areas should be redug so that roots can penetrate easily. The bed should be formed correctly (see page 91) so that water will penetrate equally over the entire surface without pooling or leaving dry spots. Compost can be added to the top of the bed prior to or after seeding, depending on the seeding depth.

Direct Seeding Methods

There are five methods of direct seeding that are helpful in the urban garden:

- hexagonal
- broadcast
- furrow
- hill
- row

The first three methods are used for plants that grow only up to 12 inches in diameter when mature. Of these first three, hexagonal and broadcast seeding produce the highest yields, while furrow seeding requires less work and is necessary in certain conditions. Hill and row seeding are used for large plants, such as squash and corn.

Hexagonal

Hexagonal seeding can be done with any size seed, but it is especially easy with larger seeds, such as beets, cilantro, beans, peas, garlic, and onion sets. We can't overemphasize the amazing yields you will get with this method.

If using small seeds you will drop as few seeds as you can manage in each hole, and if using large seeds, depending on how fresh they are, you can choose to drop one or two seeds into each hole.

- Prepare a clod-free seedbed with fine grained surface soil.
- Look on the seed packet for the correct spacing for the crop you are planting.
- Make staggered rows of holes at this spacing using your finger, a dibber, the end of a garden tool, or your seeding jig. Use a measuring stick to ensure the correct spacing until you can eyeball it. You will see that this makes a pattern of triangles and hexagons. Make the holes the correct depth for the seed.
- If you are using very small seeds that don't require a hole, you can lay chicken wire across the bed and, using the pattern of the chicken wire as a guide, drop seeds into the holes.
- Dry your hands well, and pour seeds into the palm of one hand.
- Pinch out a seed, or a few, and drop it/them into the hole or space in the chicken wire.
- Move from one end or side of the bed to the other, and wait to cover the seeds until you've seeded the entire bed (if you forget whether you seeded a hole, you can probably tell by looking closely).
- Brush soil from the bed, or sprinkle fine compost, into the holes.
- Firm the soil by pressing down firmly with your hands.
- Water daily or twice a day, depending on the temperature and the crop; see chapter 11 for best practices for watering.
- A few weeks after the seedlings have emerged, thin them or fill in as needed.

Broadcast

Broadcast seeding relies on overseeding to ensure that the bed is well filled in. If you purchase seed from good sources, the germination rate will be very high, making this unnecessary. Broadcast seeding works well with small seeds that are difficult to see. It's a great method to use if you plan to eat the thinnings, such as beet and spinach sprouts or baby carrots. If you don't plan to eat these, though, broadcast seeding is a real waste of precious seeds. These beds often don't get thinned properly and can become a pest-infested tangle of stunted veggies that often go bad before you get a chance to reap your harvest. If you plan to broadcast seed and want to avoid this mess, plan to spend hours thinning.

❖ Prepare a clod-free seedbed with fine-grained surface soil, so seeds will fall into tiny holes in the surface.

❖ Look at the bed or section of bed and visualize how you're going to cover it consistently with seed. Are you going to make swaths the length or the width of the bed? Decide on the order you will seed each section.

❖ Dry your hands well, and pour seed into the palm of one hand.

❖ Pinch out seeds and sprinkle them in your pattern, attempting to do it evenly, to the very edges of the bed and about ⅛ inch to ½ inch apart.

❖ Sprinkle fine compost over the bed, covering the seeds to the correct depth.

❖ Firm the soil by pressing down firmly with your hands.

❖ Water daily or twice a day, depending on the temperature; see chapter 11 for best practices for watering.

❖ Once the seedlings emerge, do two or three sets of thinning, separated by a few weeks, to end up with plants at the correct spacing.

❖ If there are spaces with no germination, gently transplant thinnings to fill in the space.

Furrow

Furrow seeding is a version of broadcast seeding that cuts the required thinning work in half and wastes much less seed. This method is great for small-seed crops, such as carrots, that are difficult to place exactly where you want them.

Furrow seeding is a must when using emitter-line drip irrigation, especially in dry or windy climates, where beds dry out quickly. Make a furrow either under or very close to either side of the drip line, so the seeds are close to the water source.

It is also used when planting crops you plan to trellis by making a furrow just below it. The work of thinning is cut in half with this method for this reason: Since the seeds are planted in lines spaced at the correct spacing, you only have to thin in one direction—within the furrow. An easily made furrow-seeding tool speeds the work.

❖ Prepare a good seedbed.

❖ Look on the seed packet for the correct spacing.

❖ Make furrows the length or width of the bed. Either draw a line with your hand or a stick, or press a 1 inch by 1 inch stick into the soil at an angle, making a V.

❖ Dry your hands well and pour seeds into the palm of one hand.

SEEDING FURROW TOOL

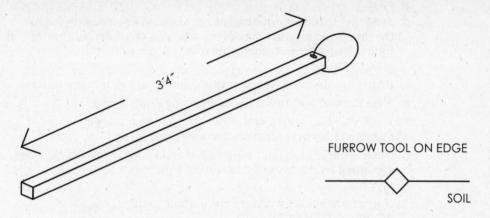

FURROW TOOL ON EDGE

SOIL

3'4"

Make this tool with a 1" x 1" stick the width of your beds. Press the tool into the soil on its edge to make furrows for seeds.

SEEDED FURROWS

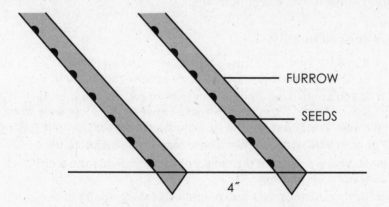

FURROW

SEEDS

4"

After making depressions with the furrow tool drop seeds in and cover with soil or compost.

- Pinch out seeds and sprinkle them in the furrows, attempt to do it evenly, to the very edges of the bed and about ⅛ inch to ½ inch apart.
- Seed the furrows consistently from one end or side of the bed to the other, and wait to cover them until you've seeded the entire bed (if you forget if you seeded a furrow, you can probably tell by looking closely).
- Brush soil from the bed or sprinkle fine compost into the furrows.
- Firm the soil over the seeds by pressing down firmly.
- Water daily or twice a day, depending on the temperature; see chapter 11 for best practices for watering.
- Once the seedlings have emerged, do two or three sets of thinning separated by a few weeks, to end up with plants at the correct spacing.
- If there are spaces with no germination, gently transplant thinnings to fill in the space.

Hill seeding

Hill seeding is a bit of a misnomer: It doesn't require hills. That's just a name given to row seeding with really big plants. This method is best for squash and vines you don't plan to vertically trellis, so it's a good method for those of you who have ample space. For the rest of you—trellis them! No hills for you! You can either direct seed or plant out greenhouse seedlings this way for cucumbers, melons, summer and winter squash, and sweet potatoes.

How to seed in hills:

- Mark out an area at least 3 feet wide in an established bed, in an underused border area or an open area of your garden.
- If you are making a number of rows or hills, make sure there is at least 2 feet to 3 feet of path space between the 3 foot wide rows.
- Cultivate the soil in a 1-foot-wide circle every 3 feet, and amend each circle with compost. These may look a bit hilled up.
- Move the soil out of the center of each circle to form a dish that water will collect in.
- Dry your hands well, and pour seeds into the palm of one hand.
- Push about five seeds into the soil 1 inch to 2 inches apart in the center area of each circle.
- Move from one end of the row of circles to the other to make sure you seed each one.
- Brush soil into the holes.
- Firm the soil over the seeds by pressing down firmly.

- Water daily by filling the bowls with water and letting it soak in at least twice.
- A few weeks after the seedlings have emerged, thin to two or three plants.
- As the plants grow, reorient vines into the growing space to keep your paths clear.

Row

Row seeding was used in the old-fashioned kitchen garden featuring neat single rows of crops with bare earth paths between them. While we don't recommend row cropping for a number of reasons, it is still the best method for a few crops, such as corn, which must be planted in at least a 10-square-foot block to pollinate correctly, or potatoes, which must be hilled up with dirt from the paths to produce significant yields. Potatoes can also be planted and hilled up in vertical tubes or bins (see chapter 7 on container planting), or in trenches (see page 226).

How to seed in rows:

- Mark out 1 foot wide rows in an established bed, in an underused border area or an open area.
- If making a number of rows, make sure there is at least 1 foot to 2 feet of path space between them. If you're planting your rows in an established 3 foot to 4 foot wide bed, you won't need paths, since you shouldn't step in the bed.
- Cultivate the soil in the 1 foot wide rows with a hoe.
- Dry your hands well, and pour seeds into the palm of one hand, or carry the bag of seed potatoes with you.
- Make holes about a foot apart and drop two corn seeds or one seed potato in.
- Move from one end of the row to the other to make sure you seed each row consistently.
- Brush soil into the holes.
- Firm the soil over the seeds by pressing down firmly with your hands or a tamper.
- Water daily until the seeds sprout; see chapter 11 for best practices for watering.
- With potatoes, dig soil from the path or the side of the bed once the sprouts are 6 inches high, and hill up to cover all but the top 3 inches of sprout. Continue hilling throughout the season.

THINNING

Removing extra seedlings is essential to growing healthy direct-seeded plants. Many methods of direct seeding overseed a bit to ensure good bed coverage, but closely spaced seedlings can become stunted, from lack of nutrients, and leggy from lack of light as they try to reach the sun above the other seedlings. Leggy seedlings produce weak-stemmed plants that can fall over or restrict the flow of water and nutrients.

The Right Time to Thin

The bigger your seedlings get, the less risk there is of the plant being completely wiped out by bugs. A larger seedling can sustain leaf damage and still grow healthy and strong. So when it comes to pests, it would make sense to thin when seedlings are quite big. However, this must be weighed against the damage done by stunting. To ensure that you will end up with the right amount of seedlings even if a few are taken out by bugs or other acts of nature, thinning is done in three successive rounds separated by a week or two, depending on how fast the crop grows.

The first thinning should be done when they have one set of true leaves (see page 184 for a description). The second should be done when the seedlings have two sets of true leaves. Your final thinning will result in seedlings with the correct amount of space between them to grow to maturity and should be done when there are three or four sets of true leaves. For crops that don't have distinguishable seed leaves, use their height and breadth as your guide. When they start to crowd each other out it's time to do a thinning. Seedlings can also be gently transplanted at the second thinning to fill in gaps. After the final thinning, it can be a real help to gently spread a layer of compost around the seedlings to cover and support leggy stems. This is especially important with root vegetables, such as radishes and beets, that can fail to form a root if the rooting part of the stem is exposed.

Thinning Methods

Thinning is a kind of selection process to help you arrive at the most vigorous plants. Try to leave the healthiest-looking seedlings and thin out the less healthy ones. (This isn't always possible, given that you're also aiming for even spacing.) It can be done by gently pulling them out, or by snipping off the unwanted seedlings with scissors just below

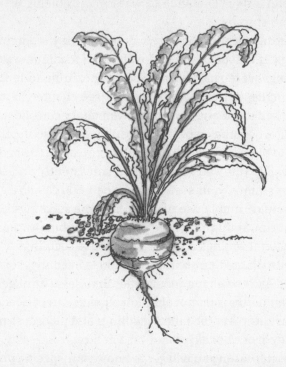

After thinning, cover the shoulders of root vegetables like this beet to ensure that they bulb up.

the soil surface. The first method can easily damage the roots, or even pull up the ones you are leaving in. The snipping method is much less invasive and yields ready-to-eat, clean sprouts. The key is to snip about ⅛-inch below the soil level, or the seedling will likely resprout.

CARING FOR TRANSPLANTED OR DIRECT-SEEDED SEEDLINGS

Reseeding Failed Seedlings and Filling in Failed Plantings

After about ten to twelve days, if you don't see any action you will need to replant the seeds, perhaps using newly bought seed. Think about how well you watered the seedbed. Did you let it dry out repeatedly

223

for more than a day? One clue to watering problems is that you see tiny, dried-up sprouts.

Look closely at the bed. You may see the tiny stumps of chewed-off seedlings. If so, your seed and watering technique were good, but your seedlings were snacked on. You may be able to tell the type of pest based on the pattern of destruction (see chapter 13 for details).

If your bed dried out too quickly, you might consider using shade cloth to protect against rapid evaporation. If you've determined that pests were to blame, you might consider protecting the bed with floating row cover. If you watered your bed consistently and see no signs of sprouts or stumps, your seeds may be bad. You can do a germination test to find out or simply use new seeds. If the seedlings germinate in a patchy pattern, at thinning time you can fill in blank areas with carefully separated thinnings or with greenhouse seedlings.

For beds planted out with greenhouse seedlings, examine your planting bed each week for about a month to see if any seedlings have died. You can do the same evaluation as with direct-seeded crops to find the cause. Replant the failed seedlings and protect as needed with floating row cover, shade cloth, or bird netting.

With each season you will get to know your specific problems and will develop methods that work for you to protect seedlings. It's a good idea to keep notes on what worked and what didn't—but only if you're the kind of person who will go back and read through them later.

This is where we remind you that one of the virtues of a farmer is patience. While the protected greenhouse produces beautiful little gems, your direct-seeded beds can be devastated in a few days by any number of pests. The same can occur when you plant out your greenhouse gems. An unexpected freeze nicely iced your spinach sprouts. Gophers gnawed the roots of three of your cauliflower plants. City birds descended en mass and ate your entire bed of lettuce. Human error is equally pesky. Hmmm. You forgot to water and now have a nice bed of fried sprouts. Perhaps you have a mental block about throwing things out and thought it would be fine to plant that seed packet labeled 1995. Nothing sprouted.

Intensive farming is successful when beds are filled with plants consistently. When plantings are spotty, not only is the harvest affected, but exposed soil loses nutrients and water, and can erode. The reality is, you will need to reseed or replant beds, sometimes more than once.

Special Techniques

Potatoes and vegetables that are typically blanched, such as leeks, endive, escarole, and celery, need special treatment as they grow. Potatoes only form new tubers off the growing stem above the seed potato; they will form roots at leaf nodes if soil is mounded around the stem covering the nodes. Hilling potatoes is essential to increase yields. Blanching keeps light from the edible portion of the plant so that it can't photosynthesize there. With leeks, this produces a longer white shaft. With bitter vegetables, this results in a milder flavor as well as, unfortunately, lower nutritional value. If you grow celery stalks or bitter greens, you might want to try them blanched and unblanched,

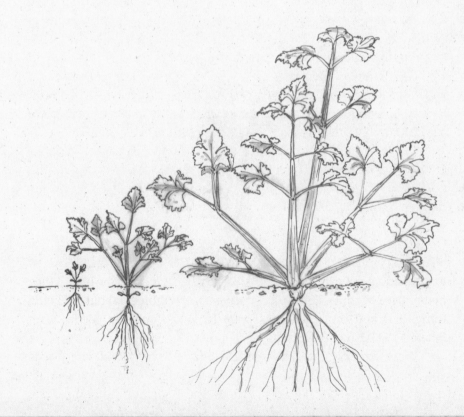

Unblanched celery stalks tend to be dark green and bitter.

so you know what flavor you like best. In addition to the hilling method, celery can be blanched by tying brown paper loosely around the stalks, leaving some of the leaves free. You can create blanched hearts on endive, escarole, and chicory by tying the outside leaves together at the top, covering the inside leaves, about a month before harvest.

Potatoes and some blanched crops can be planted in trenches to make it easier to hill dirt up over time. The entire width of an intensive bed can be trenched down 1 foot while you save the soil in bags for later hilling, or a 1-foot-wide trench can be dug with the dirt piled on the north side. Plant the seed potatoes or seedlings. Once the plants reach about 6 inches high, begin filling in the trench with soil, leaving the upper leaves exposed. Repeat the mounding until the plant stops growing taller. Be sure to leave 4 inches to 6 inches of leaves exposed.

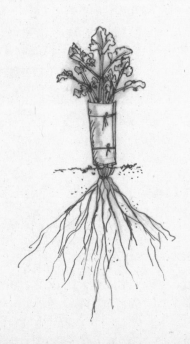

To blanch celery stalks loosely tie brown paper around the stalks, leaving the leaves exposed. This will ensure sweeter celery stalks.

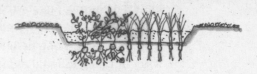

Step 1:
Dig a trench and
plant potatoes or leeks.

Step 2:
When they reach about
10 inches high, gently mound
in soil or compost, leaving
the tips exposed.

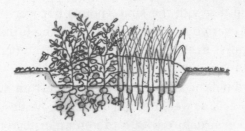

Step 3:
Repeat mounding as they continue
to grow. Harvest potatoes when
the leaves die back and leeks when
they are 1 to 2 inches thick.

CARING FOR YOUR GARDEN TO MATURITY AND HARVEST

Pest, Cold, and Heat Protection

You may have noticed we're kind of fanatical about not buying things. Floating row cover and shade cloth are exceptions. Floating row cover is a synthetic material specially formulated to let in water and light, keep out bugs, birds, cats, and mollusks, and insulate seedlings from the cold. It is usually sold in long rolls about the right width for intensive garden beds. To use correctly, you must bury or stake down the edges tightly. You can water your crops right through it. As the seedlings germinate and grow upward they will push up the lightweight material: Simply ensure that there is enough slack for them to grow as high as they need to. Remove the row cover when the seedlings have a number of sets of true leaves and can tolerate a bit of bug munching without being killed. Fold it up and save it for next time.

Shade cloth comes in similar rolls or sheets, and can block varied amounts of sunlight. Again, a synthetic, it can be reused for years if cared for. To shade tender seedlings in hot weather, drape the material over stakes, using rubber bands or string to attach it. To build a solid, reusable structure, cut 9-foot lengths of 6-6-6 cement reinforcement mesh with bolt cutters. Cut off the lateral pieces of wire at each end, leaving sharp, self-staking ends. Cover the sheets with the shade cloth using zip ties with the sharp points uncovered. Bend each panel in an arch over your planting beds with the 6-inch sharp ends of wire anchoring it in the soil at each side of the bed. (See illustration on page 49.)

Bird netting is a very effective deterrent not only of birds but of cats and deer as well. You can use the above described stakes or shade cloth hoop structure with bird netting instead if needed. However, if you decide to suspend the material over your beds, you will need to carefully stake it to the ground, making sure there are no gaps to let birds in.

Supporting and Training Plants

Vertical trellising and training are key to conserving space in the urban garden. Many people build or set up a trellis or cage, plant their seedlings, and expect them to tendril neatly up the structure. This dream rarely comes true. It's important to assist vining plants so they stay close to the trellis by tying them closely to it. Many vining plants do not produce tendrils and will not be able to find the support without your help. If you are using a fencelike trellis, tie your crops flat along it by running a horizontal string on the outside of the vines as tightly as possible without injuring the plants. Run a new string every foot or so as the plants grow in height. Loop the string behind each post as you go, and be sure to tie both sides tightly. With cages, periodically reorient the vines back into the cage.

Peppers, eggplant, okra, and many perennials need staking. Push a wood or metal stake at least 6 inches into the ground, about a foot from the plant, and tie the plant to it at intervals with natural twine, being sure not to bind the stalk from further growth.

Pruning Annual Crops

Trellising and pruning annuals go hand in hand. Pruning is often necessary in the urban, vertically oriented farm to allow nearby plants to get enough sun and airflow and to give you space to reach your harvest. You just can't have vines growing all over the place or nothing else will have room. Squash, melons, cucumbers, and especially tomatoes all can be pruned to good effect. Another benefit is that fruit production is enhanced when green growth is limited. Most vines can simply be headed back (clipped) when they get too long for your taste. This will cause it to branch out below the cut.

Tomato plants should be allowed to grow at least three main branches before pruning, with a number of subbranches each. When you are ready to prune them, begin with a careful examination. Can you distinguish between a leaf and a new branch? Leaves come out from the branches at a nearly 90 degree angle. They do not have a growing tip forming new leaves and vines. Their stems are a bit flat. Vines have cylindrical stems and tiny, new green growth at the tip.

While leaves will remain throughout the season and absorb the sun's energy for fruit production, the sprouts between the leaf and the vine will produce new vines, thus diverting the plant's energy away from fruit production. Prune off the new branches throughout the season that begin in the joint of a leaf and a branch on the main branches you initially selected to keep. You can be quite ruthless in your tomato pruning and trellising. We like to keep our tomato plants virtually flat along vertical trellises.

Removing Crops and Re-forming Beds

Vegetable gardeners sometimes procrastinate about their harvesting—everything looks so good and abundant, I want my friends to see this, etc.—but the truth is: Farming is all about ending things and starting over. For instructions on harvesting crops, please see chapter 12.

To start over after harvest, the first step is to clear the bed of debris and plant stalks. When removing old plants, you can either pull them out and knock the soil off the roots back into the bed or leave the roots in the ground and simply cut the stalk off an inch below the soil surface. New plantings will appreciate the channels made by the old roots. Move all debris and weeds to the compost pile.

Your beds will need some attention before replanting. While you should not need to double dig your beds after the initial dig, amending and re-forming them correctly is important. Over time paths tend to get wider and beds narrower. If you don't reshape beds, after a few years the ratio of bed to path may switch!

If you don't have a surveyor's eye for perspective, get out your measuring tape, stakes, and string again and reestablish the bed boundaries. Start by raking pathway mulch that may have crept up the sides into the center of the paths. If the bed soil has become compacted or a crust has formed, lightly loosen it with a regular or hand hoe. If necessary, move soil around with a bow rake or shovel to reestablish the correct shape and size. Amend the bed with 3 inches to 6 inches of compost, and rake it into the desired bed shape, as described on page 100 in chapter 6. Finally, add more mulch to the paths, making sure to keep it out of the bed. Covering your beds with straw or leaves in frosty climates is a good idea in the winter. Otherwise, mulching beds while plants are growing makes a great habitat for snails, so we don't recommend it.

BED AMENDING AND MULCHING

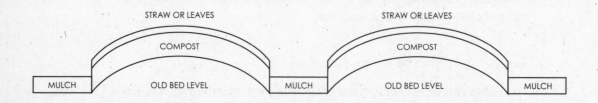

Each time a new crop is planted, add 3–6 inches of compost to the old beds. Add the same amount of mulch to your paths.

Because the seasons for planting will vary so widely due to your region's climate, we aren't going to make a month by month list of what to do. However, we have found that there are two prime periods of work. The following routine has been a support for us; farmers in the lower South will want to reverse the order of these tasks.

PREPARING FOR THE WARM SEASON

After the new year begins, the first order of business should be to review the work of the previous year and make any necessary changes to your plans. Review your seeding and planting calendars and order seeds or rootstock. Make sure your propagation setup is in working order. Plan for any necessary building or improvement projects. This is a good time to sort seed boxes, clean up sheds and work spaces, and clean and sharpen tools. As soon as it is seasonable to go outside, check on the garden and plan for any necessary bed preparation and amending. Once it's warm enough, turn your fall compost piles and see if they're ready to use or need more time (you'll know they're ready when the material is dark and crumbly and smells like earth). After the ground warms up and dries out (one should never dig saturated soil), remove any remaining crop debris or mulch. Reshape and amend emptied beds with finished compost. Lastly, double dig new beds and mulch paths.

PREPARING FOR THE COLD SEASON

We hope you will try to extend your growing season as much as possible with correct crop selection and season-extending techniques. With that said, there may still be a few months when the garden needs to be tucked in until you can plant again. It is important to cover unplanted beds with a thick mulch to protect the soil from erosion or the weight of the winter snow. Weeds are often more vigorous than crops and can tolerate some winter cold. Since they use up valuable nutrients, they should be removed before tucking the garden in. Fallen leaves work wonderfully as a winter mulch. In the fall, watch your neighbors to see when they rake leaves, then commandeer their green bin and return it empty. Cover your beds and paths with a thick layer of leaves and water them down. If you live in a very cold climate, you might want to add a top layer of straw. When you're ready to plant again the straw can be removed and the leaves can be raked into the paths.

In the fall you might want to plant a winter cover crop, such as rye (the cover crop, not the grass); these will provide a living mulch. When warmer weather comes it can be cut down, chopped up, and added to the compost pile as a source of nutrients.

Late summer to fall is also a great time to gather urban waste and thoroughly muck out your animal housing areas in order to create large compost piles or windrows to provide nutrients for the coming year. Gather materials such as manure from your own animals or local stables or farms, crop debris, fallen leaves, coffee and tea grounds, clean sawdust, or other chaff and layer them, creating as much mass as you can. We recommend using ingredients in these windrows or free piles that don't tempt rodents, such as those listed above, rather than kitchen scraps. If you've got 'em, add red wriggler worms to the top of the pile. They will find their way to comfortable locations to munch manure. Cover with tarps in case of rain. We like to leave our fall piles to compost without turning them and then add the finished compost to our garden beds the following spring or summer.

EXTENDING THE SEASON

Our goal is for your urban farm to be yielding delicious produce for as much of the year as possible. Following are instructions for extending your harvest into cold or hot weather. Please see appendix 8 for regional guidelines on extending your season.

Plants thrive in a specific temperature range.

IDEAL TEMPERATURE RANGES FOR MOST ANNUAL CROPS

Night air	55 to 75 degrees
Day air	65 to 85 degrees
Warm and hot season crop night air	Above 65 degrees
Cool season crop night air	Below 65 degrees
Night range for cool and cold season protected (greenhouse) crops	35 to 75 degrees
Night air temperature range tolerated by outdoor overwintering crops	25 to –25 degrees

DAMAGING TEMPERATURES

Frost damage possible for warm and hot season crops	Below 50 degrees
Killing frost possible for tropical plants	Below 50 degrees
Killing frost possible for warm and hot season crops	32 degrees
Growth-inhibiting high temperatures for most annual crops	Above 85 degrees
Growth-inhibiting high temperatures for cool season crops	Above 70 degrees
Heat damage	Above 105 degrees
Killing heat	115 degrees

Wintering Over

Overwintering is the process of growing crops that can survive the coldest winter months to be harvested either in the winter or early spring. Learning to overwinter crops properly is an essential urban farming skill. It can allow you to harvest in the winter months. Equally important is the fact that it can allow you to grow and harvest cool-season spring crops in the earliest spring months.

There are a number of annual vegetables and cover crops that will survive extreme winter weather, especially if protected from the wind and given the extra protection of a thick mulch. The goal in overwintering is not to have crops putting on growth in the shortest winter days but to have them to harvest in winter and to get a jump-start on spring growth. Therefore, wintering over *requires* summer or very early fall planting.

If you plan on overwintering crops, plan to have about a third to a quarter of your beds become available in high summer. If you plan to harvest crops in the winter, they must be mature or almost mature by the end of October, as they will put on very little if any growth in November, December, or January. If you plan to harvest early spring crops, they must be at least beyond their infancy by the fall. The right varieties will hold in the ground throughout the winter harvest period. If you find your winter-harvested crops are woody, you probably let them sit in the field too long and missed the best harvest window.

Successful Overwintering

✤ Select varieties that are tried and true for overwintering, often called "winter hardy."

✤ Plant crops in June, July, August, and September.

✤ Allow extra space between plants.

✤ Do not attempt succession planting after winter harvesting—the ground will be too wet and crops won't get the light they need to grow.

✤ In extreme cold, protect plants with piled-up mulch or season extenders.

Please note that garlic, the traditionally overwintered crop, is planted in the fall and harvested the following late spring or summer.

If you want to expand the list of crops you can grow into the cold or hot season successfully, use season-extending techniques. These protect your crops from killing cold or heat so that they thrive when they normally would not.

Extending the Season by Protecting Plants from Cold

Even in the most northerly latitudes, many vegetables will yield nicely when soil and air temperatures are sufficiently raised. To do this, grow your crops in the ground or in containers under plastic or glass. This increases heat by storing sun energy and protecting crops from wind. The key to extending the season is to start your seeds or seedlings for fall, winter, and early spring harvest in the warmer weather of mid-summer through early to midfall, depending on your region. This way your crops will have put on sufficient growth prior to the onset of fall and winter.

Unless the daylight hours drop below ten hours per day, the limiting factor for plant growth is heat, not lack of light. This means that most of us can indeed grow crops during the fall, winter, and spring as long as we provide the necessary heat. Even with the necessary heat, however, in most areas, the growth of crops *will* pause during December and January as daylight hours drop below ten per day (check an almanac for the dates for your latitude). After this pause plants will pick up again and give you wonderful early spring harvests.

The site for your protected crops should be a spot that will receive the most daylight hours during the winter when the angle of the sun is low. Make sure they will receive at least five hours. The site should also be protected as much as possible from the prevailing wind direction. If your garden is sloped, avoid areas of lower elevation, since cold air sinks.

We recommend you begin experimenting with protected environments by using the less expensive and less complicated methods, such as cold frames, cloches, and floating row cover. If you get hooked, you may want to invest in building high tunnels or a greenhouse.

Cloches

Cloches are the simplest method of increasing heat around plants. They are glass or plastic enclosures that fit over one plant. Traditional French cloches look like glass bells. More economical ones are gallon-size milk jugs with the bottom cut off or tubes of 6-6-6 remesh covered with polyethelene sheeting. They must be vented to allow for airflow and to release excess heat on sunny days (the screwtop lid openings on milk jugs make great vents). Propping one side just slightly up off the ground with a shim or stick does the trick.

Cold frames

Cold frames are a traditional method of conserving heat. They consist of a rectangular frame covered with glazing of some kind, typically plastic or glass. One long side is a few inches higher than the other to create a pitch to the glass or plastic roof, to ensure rain runoff. Cold frames are placed directly on the ground, and plants are grown inside either to transplant into outdoor beds or to grow to harvest. They can be fixed in place or movable, in which case they can be placed directly over garden beds to protect crops. Cold frames must have tops that can be opened easily for venting and watering.

Floating row cover and low tunnels

Floating row cover is a woven material that allows rain and sun in while protecting crops from cold, wind, and pests. Row cover can be used to extend the harvest of in-bed crops either by simply laying the material over them (they push the material up as they grow) or by installing wire supports to elevate the material a few feet above the plants. Either way, to be effective, the edges must be sealed flawlessly. Mounding soil along the edges is more effective than staking.

High tunnels

High tunnels are similar to row covers. They are simply a much larger version of hoop protection over growing beds. These are tall enough to walk into. High tunnels can be constructed out of salvaged materials or purchased as a kit. They are traditionally covered with polyethylene sheeting. If you live in a very cold climate, you can grow crops under a double layer by covering beds within a high tunnel with row covers. Neato!

Greenhouses

Greenhouses are the rich man's high tunnel. See if you can find one for free—they are out there. Often people needing to rid themselves of a greenhouse will give it away if you are willing to dismantle and move it. While glass was the traditional glazing for greenhouses, double-wall polycarbonate is more efficient and durable. If growing crops (not seedlings) in a greenhouse, there obviously mustn't be a floor of any kind, so you can plant directly in the soil. A vent is a must, and vents that open automatically will make your work much easier. Kits for automatic vents are fairly affordable.

If you are doing much growing in protected environments, you will want to purchase a thermometer to monitor outdoor and protected temperatures. Refer to the chart on page 233 for heat and cold tolerances. It is amazing how much heat can build up in high tunnels or row covers even in winter. You may need to make more vents or partially remove material from hoops or beds during the day. However, if your protected environments aren't providing enough protection and heat, you may want to create a double layer. It is rarely necessary to go so far as to provide gas heat inside your high tunnel or greenhouse, because it is so easy to capture sun energy passively, but those in the coldest environments may opt for this solution.

Extending the Season by Protecting from the Heat

Most crops begin to go into heat distress between 70 degrees and 85 degrees. When stressed, plant energy can't go toward growth but must be used biologically to help the plant survive, such as wilting. In the South, protecting crops from heat is essential. Gardeners in other regions might be surprised to learn that shading crops on hot, sunny days would increase their productivity too.

Crops that suffer from an excess of heat, such as most salad greens, should be sited in cooler parts of your garden. Choose the northern side or areas shaded for part of the day by buildings. If your site is sloped, lower elevations will be cooler.

Shade Cloth Floating Row Covers

Woven shade cloth is obtainable in a variety of shade densities. This synthetic black woven material is very durable. The greatest difficulty is in suspending it solidly so it can weather wind. To shade in-bed crops, the best method is to create hoops out of 6-6-6 remesh covered with shade cloth. (Please see the instructions and illustration on page 49.)

Shade Structures and Trees

In situations where permanent shade is desired, horizontal trellises, lath houses, or frames to support shade cloth work best. These can be designed to provide just the right shade density and constructed right over your planting beds. Planting shade trees gives the added benefit of a nice yield of fruits or nuts, but they will take a number of years to reach the necessary height. Be sure you know whether the trees you choose are deciduous or not. Depending on your conditions, you will want to choose the year-round shade of evergreens or the summer shade of deciduous trees or vines.

CHAPTER 11

IRRIGATING AND SAVING WATER

IRRIGATION

Too little water is the most common cause of crop failure we have seen at new urban farms. It's very difficult for the beginner to conceive of the relatively large amount of water it takes to soak the soil. It is very common for novices to take the watering wand, move it over a bed once or twice, and think the job is done. A mere quarter inch of the soil is moistened by this method—the water doesn't make it to the plant's roots.

On the other side, the results of too much water include air-starved roots and a higher incidence of fungal diseases, which are brought on by too much green, watery plant growth. This can also be a problem in extremely damp and humid climates. In this case, in addition to providing the right amount of water, ensuring sufficient airflow around plants by pruning, trellising, and staking will help.

Plant roots need air to perform their functions. Alternating soaking the entire root zone with a few days to a week for soil water reserves to evaporate and drain away, bringing in air from above, is required for plants to thrive. If the root zone is constantly sopping wet, your plants will appear sickly and weak—again, a condition easy to confuse with a pest or disease problem. An exception to the rule of allowing the soil to dry out for a few days is newly seeded beds. Unless it's raining, direct-seeded crops must be watered daily until the first true set of leaves emerges.

With seedlings, gently water them immediately after planting to minimize the shock of moving to a new home. Using a cup to pour water around the root zone instead of overhead spraying minimizes damage to tender leaves.

As a general rule of thumb, after plants have a few sets of true leaves you can give your garden between ¾ inch and 1½ inch of water per week split into multiple, more frequent doses in hot or windy weather. To measure the amount of water given, either through hand or sprinkler watering, using a ruler, mark a shot glass or jar at half-inch intervals and set it in your bed. Use your watch to time how long it takes to fill the glass to the desired level. This is the amount of time you will need to hold your wand over each little section of bed. If you are hand watering, ignore the cup and arc the watering wand back and forth as you normally would to soak the bed. Obviously, if you hold the wand directly over the cup it will fill in no time, but receiving droplets as you pass by will take much longer and give an accurate idea of the length of time you should water each bed. Chances are it's going to be much, much longer than you ever imagined.

So, do you give ¾ inch or 1½ inch of water? Over the growing season, the amount of water to give your plants depends on the weather, your soil structure, and the crop type.

WEATHER AND WATERING

In general, plants need more frequent water in warm and hot weather than in cool or cold weather. With higher temperatures there's more surface evaporation and the speed of plant growth increases, using more water. Signs that water is needed begin before full-on wilting sets in. Water-deprived plants can look a bit saggy and begin wilting at the tips of leaves before the entire plant wilts. As opposed to its usual glow, the water-stressed plant will have a flat, matte color.

Once you notice signs of stress, test the moisture in the soil with your fingers. Go down at least 4 inches and see if the soil feels moist or wet. If it does, your problem probably isn't water. Plants not only wilt from lack of water, but also from too much sun exposure. Plant cells can only absorb so much heat, and when the limit is reached, they wilt to slow down leaf evaporation. You will mostly see this during the hottest part of summer days. If, without extra water, plants perk back up later on, you will know the cause of the wilting was too much heat, not lack of water. And then, of course, there are pests. For example, root maggots and verticillium wilt can cause wilting. And if gophers have munched plant roots, the leaves will wilt and die. (See chapter 13 to diagnose pest and disease problems.)

As the weather heats up, increase the frequency of watering, not the duration. The reason for this is that hot weather doesn't make soil soak in water at a different rate; it just makes it dry out quicker. In other words, once the soil is soaked, additional water just runs down into the subsoil. A few examples: If a bed of chard is wilting in the heat, and you have been watering it twice a week, increase the number of watering days to three times a week and see if that solves the problem. If you have been watering each bed with a sprinkler for fifteen minutes, don't increase the length of watering, simply the number of watering times per week or day. In extreme heat, twice daily watering can be necessary for some crops.

SOIL STRUCTURE

Your soil will retain more or less water depending on the soil structure. Water soaks easily into sandy soil, but sandy soil retains the least amount of water over time. Shortly after that soil is soaked, moisture begins to drain away. Clay soil takes longer to become soaked with water but retains moisture for a much longer time.

Water requirements for different types of soil

Sandy soil: less water more often $\longrightarrow$ Silt soil $\longrightarrow$ Loam soil $\longrightarrow$ Clay soil: more water less often

The speed at which you deliver the water should also be determined by your soil type. Sandy soil generally absorbs water quickly, so you can use a harder spray. Clay soil can take a long time to absorb a small amount of water, so you will need to use a gentler spray, or a bubbler. You will know you're watering too fast if water is running off your beds into paths, or downhill.

CROP TYPE

The majority of garden vegetables produce large root systems that penetrate from 4 feet to 10 feet. If you split the root zone of any plant into quarters, 40 percent of the water the plant uses is absorbed in the quarter closest to the soil surface, 30 percent in the second quarter, 20 percent in the third quarter, and 10 percent in the deepest quarter. If you only provide enough water to soak the top quarter, the plant will develop most of its roots in this shallow zone and will be more susceptible to stress from lack of water. Overwatering has the same result, since roots don't need to search for water as the soil dries between waterings. These conditions counterproductively produce plants that

AVERAGE WATER EXTRACTION PATTERN

Plants Growing in a Soil Without Restrictive Layers and with an Adequate Supply of Available Water Throughout the Root Zone

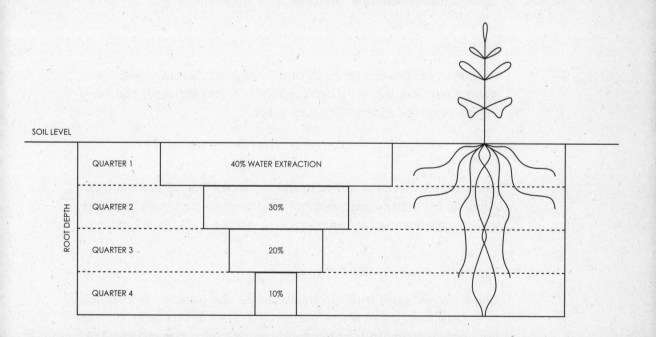

SOIL LEVEL

ROOT DEPTH

QUARTER 1 — 40% WATER EXTRACTION

QUARTER 2 — 30%

QUARTER 3 — 20%

QUARTER 4 — 10%

Most people give too little water, only soaking the first quarter. This encourages the plant to grow a shallow root system that needs more frequent watering.

need water more often to avoid stress, since they don't have the deep roots that would enable them to scavenge water. This is precisely why container gardens require much more frequent watering than those planted in the ground. Less frequent deeper waterings encourage deep roots that can forage for water and nutrients.

While it's okay for most waterings to soak the top half of the root zone, you will want to periodically give a deep soaking to the deeper-rooted crops. These can take less frequent watering as well, since their deep roots can forage for water below.

The best thing to do if you are growing a wide variety of crops is to group plants with the same water needs together. This is especially important with drip irrigation.

While crops need sufficient water throughout their life cycle, most crops are particularly sensitive to water stress during specific times. Erratic watering causes plants to want to go to seed *now* to ensure reproduction. For this reason, crops that are grown for leaves and heads, such as mustard greens or cabbage, need the most uniform water in order to delay seed formation. Crops that are grown for fruits or seeds can be tricked into producing quicker and higher yields through erratic watering. Go figure.

Leaf Crops

Leaf crops require uniform watering throughout their life cycle. Water stress can cause leaves to become bitter or pest-infested, and early bolting (seed stalk formation) is common.

Root Crops

Root crops are most sensitive to water stress when the roots are expanding, but they do well with uniform watering throughout their life cycle. Uneven water can cause cracking, off flavors, and early bolting.

Tomatoes, Melons, and Peppers

Correct watering of fruiting crops can test the patience of even the most seasoned farmer. Many fruiting crops send down deep roots and are good foragers, so less frequent watering once plants are established can work well. However, frequent watering in the second growth period, when the plant is forming stems and leaves, can encourage the plant to put on a lot of unnecessary succulent green growth and delay flowering. You've seen this happen with tomato vines that are huge and lush for months with nary a flower or fruit. So depriving them a bit until flowering begins can be good. During the initial flowering and fruit set time, sufficient water should be given for the fruits to swell. Lack of water during this phase can cause cracking, small, dry fruits, or fruit drop. Finally, once the fruits have swelled to about two-thirds of their final size, slackening off on water can improve flavor and avoid mealy texture. Whew!

Summer Squash, Eggplant, and Cucumbers

The above instructions apply—except for the last phase. Be a bit stingy with water prior to the commencement of flowering, but after this point continue uniform watering till harvest.

*Allium*s (Onion Family), Potatoes, and Winter Squash

These crops require uniform watering until the very end of their life cycle. Once leaf growth begins to brown, cease watering entirely, or the crop may begin to rot.

Heading *Brassica*s

Broccoli, cauliflower, and cabbage are most susceptible to water stress when they're heading up. Insufficient water can cause poor head formation and early bolting.

Pod Crops

Fresh beans and peas are most susceptible to water stress during flowering.

WATERING METHODS

Although it's less water efficient than using drip irrigation, it's very useful to hand water your crops at the beginning of your gardening endeavor. Hand watering will give you time to develop your observation skills, and your observations, in turn, will help you learn gardening skills faster and help you learn to diagnose plant problems correctly.

Who knows, an emotion of care and love toward your plants may even develop.

If done correctly, the following watering methods can all be effective. They vary in efficiency of water usage.

WATERING METHODS

Types of Overhead Watering	Types of Drip Irrigation
Dipping water out of a bucket—allows you to use rainwater or grey water	Emitter lines
Overhead sprinkler	Soaker hose—allows you to use rainwater or grey water
Watering can—allows you to use rainwater or grey water	Spot emitters
Watering wand	Spray emitters
	T-tape

Overhead Watering

Overhead watering of any kind is less efficient than drip irrigation, because there is more surface evaporation. With that said, to grow healthy plants and not waste your time and effort getting them in the ground, what's most important is watering your garden, however possible and whenever convenient.

Opinions vary as to the best time of day to water. In many cases it's really just splitting hairs. The important thing is to water. With your busy schedule, minute differences between surface evaporation rates at different times of the day don't really figure in. Evaporation is highest on warm or hot days at midday, so don't water then. Otherwise, water at a convenient time for you.

When hand watering, it's very important to develop a mental picture of the pattern you will use. This will ensure even and sufficient water distribution to all parts of the bed. Develop a rhythm by counting to yourself in your head. Hold the watering wand, and with a slow, inner metronome sweep it across the bed at an even speed, counting: one . . . two . . . three . . . four. Then move to the next little strip of bed. After doing a few strips, go back and give each strip a second and a third soaking. When approaching the edge of a bed it's important to actually pass beyond it, so as to give equal water to the plants there. If plants are in hills (doughnut dishes), boxed beds, or beds with a lip to hold water, flood each area in turn, and then return two or more times.

Be systematic within the entire garden. Count how many times you've watered each bed to make sure you're consistent. As you can see, if you have three or four planter boxes, watering your garden can take an hour. Be sure to allow enough time.

Drip Irrigation

Once your urban farm is established, you may opt to save time and water by installing a drip irrigation system. We highly recommend drip systems, since they can conserve as much as 50 percent of the water you would use with other watering methods. Drip components have improved a great deal over the last quarter century, and what was once a confusing and technical quagmire has become easy as pie. Don't be intimidated by the prospect of installing your own system—it's just like playing with Tinkertoys.

Choosing the right system for your needs

Drip irrigation systems for ornamental plantings often consist of solid polyethylene (poly) tubing supply lines with single emitters placed at each plant. This works well for perennial plants, but for annual veggie beds, an emitter line (a semirigid plastic tube with in-line emitters that drip along its length), soaker hose (a hose that secretes water through pores), or T-tape (a flat plastic tube with multiple holes) are more appropriate. These components slowly drip water from tiny openings in poly tubing. A number of lines of tubing are run along each bed, ensuring even coverage of the entire growing area. The tubes are on the ground releasing one tiny drop of water at each opening every moment, so very little water is lost to wind drift, evaporation, or runoff. As the droplets soak into the ground, the water spreads out, so while beds irrigated with drip lines may appear dry on the surface, the soil is moist throughout below.

If you are using city water, we recommend using either emitter line or T-tape. If you are using rainwater or grey water, use soaker hose, as it requires less water pressure to function. Both T-tape and emitter lines are attached to solid ½-inch poly supply lines with special fittings (your drip irrigation supply company can let you know what you will need). When designing your system, we recommend investing in on/off valves that are placed at each bed. That way, if a bed is not being cultivated, or you need to suspend irrigation at a certain phase of growth, you will be able to do so.

Because of its affordability, T-tape is especially appropriate for larger urban farms. T-tape is traditionally used in field farming, in which beds are tilled each season, because it's easy to roll up and set aside. It comes in a few grades that each last a certain amount of time before degrading, at which point it must be disposed of. If you use T-tape, invest in the most highly rated grade, so it will last longer. It has a seam where the emitter holes are located that should be placed facing up to avoid plugged emitters.

Emitter lines are much more durable and long-lived than T-tape, and easy to patch if you stab them with a digging fork, but they are more expensive and a bit more cumbersome to move around after they're installed. Emitter lines come in ½-inch and ¼-inch diameters, with varied rates of flow (talking to your drip irrigation supply company will help you decide which flow rate is right for you). The emitters are inside the polyethylene tubing. If you opt for emitter lines, we recommend using ¼ inch instead of ½ inch, since it takes up less space in the bed and can be moved aside temporarily more easily during cultivation. The fittings are also much cheaper.

Drip irrigation supply company employees are extremely helpful when it comes to designing these systems. They have an interest in helping you—you're buying their stuff. Many of them have parts catalogs that explain in detail how to design and install systems. The most complex aspects of the project will be measuring your water pressure, determining how many zones (areas set to turn on at different times) you need, installing multizone timers, and installing valves (plumbing parts that contain backflow on/off valves and filters and are connected to the timer). These aspects require some plumbing and electrical know-how, and if you haven't got it or don't want to learn it, hire someone.

Except for the valves, installing drip components is very easy, especially if your parts supply store has gone over your layout with you and helped you determine the maximum length of the individual lines that will work with your water pressure.

Installation Tips

Poly-tubing fittings are compression fittings, meaning that the fitting attaches so tightly around the tube that water can't seep out. In other words, the tube is compressed into a slightly smaller opening on the fitting. So how do you get the wider ½-inch line into the narrower fitting? When connecting a ½-inch emitter or supply line with fittings,

DRIP IRRIGATION LAYOUT ZONE 1: EMITTER LINE

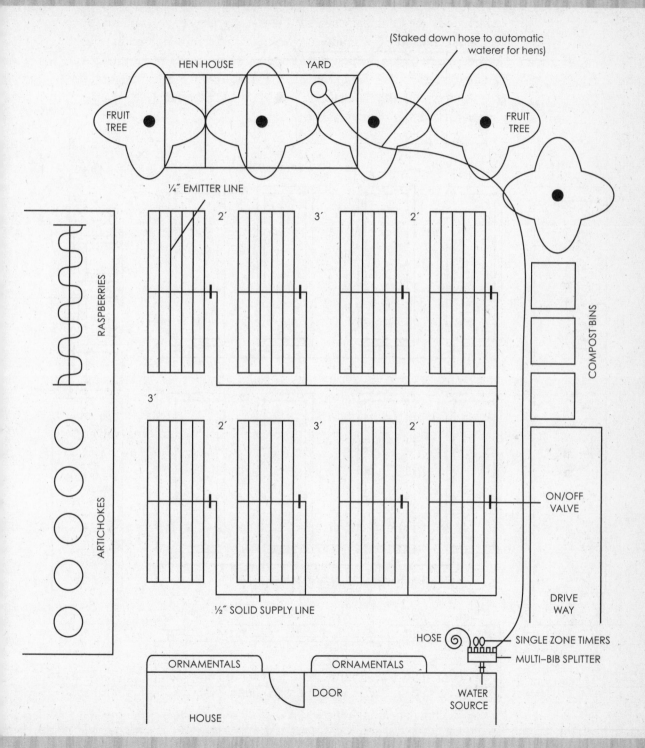

Example of how you can lay out a drip system using in-line emitter tubing

DRIP IRRIGATION LAYOUT ZONE 2: SPRAY EMITTERS AND DRIPPERS

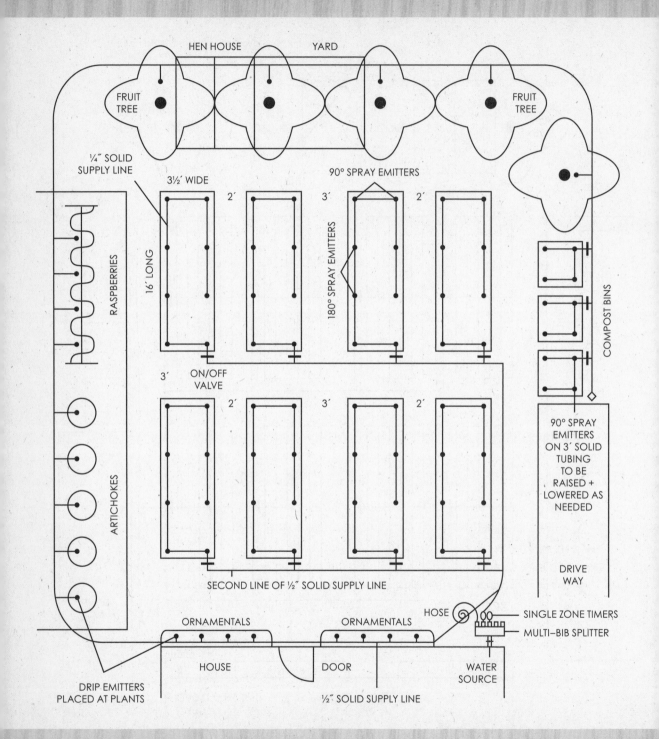

Example of how you can lay out a drip system using spray emitters and drippers

ATTACHING ½-INCH POLY TUBING LINE TO FITTINGS

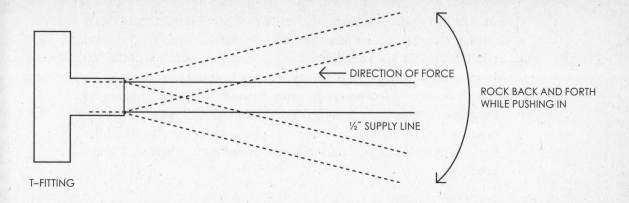

DIRECTION OF FORCE

ROCK BACK AND FORTH
WHILE PUSHING IN

½" SUPPLY LINE

T–FITTING

To attach ½-inch poly tubing to fittings, rock back and forth while pushing. Don't twist!

the trick is to use a rocking, back-and-forth motion while pushing the tube hard into the fitting, not using a more intuitive twisting motion. This rocking and pushing motion can be very tiring on the fingers. A secret of the trade is to keep a thermos of hot water on hand. To soften the poly material and make it slide in more easily, dip the tube ends in the hot water for a minute before pushing them into the fittings.

Whichever type or diameter of emitter line you use, you will space the lines in your growing beds at specific intervals; for instance, every 6 inches or 12 inches (your irrigation supply company will determine this for you in the design process). It is important to correctly space the lines at the edges of the beds. The spacing interval is selected so the last line before the edge of the bed should be half the interval from the edge. Otherwise, the edges of the beds will get half the amount of water as the rest of it.

Once attached to the supply line, the emitter lines are set at the correct interval and staked in place along their length. To properly stake a line, go to the end, space it correctly, pull it taut, and work your way back to the supply line, staking the latter every 12 inches to 18 inches. Otherwise, lines can snake through the bed in an irregular manner and make for spotty watering.

Naturally, in the course of working in your beds, and over time, emitter lines will be displaced. When you want to add compost or re-cultivate the top few inches of soil, remove all the stakes and flip the lines into the paths or into another section of the bed.

The Challenge of Misting Direct-Seeded Crops

As a primary watering method, overhead misting is extremely inefficient and doesn't soak the soil sufficiently for deep-rooted crops. But
because emitter lines deposit water just below the surface of the soil,
seeds tend to sit just above the moist zone and fail to germinate. For
the production-oriented urban farmer, one of the most challenging
aspects of drip-system design is providing the necessary, even overhead misting to enable direct-seeded beds to germinate. We've fiddled
with various setups and find that no one solution is perfect. Choose
your poison.

Hand-Watering Seedbeds

You can choose to hand water seedbeds or put a sprinkler on them
until they germinate. It's really easy to miss a day or waste water, but
if you're organized enough it can work fine.

Spray Emitters

Run two separate supply lines to each bed when installing your system.
Put in drip lines on one supply line, and attach a second system, of ¼
inch solid-line with spray emitters on riser stakes every few feet and
at corners, to another supply line. Depending on the width of the bed
and the reach of the sprayers, the spray-emitter supply line in the bed
can either go down the middle or around the perimeter. Sprayers on
riser stakes (stakes that poke into the ground with a connector to the
supply line and a ¼ inch tube with a spray emitter at the top) are attached to ¼-inch supply lines at intervals that allow for a slight overlap of spray. There are emitters available that spray 360 degrees for
center-bed lines, and 180 degrees for perimeter lines, and 90 degrees
for the corners of perimeter lines. Be sure to install ¼-inch on/off
valves to the supply lines so you can turn the sprayers off when not in
use. When you want to use the sprayers, turn off the emitter lines, and
vice versa. Sprayers usually come on 12-inch risers, which can be too
high for beds (6 inches is better). Cut the risers in half, and be sure to
set each mister stake carefully at the right angle so the bed will be actually misted and not the air or the path.

The fineness of the spray is also an issue in seed germination.
Several companies produce spray emitters of various types and qualities. Because their biggest market is ornamental landscapers, products are geared toward perennials. What veggies need is a spray that

will mist the bed evenly with a fairly gross spray. Very fine mist, while fine inside a greenhouse, is just blown away out-of-doors. Harder sprays that shoot to the edge of the range of the sprayer, missing the space in between, are useless. Will someone please design a perfect product for the needs of urban farmers?

Movable System

If you have a backyard-size garden, chances are you will only have newly direct-seeded crops in one bed at a time. A line of spray emitters can be attached to a long garden hose with special fittings, and to a relatively inexpensive single-zone timer that would be attached to your hose bib (the outdoor hose faucet connection) (you'll need a metal hose splitter to create two faucet connections—one to allow for a regular hose; another for the drip system). When you're ready to mist, turn off the drip lines to the bed. Carefully stake the line of spray emitters, and turn it on to test for coverage. Set the timer for one or two watering times per day. When not needed, the hose and poly tubing line, with spray emitters attached, can be carefully coiled up and stored in a tote. This system doesn't work well in larger gardens. Even in small gardens, two beds can easily be direct-seeded at the same time. The other downside is that care must be taken when taking it all out, setting up, and storing, or the parts can easily be damaged.

SAVING WATER

Farming requires a lot of water. Seeds need moisture to germinate, crops require water to grow and produce fruit, and your harvest must be washed before you can cook it. While rural farmers get theirs from wells or at discounted agricultural metering rates, as an urban farmer, you will most likely be paying for it at the relatively expensive residential rate. And conserving water is not only important for your pocketbook; it's important for the community. Clean drinking water is becoming scarcer as demand increases. Given these pressures, it makes sense to conserve water as much as possible on your urban farm.

Many regions of the United States deal with water shortages, recurring droughts, and a lack of rain during the main growing season. In similar climates in other countries, the use of water catchment systems, cistern water storage, and water-conserving irrigation techniques is nearly universal.

There are many ways to conserve water on your farm, and, indeed, an entire book could be devoted to this issue alone. Here we offer a number of ways to save it, to use your water wisely, and to decrease your farm's thirst while growing the healthiest and most productive crops.

Mulching

The intensive-farming method is, in itself, a water-conserving farming method. Mulching the ground to decrease evaporation from the soil can be done in two ways. The first is called a "living mulch," and it is accomplished through the intensive-planting patterns previously described. Rather than having areas of bare ground between them, the outer leaves of the crops touch as they grow. The soil below is shielded from the sun's rays, and water-hogging weeds are also shaded out. It's a big no-no in organic and intensive farming to leave soil bare, as nutrients are more easily lost and erosion can set in.

If the living mulch method can't be done (in the case of eggplants, which are heavy feeders and need to be spaced apart, for example, simply apply a couple inches of dry leaves, compost, or straw in between plants. This duff of carboneous material will suppress weeds and retain moisture in the soil. In extremely dry areas, one option is to create a "waffle garden," which consists of plants in deep wells formed by piling up to a foot of mulch. You would need to space your crops farther apart to allow for piles of mulch in-between individual plants or rows.

Other areas of the garden should be mulched as well to suppress weeds (weeds use a lot of water!) and to slow surface evaporation. Paths, trees, and perennials can be mulched with any organic matter, such as cardboard and wood chips, straw, solid wood sawdust, clover turf, gravel, or crushed rock path fines. Paths and other areas can take as much mulch as you can practically add.

Dry Farming

Do you want the most fabulous, flavorful tomatoes, potatoes, or winter squash? Try dry farming. Dry farming is a popular technique that works well in dry summer areas. In climates where there is summer rain,

WAFFLE GARDEN BED FOR ARID CLIMATES

Plant plants in wells and cover bed surface with mulch of straw or wood chips.

Waffle bed gardens work well in arid climates.

modified dry farming methods can still be applied. On a small scale, in urban environments, dry farming will significantly reduce your water use. This method can be very effective for fruiting crops, such as tomatoes, potatoes, corn, squash, and melons but doesn't work well for leaf crops or *Brassica*s.

The main principle of dry farming is to provide a deeply cultivated soil that acts as a sponge, storing water during dry times. The goal is to keep the lower regions of the soil moist so that plants will send their roots down deep. The first step is to amend your vegetable beds heavily with compost during the rainy season. The compost will sop up the moisture and create a sponge that will retain water throughout the summer. In the spring, beds are planted directly with seed or transplants. As the plants grow during spring rains, add a light, 1-inch dusting of mulch (shredded straw, leaves, or wood chips) across any bare soil. A doughnut-shaped depression should be left around the plants to collect water. If it doesn't rain while plants are putting on green growth, you'll need to irrigate. Once the summer begins, place a deep layer of mulch—up to 6 inches—around the plants. Evaporation of soil moisture is discouraged by this addition. Water sparingly, if at all, into the depression, until the plant begins to set fruit. Then stop watering entirely.

Dry-farmed fruiting crops are usually tastier because they contain less water. Yields are decreased, however, often by one-third. But you are less impacted by lower yields on an urban farm, since your livelihood doesn't necessarily depend on it, and the cost savings from reduced water use could make it worthwhile.

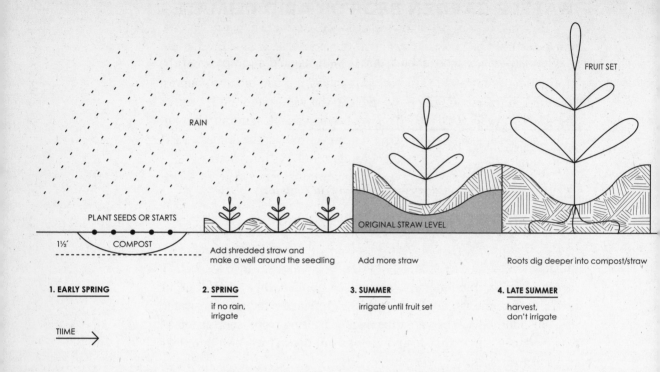

RAIN

PLANT SEEDS OR STARTS

1½' COMPOST

Add shredded straw and
make a well around the seedling

ORIGINAL STRAW LEVEL

Add more straw

FRUIT SET

Roots dig deeper into compost/straw

1. EARLY SPRING

2. SPRING
if no rain,
irrigate

3. SUMMER
irrigate until fruit set

4. LATE SUMMER
harvest,
don't irrigate

TIIME

Dry farming method for potatoes, tomatoes, melons, and corn

Grey Water

By now you know that one of the greatest benefits of farming in the city is the amazing waste stream just waiting to be captured by an enterprising urban farmer. Used household wastewater—known as grey water—is another way to channel a valuable waste product into growing crops. Think about how much water you and your neighbors use in your houses: washing veggies or dishes; runoff from a quick shower; or rinse water from the washing machine. Though we wouldn't want to drink it, plants in the garden love to soak it up.

If you plan on applying grey water directly to your garden, observe the following rule: Never apply raw grey water directly to plants with low-lying fruits or on leaves of plants that you plan on eating raw (e.g., lettuce leaves, basil). Instead, use it to water fruit trees, shrubs, and your compost pile. By diverting your household water this way, you reduce their need for fresh water and can "spend" your metered water on your other plants.

How to go about creating your grey water system depends on your house and circumstances. If you own your house, it's a good investment to have a grey water system installed by a knowledgeable expert. A professionally installed setup often involves the use of sump pumps, biofilters, and a circulating pond to clean the water before it is used in the garden. These are beautiful, permanent systems befitting a person with an ecohome. There are detailed books about installing them if you'd like to do the work yourself (see appendix 11 for resources).

If, however, you're a renter, your landlord may not want to re-plumb their building to save water. Luckily there are some low-tech, low-budget systems that you can install with ease. These simple systems can also be uninstalled quickly if you move or if your landlord objects. These systems are sanitary and save water. Here are some idealized, low-cost scenarios that might apply to your urban farm.

Washing Machine

By modifying your washing machine hose, you can easily route your wash water into your garden. Simply attach a long garden hose to the outflow pipe on your washing machine and run it out a window or door and into your garden. In addition to a long hose, you'll need to buy biodegradable or biocompatible laundry soap. Our favorite brand is Oasis, because in addition to being a cleaning solvent, it actually contains plant nutrients. Biodegradable soaps are second best. With these you will need to move the hose to different locations in the garden to avoid excessive salt buildup over time.

Once your washing machine system is set up, place the end of the hose in a wood chip pile or mulch basin that surrounds a fruit tree or shrub, or directly into your compost pile. The carbon-rich mulch acts as a natural biofilter that will provide a holding area for particles that come out with the wash water. Quietly, and more importantly, unnoticed by the nose, the particles will decompose. The wood chip pile also prevents overwhelming an area with a torrent of wash water. Alternately, you can route your hose into a fifty-five-gallon drum with a spigot on the bottom if you want to hold the water and use it in the garden later. But don't let it sit for more than a day or so—it will begin to fester.

Bathtub

Bathing is the prime water need in most households, especially if you have a chronic bath taker (showers use much less water) in your

WASHING MACHINE GREY-WATER SYSTEM

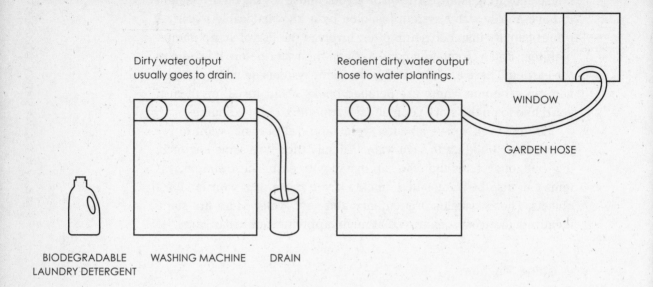

Dirty water output usually goes to drain.

Reorient dirty water output hose to water plantings.

WINDOW

GARDEN HOSE

BIODEGRADABLE LAUNDRY DETERGENT

WASHING MACHINE

DRAIN

Washing machine water usually drains into a drainpipe. To use grey water, simply remove the hose from the drainpipe and redirect it to the garden. Attach a garden hose if you need more length.

midst. Put that fairly clean wash water to good use by keeping the drain stop closed, leaving the bathtub full and reusing this water around the house; this will offset increased garden water use. One use, is instead of flushing your toilet with the lever, keep a bucket near the side of the tub. When it is time to flush, scoop up a bucketful of water from the tub and pour it into the toilet bowl. As the water goes into the bowl, the toilet will automatically flush out to the sewer, just like it does with the water in the toilet tank.

Alternately, you can scoop water out of the tub, carry it out to the garden, and use it to water your fruit trees, shrubs, and vines—just be sure to use natural soap instead of detergent-based body products. If that seems like too much of a schlep, buy an inexpensive sump pump with a long hose; place it into the water-filled bathtub, put the end of the hose out in the garden in a place covered with wood chips (to filter

BATHTUB GREY-WATER SYSTEM

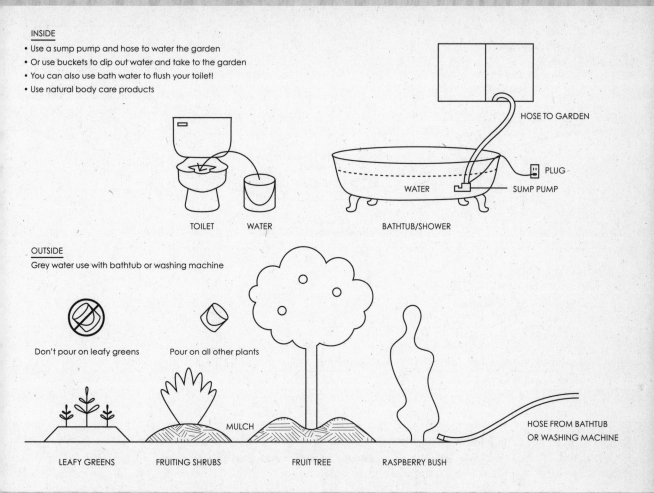

INSIDE
- Use a sump pump and hose to water the garden
- Or use buckets to dip out water and take to the garden
- You can also use bath water to flush your toilet!
- Use natural body care products

HOSE TO GARDEN

PLUG

WATER   SUMP PUMP

TOILET   WATER   BATHTUB/SHOWER

OUTSIDE
Grey water use with bathtub or washing machine

Don't pour on leafy greens   Pour on all other plants

MULCH

HOSE FROM BATHTUB
OR WASHING MACHINE

LEAFY GREENS   FRUITING SHRUBS   FRUIT TREE   RASPBERRY BUSH

Using bathtub grey water to flush your toilet or to water your garden

out particles and dirt), and pump the grey water directly into your orchard. Your back—and trees—will thank you.

Rainwater Catchment

Catching rainwater is most popular in areas that get summer rain, like the East Coast, Midwest, and Pacific Northwest. It can be used for irrigating the garden, and rainwater is considered to be the best for your garden, because it has low salts and additives like chlorine. The

SWALE IN TERRAIN AND SWALES FOR TREES AND SHRUBS

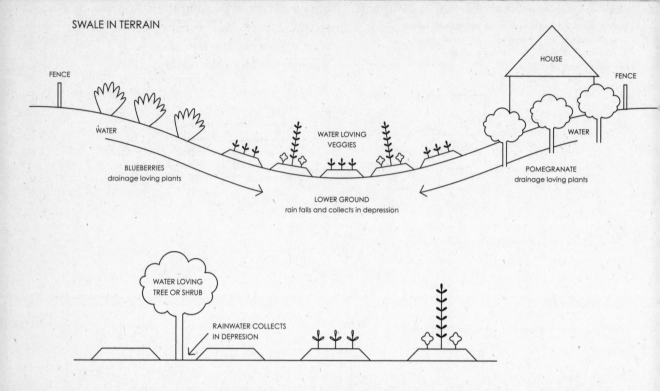

SWALE IN TERRAIN

FENCE

HOUSE

FENCE

WATER

WATER LOVING VEGGIES

WATER

BLUEBERRIES
drainage loving plants

POMEGRANATE
drainage loving plants

LOWER GROUND
rain falls and collects in depression

WATER LOVING
TREE OR SHRUB

RAINWATER COLLECTS
IN DEPRESION

Above: Ideal planting plan for a natural swale in the landscape
Below: How to build swales to direct water to thirsty trees

simplest way to catch water is to build what are called "swales" in the garden: You just create low-lying areas where you want water to collect naturally during a rainstorm. Swales create moist zones where water is stored in the soil.

Another method is to buy or make rain barrels. These attach to the rain gutter downspouts of your home, to catch summer rain as it happens. This water can then be directed toward the garden when needed. If you have space for a large cistern (two hundred gallons or more), it makes sense to store winter water for summer use. The Australians have almost perfected rainwater catchment, with narrow yet high-capacity containers that are placed along the sides of their homes. All catchment systems storing water for longer than three days should be closed, so insects, sunlight, and leaves can't enter. It should have an automatic outflow system, in case you get a sudden torrent that overflows the barrels. It should also have a vent to prevent the container from collapsing when you water your garden. The best systems situate the containment so that gravity will do the work of distributing the water into the garden (see appendix 11 for information and resources).

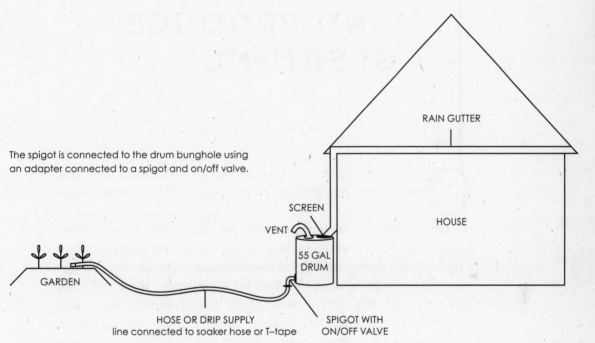

The spigot is connected to the drum bunghole using an adapter connected to a spigot and on/off valve.

RAIN GUTTER

HOUSE

SCREEN

VENT

55 GAL DRUM

GARDEN

HOSE OR DRIP SUPPLY
line connected to soaker hose or T–tape

SPIGOT WITH
ON/OFF VALVE

Rainwater catchment system. Rain flows from the gutter into a drum with an on/off valve spigot.

CHAPTER 12

HARVESTING AND PRODUCE GLEANING

One would think that harvesting from your urban farm—pulling up carrots, plucking ripe tomatoes, clipping fresh lettuce leaves—would be the simplest and most satisfying task on the farm. In fact, we've found it to be the exact opposite, with many urban farmers resistant to picking their produce. We've even done it ourselves—let the broccoli go to flower, whale-sized zucchini escape our notice, and sometimes tomatoes rot on the vine. There are many reasons for this reluctance; one is simply time. In general, we all lead incredibly busy lives, and harvesting, cleaning, and preparing garden produce adds one more chore to a long list. The fact that you may have invested a ton of time and money into your gardening endeavor doesn't translate automatically into efficient use of your harvest. It's important to note that only in the affluent West, where cheap calories abound, would this even be thinkable. Another reason is most people don't know how to harvest from a high-producing garden. If you want to eat out of your garden, if you want crops grown in your community gardens to go home and actually be eaten by community members, if you want the children to eat the food coming out of your kid's school garden, you will have to muster great strength to go against ingrained habits.

In the interest of your success, we offer here not only nuts-and-bolts information on how to know when your veggies are ready to harvest and how to harvest them, but suggestions to make it as easy as possible for that food to end up in your belly. Know that you will miss the mark on harvesting or eating some things. Move on. You can only build new habits by having successes, so celebrate them, even if they're small. In time it will be possible to change the way you eat. One day, when you're hungry you will open the garden gate rather than the door to the refrigerator without even thinking about it. But it's going to take some time to develop those habits.

There's another reason for not neglecting the harvest. Harvesting or the lack thereof also affects the health of your garden. Neglected plants can breed insect pests. Crops left to rot bring mildew that can be hard to eradicate. And you're certainly not going to win over your neighbors when they see you letting food fall to the ground and broccoli bloom. To avoid these issues, this chapter will detail ways to make harvesting a pragmatic, pleasurable activity that will become second nature.

ROUTINE: THE KEY TO HARVESTING AND EATING YOUR BOUNTY

You already have fully formed food preparation and eating habits, and eating out of your garden is not a part of them. The easiest way to incorporate your garden produce into your diet is to act as if you're going to the grocery store. We recommend that you make a real effort to create a "market day," when you harvest, wash, and bunch all the mature produce in your garden, so it will be sitting in your refrigerator ready to use during the rest of the week. If you are a home gardener this will probably take under an hour on one day, with a few extra minutes of squash, bean, or pea picking on two other days at the height of production. A few hours to an entire workday may be necessary if you're managing a larger garden.

Remember that rhythmical schedules eliminate the need for strength of will. Creating a routine will mean you don't have to muster energy and fight old habits each time you think to harvest. You will feel great when you are using all the bounty from your garden. If you go out and see a whole bed of bolted lettuce . . . well, you're not going to feel great. Use the pleasure principle, not guilt, to reinforce new efforts in your life. The experience of wolfing down a freshly picked bowl of salad greens drizzled with oil and vinegar like you've never eaten salad before is the pinnacle and culmination of your work. It feels and tastes good, and you'll want to repeat the experience again and again. Pick one day and time in your week that will be your harvest time and try to stick to it even in times of scanty harvest. We recommend you don't pick a weekend day, since many travel plans and special occasions will interfere.

WHEN IS IT READY?

You already know when most vegetables are at harvesting size. You've seen them all your life in the grocery store. What confuses us in the garden is that there's all this other stuff attached to the edible portion. It throws us off. Try to zero in on the edible part of the plant and ignore all those leaves or vines. Does it look like what you buy at the store? If so, it's probably time to eat it.

Be aware, however, that we have become used to eating certain vegetables in bunches, not knowing what they look like singly. Spinach and lettuce heads in the store are often bunched together to make one big head. Your garden-grown lettuce and spinach may never look like those, so you're in danger of missing the window and having bolting crops on your hands.

Your farming methods are also different from most commercial growers' methods, and this will make for some important differences in the size at harvest. Many commercial growers pump their crops full of water to make them grow as quickly and as large as possible. They also choose varieties that produce the biggest heads over the most flavorful varieties. Those who value flavor and water conservation, on the other hand, attempt to grow varieties that may be smaller at maturity and give just the right amount of water. As you gain experience you will be able to tell when radishes or lettuce are getting ready to go to seed or when a head of broccoli is about to flower, and you will know to harvest well ahead of this for peak flavor.

There are plants that are removed entirely at harvest time, like cauliflower, and plants that yield over a period of time, like parsley. For the former, individual plants often mature at different rates. From germination on, the vigor of individual plants in a planting is not the same. Home growers actually prefer to have veggies that can be harvested over a longer window, but it makes for more observation work. Keep abreast of what's going on in your cauliflower bed. When you harvest your first mature head, note the stage of the others, so you can

plan to harvest them as needed, or if they're all coming on at the same time, to preserve or share them.

Also important is keeping up on the specifics of the varieties you planted. If you ordered seeds, you knew at that time what they would look like from the beautiful glossy pictures. When the packets arrive, often none of this information is listed on the packet. It's good at this time to go back to the Web site or catalog, read the plant culture information provided, and make notes on the packet or on a piece of paper, sticking both packet and notes in a baggie. This information will help you in the greenhouse when labeling your seedlings, and also in the garden when harvesting. When you plant seedlings or seeds be sure to write what you will want to know on the plant label. If you are waiting and waiting for your broccoli to form heads, you may have a sprouting variety on your hands. If you're not sure what those yellow tennis balls are doing on your cucumber plant, you may have planted lemon cukes without knowing what they look like.

WHEN AND HOW TO HARVEST AND REMOVE PLANTS

Depending on the type of crop, we harvest into buckets half full of water, empty buckets, baskets, or shallow trays. Root vegetables, greens, herbs, and broccoli or cauliflower heads will continue growing and stay fresh and crisp if harvested into water. Just make sure the cut ends or roots are submerged. Fruiting crops are often damaged by water and excessive weight. Unless you have time to let them dry in the air, don't expose them to water until you're ready to eat them, and if they're heavy, harvest in a single layer to avoid bruising. Onions and storage vegetables can be washed briefly but shouldn't soak in water.

In the heat of summer you can place a wet sheet over your buckets, baskets, and trays. As the water evaporates it will create a cooler environment below.

The same harvesting practices generally apply to the following broad categories of crops. Unless you plan to save seed, take out the plants as soon as they're not productive anymore.

Roots

Root vegetables are often overseeded, so the first phase of harvesting them may be thinning. If you wait until small but edible roots have formed, you can eat your thinnings as baby veggies. Alternatively, use the leaves in salad mix.

These vegetables send down long taproots with extensive root systems growing off of them. You can brush dirt away from the top of the root to see if it's big enough to eat without damaging this system. The size of the "shoulders" of the root will let you know how wide it is and give you a guess as to its level of maturity. There is a cost/benefit analysis to be done with the concept of baby root vegetables. Since growth is exponential rather than linear, harvesting roots before they reach their full size can unnecessarily lower your yields. Some will grow woody or lose sweetness if allowed to get too big, but most open-pollinated varieties have been bred to have good qualities at standard size, and good holding ability in the ground once they've reached maximum size. The best way to determine the optimum harvest moment is to taste test over a few weeks.

Root vegetables vary in their ability to hold in the ground. You will want to know up front by reading the description in the seed catalog whether the variety you've grown will hold its flavor over a number of weeks or months if left in the ground and harvested in stages. If not, when maturity arrives it's time to can some beets or freeze some carrots.

Be sure to use a digging tool when harvesting longer roots, as it is quite easy (and frustrating) to break them in half. To keep them fresh in the field, harvest them into a bucket of water. If you like you can bunch them up with rubber bands or twist ties as you go along. Once dug, wash the roots, trim the leaves and taproots, then refrigerate.

Like other life forms, all plants have a biological imperative to reproduce themselves. When root vegetables prepare to go to seed, all the stored energy transfers to the seed stalk, and the root actually becomes quite thin! If the leaves have begun to form a thickened, round stem in the center, you've probably already lost some of the bulb, and what's left may be spongy. Even if small, be sure to harvest at the first tiny sign of bolting.

Cooking Greens and Herbs

Cooking greens and annual herbs produce leaves that can be harvested over a period of time. They either grow in a rosette of leaves, with larger, more mature leaves on the outside and newly forming leaves on the inside, as with chard, or in the form of a stalk with staggered leaves, as with some types of kale and collards. Harvest the more mature, outer leaves when they reach the size you are used to seeing in the grocery store. Again, to maximize your yields you will want to harvest mature leaves, not immature ones. If left too long, bugs will eat some leaves and/or others will become mildewy. Many people believe that collards and kale become bitter if too big, but it's actually heat rather than size that produces bitterness.

Use scissors, a knife, or your fingers to harvest the leaves of bunched greens or herbs around the outside of the rosette; attempt to leave as little stem on the plant as possible (long stems can introduce rot). If removing leaves with your fingers, use a side to side or twisting motion. Pulling may damage roots or actually uproot the plant. To maximize the plant's ability to generate new leaves through photosynthesis, leave two thirds to half of the leaves. As you move from plant to plant, gather enough leaves for a bunch, line up the ends, and wrap them with a rubber band. Place the bunch cut ends down in a bucket with a few inches of water.

For a braising mix, harvest cooking greens with the cut-and-come-again method. Some varieties are more suited to this than others, so you might want to start with a mix intended for this method. Your plants will be more closely spaced together. When the seedlings have reached 3 inches to 4 inches high, gather up all the leaves of each plant with one hand, and, with scissors, cut all the leaves on the plant about 1½ inch above the soil surface. This leaves immature leaves unscathed. Put the leaves in a bucket of water to keep them fresh until you go wash and spin them dry before refrigerating.

Occasionally, especially with some varieties of spinach, mustard greens, and Asian greens, you may choose to harvest whole plants rather than leaves. Cut the plant off at the base of the rosette, leaving the stem and roots to deal with later. This will save you a lot of work in the washing department, as you won't spread dirt to the leaves.

If you notice that plants are beginning to form a flower spike or thick, round central stem, cut this stem or spike off to encourage leaf production for a few more weeks. The earlier you catch it the better. At some point the plant will get ahead of you and make one or more flower spikes. Harvest any edible leaves (be sure to taste test them, as they might be bitter or too spicy), and remove the plant.

Salad Greens

Lettuce, bitter greens, some mustards and Asian greens, spinach, arugula, and a host of other unrelated greens and herbs have all become a part of the famous mesclun or spring mix now to be found even at Walmart. Vive le mesclun! These delicious, vital, complexly flavored greens are especially tasty and healthy eaten fresh from the garden. The above instructions for harvesting greens and herbs apply in the case of salad mixes. Here we will outline three basic methods for harvesting lettuce and salad mix that extend the harvest and yields of your garden.

Whole heads last longer in the fridge, since their leaves are still connected to the stalk. In spite of the popularity of prewashed mix, it's okay to like heads of lettuce! If you have a lettuce bed that was planted all at once, and you wish to harvest whole heads, start when they are slightly immature, so you won't be stuck with twenty heads at once. Remember that garden-grown heads may look smaller than what you're used to in the grocery store. It is also possible to stimulate lettuce plants to produce a second head: Harvest a smaller inner head, leaving a ring of mature leaves around the outside; these leaves will photosynthesize and stimulate the plant to try again.

Harvesting outer leaves is a great way to harvest lettuce if you want mid- to large-size leaves but don't want a whole head. It's also an appropriate way of harvesting mesclun with lettuce varieties that tend to bolt when using cut-and-come-again culture. This is a great way to get a mix and stimulate the plant to be as productive as possible. As

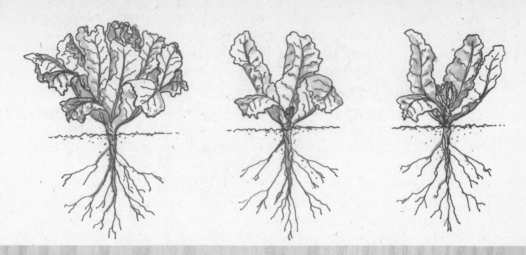

Harvesting the inner lettuce heart to stimulate regrowth

with cooking greens, break off outer leaves cleanly and close to the stalk, and toss them in a bucket of water to keep them fresh as you go. If you notice a thickening stem in the center, break it off to encourage more leafing.

Cut-and-come-again salad greens are so fun to grow! You are basically growing and then shearing a carpet of greens when the leaves are immature and tender. Special blends for summer and winter are seeded or planted with a final spacing of 1 inch to 3 inches apart. When the seedlings have reached 3 inches to 4 inches high, it's time to harvest. Gather up all the leaves of each seedling with one hand, and with scissors, cut all the leaves about 1½ inches above the soil surface. Immature leaves go unscathed. Put the leaves in a bucket of water to keep fresh as you go, but don't leave too long or they will become waterlogged. You can also harvest dry into a basket or tote if it's not too hot. Wash and spin dry before refrigerating. It's lovely to top off your bag of greens with nasturtium, calendula, and herb flower petals.

This method takes a bit of experience to get used to, so be sure to observe the results a week after your first harvest. If you cut too low, you can damage or destroy the inner growing leaves. If you cut too high, you can waste a lot of leaf material.

Eventually, all lettuces and salad greens must produce seeds or forfeit their lives. Once you begin to notice thick, round center stalks forming, your days of cutting and coming again are numbered.

To grow cut-and-come-again greens, simply shear the bed like a lawn and the plants will regrow for a number of additional harvests.

Fruiting Crops

Fruiting crops, such as tomatoes, are some of the most difficult in the harvesting department. Pick them too immature and they will never ripen. Pick them too ripe and they will begin to mold by the end of the day. Learning how to pick a ripe melon in the grocery store practically requires a degree in agriculture. Sweet corn can be ambrosia one day and mealy starch the next.

Because of the extremely early harvesting of store-bought fruiting vegetables, notably tomatoes, many home growers overdo it in the opposite direction. Overripe veggies have little to no shelf life in your kitchen. There is a point at which the carbohydrates are fully developed. At that point, your harvest can finish ripening off the plant without harm—it produces its own ethylene ripening hormone naturally. Changes in color and giving slightly under the pressure of your thumb are indications of maturity.

A good way to see this visually is to harvest a truly green tomato, and then harvest a green tomato with the beginnings of color on its skin, and let them sit in your kitchen for a week or two. The second will become ripe while the first will remain hard and green. The flavor of the tomato ripened from a green state won't be as sweet as a vine-ripened tomato, but it sure beats a mushy rotten one. There's a happy middle that you're shooting for.

With other, nonstorage fruiting crops (see storage crops, below), use the plant culture information on the Internet and pictures in the seed catalog to know what to look for at maturity (size, shape, color), and taste test each variety to determine when to harvest the remaining fruits. Some examples: While English cucumbers are delicious and seed-free when quite large, many slicing cucumbers are bitter if harvested too young but seedy if harvested too large. Lemon cucumbers are perfect when about the size of a tennis ball and still a bit green, but they can be bitter and tough-skinned if allowed to become dark yellow. Many peppers are edible immature or mature. And woe be to she who lets summer squash grow unfettered. To catch 'em when they're small and tender requires twice-weekly harvests. The flowers are edible too. So you can see, there is a great variety of stages when different crops are ripe. You will need to educate yourself and taste crops at various stages of growth to get a feel for the best time to harvest.

Storage crops, such as winter squash and pumpkins, should be left on the vine, perhaps with some straw or a piece of cardboard eased underneath, until the vines begin to die back and turn yellow, then brown. Starchy vegetables, such as winter squash and pumpkins, should then be dried in the open air. This process allows carbohydrates to develop fully before storage and prevents premature rotting. If weather permits, cure, or dry, your pumpkins and winter squash in a shady area on a bed of straw or paper. If rain threatens, be sure to bring them in. Once cured, keep them in a cool, dry cupboard, preferably out of the humid kitchen.

Tomatoes, peppers, and eggplant can be harvested with an upward twist—they will separate from the main stem. Other crops should be taken with a knife. As you look for summer squash to harvest, move the vines where you want them (since they can trail into paths). The fruits will be at the ends of the vines. You can wash and air-dry fruiting crops, but don't refrigerate them until dry. Tomatoes should not be refrigerated. When the tomatoes really come on, it can be hard to keep up with them, and some ripe ones even fall to the ground. Again, be sure to remove all ground-fall each week during harvest, since mold may breed and spread to your healthy crop.

To determine when to remove your plants, keep track of new flower formation. Once it slows to a crawl, feel free to remove the plants. You can pickle immature tomatoes and small peppers. You'll probably be needing the space for your fall garden anyway.

Pods

Fresh beans and peas are perhaps the most time-consuming crops to harvest. We think they're worth the effort, though. They are versatile, well liked by children, and have high protein levels. Pea and bean pickin' time is a time to be with your plants, look up at the sky, notice the weather, listen to the birds. It's going to take a long time. Different bean varieties are so variable in their perfect harvest moment that there's nothing much we can say here except test your produce. Nowadays many specialty fresh bean varieties (notably from Renee's Garden Seeds) are crunchy, moist, and starch- and string-free when quite large, so size is not necessarily the determining factor. One thing that's certain is that to keep up with the edible pods, you're going to have to harvest three times a week at peak season. You need to know what type you are growing to know when to harvest. Snow pea pods should be at least 3 inches long, but flat (if bumps form they're getting old). Edible pod peas (sugar snap peas) should be at least 2 inches long and have peas inside. If left too long however, these peas can become starchy. Shelling peas should be filled out with peas, but they should taste sweet, not starchy.

To extend the harvest it's important to pick every single mature bean and pea. As long as you do so the plant is spurred to produce more flowers in an attempt to produce seed. You can always spot people with attention deficit disorder in a bean row. Picking beans and peas requires sustained, systematic attention. Take it as a test of wills, yours against the plants' ability to make more pods. You want to win, don't you? Keep at it. Don't go sit in that lounge chair yet!

Wearing a smock with big pockets can be a real help when picking pods. Start at one end of the row and make a mental picture of how you're going to do the job. Will you get all the high ones first and then come back for the low ones? Or will you pick each vine from bottom to top? You need a plan. As you go, be sure to look under leaves, where many pods hide. To pick the pods, hold the stem with one hand and grasp the pod close to the stem with the other and snap it off. This is why you need two hands free. You can also use small scissors or clippers. It's important not to damage the plant, as this will increase the likelihood of pest and disease attacks. If you do break stems by accident, remove the broken part.

As you pick, transfer pods from your smock to a basket or bucket. You can wash them and allow them to air-dry before bagging them for the fridge. As with other fruiting crops, remove the vines when flower formation slows to a crawl or when you're simply ready to move on.

Alliums (Onion Family)

Onions and even garlic are generally grown for fresh green or bulb eating. You will know what you're dealing with because you will have labeled what you grew. Otherwise it's hard to tell the difference at first between a green onion and a bulb onion.

There are two methods for harvesting green or spring onions. You can dig them up or you can cut the top off at about half an inch above the soil, and the top will regrow one or two times. When digging green onions, use an implement so that the white part doesn't break off. For storage onions, harvest when one-third or two-thirds of the tops have died back, then pull off any outside green leaves that are rotting or mildewing and set the bulbs in a single layer in a tray outdoors to cure for a few weeks. After curing, trim the roots and cut off the now dry tops about half an inch above the bulb. Take note that some varieties of bulb onions are not intended for storage but for fresh eating and do not need to be cured. If you see a flower pod beginning to form on an onion it's time to harvest it, even if it's small.

Garlic is ready to harvest when one third to two thirds of the tops have died back and a head with individual cloves has formed belowground. Hardneck garlic will have a hard, cylindrical stem that doesn't allow you to make it into the classic garlic braid. Softneck garlic can be braided after harvest.

Storage *Alliums* should be kept in bags or boxes that allow airflow, in a cool, dry cupboard, preferably out of the humid kitchen.

Potatoes

Allow the potato vines to flower and die back, which allows them to form their skins. Use a pitchfork to gently unearth the potatoes and place them in a bucket. Keep them in a box in a cool, dry place, and don't wash them until you are ready to cook, as the dirt prevents them from sprouting. New or spring potatoes are harvested just after flowering and should be eaten soon after harvest, or stored in the refrigerator.

HARVESTING MATERIALS, WASHING STATION, AND SANITATION

Having everything you need at your fingertips will make harvesting much easier.

The first thing to do is gather the necessary materials. Creating a special basket or tote can help. Second is deciding where and how to wash and prep your produce. If possible, setting up a washing station in your garden is the best option. Washing and prepping in your kitchen is fine too. It will be help if your kitchen is clean and you have ample clear counter space.

Harvesting materials:

- rubber bands or twist ties
- plastic or muslin refrigerator bags
- hand pruners
- a sharp harvest knife
- scissors
- harvest buckets, baskets, and trays
- egg cartons, if you have fowl
- berry baskets on occasion
- an old, light-colored sheet
- a bucket for produce waste
- a garbage receptacle
- chlorine bleach (sodium hypochlorite)
- biocompatible or biodegradable soap
- scrub brushes (one hard and one soft)
- salad spinner
- drying rack (old dish drainers or tiered hanging veggie baskets can work well)
- sink or bathtub drain stoppers

Washing Station

There's nothing nicer than washing your freshly picked fruits and vegetables directly in the garden. It avoids soiling your kitchen, and you can send the wash water directly back to the garden.

If you choose to create a wash station, site it close to the house in a shady area. Find or buy one or two sinklike receptacles: laundry, mop, or regular kitchen sinks, or bathtubs. In a pinch, large plastic bins can work too. Having two will allow you to wash veggies in one tub with small quantities of biodegradable soap like Oasis or Dr. Bronner's, white distilled vinegar, or small amounts of chlorinated vegetable wash (see table on page 275 for exact amounts). You will rinse them in the second tub. If you are managing a larger garden, two old bathtubs elevated on blocks with tables on either side make a great setup.

Build or find a simple table, use a jigsaw to make a cutout in the top, and drop the sink in. You can also use salvaged kitchen cabinets if your area is sheltered from the rain. It's nice to have as much counter space as possible, so you can deposit dirty veggies on one side and drain clean ones on the other. If you lack such space, add other tables to either side of the sink. Covering your tables with oilcloth stapled to the underside will help you maintain sanitation.

OUTDOOR VEGETABLE WASH STATION

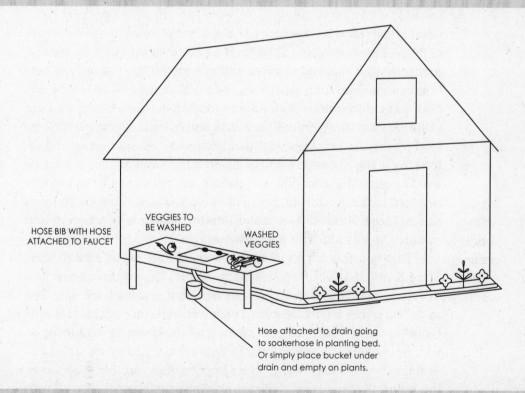

HOSE BIB WITH HOSE
ATTACHED TO FAUCET

VEGGIES TO
BE WASHED

WASHED
VEGGIES

Hose attached to drain going
to soakerhose in planting bed.
Or simply place bucket under
drain and empty on plants.

An ideal way to wash and prep your produce for the kitchen, outside

Buy a hose that is lead-free and maintains water as safe for drinking. Attach the hose to your garden hose bib and run it to the cold faucet on the sink or bathtub. Use splitters as needed at the hose bib or if you have more than one sink. Since some plumbing know-how is needed to reduce a drainpipe to a garden hose, the easiest thing to do for the used water is simply to place a bucket beneath the sink drain opening to collect the waste water. In our garden sink setup we use a plastic pitcher to dip out carryable amounts and water perennials or other veggies not on our drip system. If you're more sophisticated, run another hose from the sink drain and place it wherever you want to water, for instance, on fruit trees.

Purchase produce- and sink-washing products that are biodegradable or biocompatible, so that your garden plants aren't poisoned. The exception to this is chlorine, which you will need if you want to ensure sanitation. Chlorinated water can be dumped on wood-chip piles or the ground, where it will degrade.

Sanitation

Of late, food safety has become a hot topic in the news, due to outbreaks of *E. coli*, *Salmonella*, and *Shigella*. Most outbreaks occur with factory-processed and -packed produce and raw juices. The majority of outbreaks occur with fruit, salad bars, and salad mixes, or with lettuce that have come into contact with bacteria during processing and packing. Outbreaks are rarely traced back to practices on the farm itself. Some food processing practices lend themselves to such outbreaks, and efforts to pin the blame on small-scale producers and organic growers is a red herring. The trend toward washing and packaging fresh produce is linked to the greater incidence of outbreaks. A notable exception is factory farms situated near major livestock operations, where runoff contaminates produce on the farm itself.

That said, it is not impossible to acquire bacteria on a small farm. Since *E. coli* and *Salmonella* can be deadly, it's important to know how to ensure that your harvesting and handling practices are safe. The main risk comes from raw animal manures, and since organic farming relies on manure for nutrients, there is a risk. If you have animals on your farm, there is a possibility raw manure can end up on tables and in sinks used for harvesting, so it's important to sanitize these areas. It is especially important if you are sharing or selling produce off your farm. The following practices will help maintain safety.

- Only use fully composted or aged manure on your growing beds.
- Do not allow pets or farm animals to defecate in growing beds.
- Always wash your hands, fingernails, and tools after working with animals.
- Always wash your hands and fingernails before harvesting and after using the restroom.
- Wash with soap, and then sanitize harvest containers prior to use.
- Wash with soap, and then sanitize harvest tables, sinks, and/or tubs prior to use.
- Use chlorinated vegetable wash water in the correct solution (many crops are sensitive to chlorine so don't be tempted to use more than the recommended amount).
- Change the veggie wash water when it becomes dirty.
- Do not put harvested produce directly on the ground or put containers that have been on the ground on your produce preparation surfaces.

A good protocol for sanitizing and washing is to first fill a bucket with soapy water and scrub your sinks, countertops, buckets, and harvest containers, and then rinse them all with clean water. Next, fill your sink with correctly diluted sanitation solution and use a bit to sanitize your countertops, buckets, and harvest containers and tools before you begin washing your harvest. Dump your soapy and sanitized water on a brush pile, gravel area, or wood-chip pile.

Mixing Sanitation Solution

Proportion recommendations for different vegetables vary from 50 to 500 parts per million of chlorine to water. Mixing various solutions is unworkable and probably unnecessary for home or small-scale farmers, so we recommend using a 50 to 150 parts per million solution for both sanitizing surfaces and washing produce.

RECOMMENDED PRODUCE SANITATION SOLUTION RATIOS

Chlorine Bleach (target parts per million)	Sodium Hypoclorite (5 percent to 6 percent) (amount per 5 gallons of water)
50	3½ teaspoons
100	1 tablespoon + 1 teaspoon (7 teaspoons)
150	3 tablespoons + 1½ teaspoons (10½ teaspoons)

WASHING AND PREPARING PRODUCE
FOR THE FRIDGE

Start with the generally cleaner produce, such as salad mix, leafy greens, and fruiting crops and end with root veggies or other dirty produce. Submerge your produce, giving a good scrub down if needed. Drain or spin as needed, and place produce in plastic bags ready for the fridge (don't close until ready to put away, as evaporation will continue). To dry bunched greens or herbs, take one bunch in each hand and arc your arms down to fling the excess water off. If you prefer, you can even prechop veggies, so they will be ready to throw in the pan during the week. Be sure they are relatively dry when bagged.

If you have heavy-metal soil contaminants, you will need to do an extra careful wash of leafy greens and root vegetables. Most heavy metal contamination from the garden comes in the form of dust on leaves rather than lead absorbed into the produce. It's difficult to remove fine dirt and dust because of the surface tension of the water. You will need to use a few dashes of soap or a 1 percent vinegar solution to break the surface tension. Root vegetables should be scrubbed vigorously and, for extra safety, peeled when eaten.

EATING PEST-DAMAGED PRODUCE

Just because another creature munched on your produce doesn't mean you can't eat it. They didn't give your produce cooties, okay? If you eat canned food, you eat a certain amount of pests, so keep that in mind if you find a worm in your tomato. Just cut around it and enjoy. How much pest damage is too much? Just use your common sense. If mold has begun to set in as a result of pest munching, don't eat it.

If you are particularly creeped-out by bugs, it may take some time and fortitude to learn to live with them. Keep in mind the alternative: We can either live with visible bugs that don't pose any real health risks or invisible poisons that pose absolutely real health risks.

GLEANING & GATHERING WILD FOODS

Gleaning, an old practice recommended in the Bible, is simply harvesting leftover or neglected fruits and vegetables from farm fields.

Gleaning can happen in grape fields post–wine crush, or in your neighbor's forgotten plum tree. Wild foods can be found in wilderness areas and parks, where delicious weeds, nuts, and mushrooms might lurk.

To become a master gleaner and gatherer, you'll have to become organized over the years. Cultivate relationships with farmers who will allow you to collect what's left over after the main harvest. Have a calendar that is marked with a list of fruits and nuts in your neighborhood. Here in California we know that morel mushrooms spring up in May; green walnut and feral plum season is in early July; olives are in November. In the Pacific Northwest there are wild blackberries in summer; in the South, wild mulberries. Every region has their gleanable crops! Be sure to take notes on where your favorite gleaning and gathering spots are, and remember to share, and leave some of the crop behind for others.

CHAPTER 13

MANAGING WEEDS, PESTS, AND DISEASES

Before we tell you how to prevent and destroy weeds, a few kind words. Many plants considered weeds are edible, if not downright delicious. Chickweed and purslane make nice additions to spring salads, for example. Others, like nettles and plantain, are important medicinals. Still others—pellitory, *Malva*—make outstanding fodder for animals on the farm. Because of these benefits we have been known to allow some of them to mature as "accidental" crops. We have done this with a plan, though, not just as a way to put a happy face on laziness. No matter how nutritious chickweed is, a whole gardenful would never be used. By the same token, many crops started intentionally can become pesky weeds. You may rue the day you decided to experiment with amaranth, for example.

The patchwork layout of the land in the city makes for different weeding issues. Fields full of wild plants and weeds in the countryside can make elimination nearly impossible. Rural farmers often focus on mowing and tilling, expecting weeds to come back season after season. Urban areas are not subject to windblown weed seeds from agricultural or wild lands. Our weeds tend toward the imported type—lawn grasses, such as kikuyu or Bermuda grass, and ivy and berry vines—and those that thrive in soils compacted by cars, foot traffic, and construction: wild or ornamental morning glory, dock, dandelion, mallow. We city dwellers are lucky: we *can* hope to eradicate weeds from our relatively small pieces of land, though it may take years.

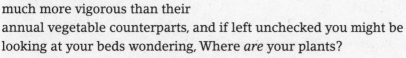

An integral part of intensive farming is maintaining weed-free planting beds and paths. Weeds rob your crops of water, nutrients, and light and can multiply insect infestations by providing habitat. Maintaining the fertility necessary to grow closely spaced crops from the very beginning to the utmost end of the growing season takes time and energy, and you certainly don't want to spend it feeding weeds. They are often much more vigorous than their annual vegetable counterparts, and if left unchecked you might be looking at your beds wondering, Where *are* your plants?

Here's a quick look at the best plan of attack:

1. Remove as much of the pernicious and poisonous weed material as you can when you initially prepare your garden.

2. Learn about the common poisonous weeds in your area. Keep an eye out for them and remove them before they produce seed.

3. Once a year, intentionally sprout and eliminate annual weeds in each bed prior to planting.

4. Try not to let weeds go to seed. In maintenance weeding, start with any that are getting close to producing seed, then work your way to smaller ones.

5. Keep an eye out for the reemergence of pernicious weeds and dig them out by the roots before they spread.

6. Take care not to import weeds into your garden by avoiding cover crop mixes with invasive species, keeping straw and hay contained (mulching with straw is dangerous), and know which intentionally grown crops can become invasive.

PERNICIOUS, POISONOUS, AND NONPERNICIOUS WEEDS

Make eradicating pernicious weeds your priority. Most are perennial and spread vegetatively by roots, rhizomes, bulbs, and taproots. Because they live on season after season, and can reestablish themselves from one tiny piece of root, they are of the most concern.

Pernicious weeds include:

- grasses
- tubers and bulbs
- vines

Some common pernicious weeds to look for in urban areas:

- berries
- Bermuda grass
- bindweed (wild morning glory)
- ground elder (herb gerard, bishop's weed, goutweed, or snow-in-the-mountain)
- ivy
- kikuyu grass
- kudzu
- quack grass (couch grass or devil's grass)
- oxalis (wood sorrel)
- vines

Poisonous weeds are also worrisome, especially where children and animals are concerned (see part III for information on farm animals; we regret we are not including information regarding plants poisonous to pets). If you are unsure, you can look up the names of poisonous weeds on an Internet image search to see what they look like. Anything with solid, round, black or purple berries is likely a poisonous nightshade, and since children think berries are delicious, these should be carefully eradicated.

Common poisonous weeds:

- poison hemlock (wild carrot)
- poisonous nightshades include:
  - bittersweet nightshade
  - black nightshade

- datura (jimson weed)
- horsenettle
- Jerusalem cherry
- poison oak/ivy/sumac
- stinging nettle (actually edible and nonstinging when cooked)

Nonpernicious weeds are perennials that spread less aggressively and annual ones that reproduce by seed. Because annuals can produce prodigious numbers of seeds, they are not to be taken lightly just because they aren't pernicious. Even if you don't allow annual weeds to set seed, old seeds that were in the ground before you started your garden can lie dormant for years, some sprouting each season. Exposure to light is what allows latent weed seeds to sprout, another reason the low-cultivation methods of intensive gardening are great. They don't allow those seeds to see the light of day.

THE INITIAL REMOVAL OF PERNICIOUS WEEDS

If you are starting a garden from scratch and have a healthy cover of pernicious weeds, in the late spring or early summer start by using an electric weed whip machine and/or loppers to cut the top growth as close to the ground as possible. If you unearth tangled aboveground roots, as often happens with blackberries and ivy, cut them up. Unfortunately, you will most likely need to throw all the material away, instead of composting it, since new plants can likely sprout from pieces. Plan to solarize the ground for the rest of the summer. This will kill some or all of the weed roots. You have to carefully follow the procedure below for it to work.

Solarizing

Solarizing heats the soil to over 100 degrees to a depth of about 1 foot, killing many pests, diseases, and weed roots. Unfortunately, good soil organisms may be killed as well, but you will bring them back with your soil-building activities. If you solarize your garden land in batches throughout the warm season, you can also be digging up solarized roots and planting cover crop in finished areas to enrich the soil for the following season.

1. Even out the ground, filling in any holes and leveling out lumps so you have a fairly flat surface.

2. Set a sprinkler or sprinklers on the ground and soak for a few hours. The water will help to conduct the heat into the ground.

3. Spread clear polyethylene plastic sheeting over the area.

4. Dig shallow trenches around the edges of the sheet or sheets and carefully bury the edges, covering them all the way around with dirt. The plastic must be shored up like this because if any air can get in, the temperature will drop significantly, and the method won't be effective.

TIME REQUIRED

Outdoor Air Temperature (in degrees)	Duration of Solarization (in weeks)
Above 90	4
80–90	6
70–80	8

After each section is solarized you may want to dig out large weed-root stumps and rake up any vines. At this point you can either begin preparing your beds and paths for fall crops, plant the entire area in a nitrogen-fixing cover crop, or sheet mulch (see page 285 for details) the entire area and form your beds the following season.

MAINTENANCE WEEDING METHODS

We have found that in intensive urban gardens with permanently established beds, where serious time was spent at the outset to eliminate both perennial and annual weeds and establish well-mulched paths, maintenance weeding is a lighter chore. Weeding happens quite naturally with the changeover from one crop to the next. For example, if you've just harvested the last of the broccoli in a bed, the preparation of the bed for the next crop will in itself eliminate weeds. After cutting the broccoli plants off an inch below the soil, quickly remove any large weeds and cover the bed with new compost, effectively smothering small weeds.

Beds with crops that stay in the ground a long time, such as some cooking greens, onions, or garlic are cultivated when interplanting. The top inch or so of soil is gently cultivated and weeds removed on planting the interplanted crop. A good example is chard (productive for up to a year) interplanted with a succession of radishes, lettuce, and green onions over the course of a year.

If you live in an area with wet summers, or started out with certain uberweeds, this idyllic description may make your eyes roll. Yes, some are beset with hoards of vigorous weeds that can choke out plantings. Oxalis and purslane come to mind. Under these conditions, you must schedule serious time, when your crops are under a month old, to weed. Consistent effort to not allow photosynthesis to build up energy in the weed roots will pay off after a number of years.

In general, before weeding, make sure that the ground is moist for easy removal and that you have two buckets: one for pernicious weeds and garbage that must be thrown away, and one for weeds that can be composted. Weeds left in place in beds or paths can easily resprout or become attractors for mildew and pests. Nonpernicious weeds make a great green manure for your compost bin.

Presprouting Weed Seeds Prior to Planting

An amazingly effective way of controlling annual weeds is to sprout and eliminate them in their young tender form prior to planting or seeding crops. To do this, water the bed thoroughly every other day after you have prepared a bed to be planted, until weed seeds sprout. Allow them to grow to a height of 1 inch to 4 inches, and then either pull them out by hand or hoe them. If you use a hoe, return a few days later to ensure that none have resprouted. Then plant your crops. An alternate method is to take advantage of the fact that many weed seeds are more vigorous and quick sprouting than vegetable seeds. If you know what your intended crop sprouts will look like, direct seed a bed and then hand weed all the tiny weed plants that sprout prior to your seeded crop.

Hoeing

Hoeing can remove many weeds quickly. A good, back-saving technique starts with the right tool. The hoe should be sharp and skim about a half an inch below the soil surface, cutting off weeds at the roots, so they won't sprout again. The blades of many garden hoes are at an angle that forces the user to bend over. These really aren't for weeding but for breaking up soil with an up and down motion. A weeding hoe should have a blade at a 70 degree angle to the handle so that the blade is parallel to the ground when the handle is angled toward the user. Circle or scuffle hoes also work well, since they can be skimmed half an inch below the surface without disturbing the soil, and they cut on both sides. Once beds are planted it's easier to work from a low stool or on your knees with a Japanese hand hoe, since the angle of the blade is just right for closely spaced intensive crops.

Hand Weeding

Hand weeding is often the most effective upkeep method. Even with the best hoeing technique, weeds can resprout: Nothing is more effective than pulling them up by their roots. With direct-seeded beds, be sure your crops are well-established enough that you can distinguish them easily and won't pull them up in clumps of weeds. We find that we rarely use trowels for weeding, as they disturb the soil structure too much, but a digging-stick tool is essential to have strapped to your gardening belt to loosen taproots and pernicious weed roots. However, hand weeding a site bigger than a tenth of an acre can be a ridiculous proposition; read on for other methods.

Malva, commonly known as cheeseweed, has a very deep taproot. Weeding them when young is ideal; if allowed to grow to maturity, a shovel will be required.

Flame Weeding

Flame weeders are not cheap, but if you have a large farm site and a particularly invasive weed, such as oxalis (wood sorrel), purslane, or crabgrass they can save your sanity. Flame weeders run on a propane tank that can either be wheeled around as you go or be worn as a backpack. A nozzle on a long handle is held close to the ground. Rather than actually torching weeds, you just heat them till they wilt. When you come back later they will be crisp and dead. We see flame weeders in the same category as weed whips. They use fuel—a dwindling resource—but unless you've perfected the lost art of scything, hand working larger pieces of land probably won't fit with your other life responsibilities. If you do purchase a flame weeder, opt for a model intended for professional farmers, not for hobbyists.

USING CHICKENS, DUCKS, OR GEESE FOR WEED CONTROL

If someone suggests keeping poultry in the vegetable garden for weed control, be very suspicious. Chickens, ducks, and geese will eat your vegetables and destroy your garden. We do think it's a good idea to set your fowl on your beds *after* you have fully harvested a crop. Some people will build a "chicken tractor" to do this, which involves making a movable cage with an open bottom. Chicken tractors are too big and cumbersome for most urban gardens. Instead, we highly recommend putting up temporary stakes (4 feet to 5 feet high) around a recently cleared bed, then tying plastic netting, or whatever you have on hand, to them. Stake the fencing around the edges with irrigation stakes to prevent escapees. Place the birds into your improvised corral, and they will go to work, scratching up weeds and weed seeds and simultaneously fertilizing the bed with their droppings. To make the best use of this method, presprout the latent weed seeds in the bed a few weeks before setting out the fowl by watering and letting weeds grow to a few inches high.

Sheet Mulching

Sheet mulching mimics the natural processes of nature in wild lands, especially forests, where a lot of organic matter breaks down continually, providing needed nutrients for plant growth. With the addition of an organic weed-blocking layer (usually cardboard), the process of mulching also acts as a weed suppressor. Sheet mulching is our favorite method for suppressing weeds in paths. Not only does it save weeding labor, it enriches the soil by adding organic matter; greatly multiplies soil microbial and worm life; conserves soil moisture; and recycles waste materials. A word of warning: The effectiveness of sheet mulching to suppress pernicious weeds, especially grasses, is totally dependent on meticulously removing weeds and roots prior to applying the sheet mulch. Many people think sheet mulching is a way to get out of hand-digging Bermuda grass or blackberry roots. Unfortunately, this fantasy rarely comes true: Weeds store energy in roots for a number of years and will emerge in the first chink in the sheet mulch.

Finding free materials for this project is a fun adventure and a great way to meet people in your community. Unlike our rural counterparts,

SHEET MULCHING PATHWAYS

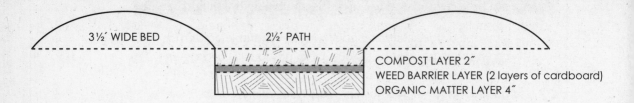

3½´ WIDE BED   2½´ PATH

COMPOST LAYER 2˝
WEED BARRIER LAYER (2 layers of cardboard)
ORGANIC MATTER LAYER 4˝

Sheet mulching pathways is a natural way to keep down weeds and build soil

urban farmers have easy access to unimaginable amounts of free cardboard boxes, coffee sacks, wood chips, sawdust—the makings of sheet mulch. Many business owners believe in recycling, and you can help them out by offering to take some of their waste and put it to use. Be creative: Coffee roasters may be throwing away burlap sacks; food processors may be throwing away various types of chaff; tree trimmers may be taking wood waste to the dump; contractors may be throwing away used carpet. Call local businesses owners or managers to see if you can make an arrangement, and if they agree, be dependable. Share the love and bring a gift of some garden produce to show gratitude.

All you do to sheet mulch is create layers, or sheets, of various organic materials over areas of your garden, especially paths and permanent perennial plantings. The sheet mulch is replaced from time to time as it breaks down and turns into soil. For this reason the materials must be free of plastic and any other material that either does not compose or is poisonous. Be especially wary of donated lawn clippings and sawdust. Lawns, not agricultural areas, receive the greatest volume of herbicides in the United States. Only take lawn clippings from trusted sources. Sawdust often contains glue from laminate lumber and poisons from treated lumber. What the woodworker does is a clue to safety. Seek out fine woodworkers, such as furniture, window, or cabinetmakers and ask them if they are willing to separate their sawdust for you if they do use laminates or treated wood.

The layers of the sheet mulch sandwich are:

1. a compost or manure layer, to encourage microbial growth (optional)

2. a *light-blocking* organic weed barrier

3. a *weed-seed, poison-free* organic matter top layer

The layers of material can be as numerous and thick as you wish. · To achieve even more weed-blocking action, the second and third layers can be repeated. If used in paths, obviously the height can't be greater than the beds. At a minimum, the first layer should be at least 1 inch thick and the third layer should be at least 3 inches thick.

How you apply the second layer will greatly affect its effectiveness in blocking weeds. Here are a few tips:

1. Overlap all edges 3 or more inches.

2. Run the material up the sides of beds, boxes, or fences a few inches to block weeds growing into these edges.

3. If using paper or cardboard, wet it as you go to help flatten it and adhere to the ground or first layer.

4. Use as many layers as are obtainable and practical.

COMMON SHEET MULCHING MATERIALS

Compost Layer (1 inch or more)	Weed Barrier Layer (At least one layer)	Organic Matter Layer (3 inches or more)
Compost	Cardboard*	Wood chips and bark
Manure	Newspaper or other poison-free paper (0.5 inches thick)	Sawdust (solid wood only)
Worm castings	Burlap (two layers is best)	Straw
Green grass clippings	Recycled natural fiber rugs	Leaves
Waste pomace	Recycled natural fiber clothing (jeans are especially good)	Chopped-up garden brush
Coffee and tea grounds	Recycled gypsum board (sheetrock)	Dried-grass clippings
Brewery mash		Animal bedding
		Grain or coffee chaff
		Nut shells
		Ground corn cobs
		Pine needles (be careful of high acidity)
		Seaweed

*If using cardboard, prepare it by removing plastic tape and staples.

CHICKWEED

A tender spring green, chickweed is considered a general tonic herb. The edible leaves and tiny vines are the lightest green and great chopped up in salads or sautés.

LAMBS QUARTERS

These slightly tannic-tasting, soft, arrow-shaped leaves are great in salads or steamed.

MINER'S LETTUCE

These succulent, perfectly circular leaves, suspended on a stalk, make great salads.

PURSLANE

Purslane's tiny but fat leaves resemble those of succulents such as jade plant. Their tart flavor belies their high nutritional content. They are a delicious snack while weeding and a good addition to salads. Some find the slightly slimy texture off-putting.

NETTLES

In the spring, get out your gloves and pull up any pesky stinging nettles to steam for a spring health tonic. Once cooked, the sting disappears.

YOUNG AMARANTH

Similar in taste to lambs quarters, young amaranth sprouts often sport dual-colored, silver-and-deep-red leaves. They are good in salad or steamed.

YOUNG MILKWEED PODS

If picked at the right moment—spring or early summer—these horn-shaped seedpods are delicious when steamed. Kind of like broccoli with a fun texture.

YOUNG MALLOW

Mature mallow has a huge, fibrous taproot that can be nearly impossible to dig out. Picked young, mallow leaves are a lovely, mild version of raw or cooked greens. The seedpods, which look like hollyhock pods, can be steamed, and they taste like okra.

For more information about eating weeds in the city, consult *Identifying and Harvesting Edible and Medicinal Plants in Wild (and Not So Wild) Places* by the famous urban forager "Wildman" Steve Brill, with Evelyn Dean.

Maintaining your sheet-mulched areas over time will increase their longevity. Keep an eye out for weeds that pop up along the edges of paths and fences, and dig these out. Left alone, these can creep into the mulch above the weed-blocking layer and find plenty of nutrients to help them grow there, sending new roots down through chinks. If the weed-blocking layer was well done, it can last for a number of seasons, and your chore will simply be replenishing the top layer as it decomposes. If it is time to redo the sheet mulch, your compacted paths can be especially difficult to weed. Water them the day before so roots can release from the highly compacted soil. Then use a hoe, in conjunction with a digging fork where needed, to remove weeds. Be sure to rake them up and dispose of them. Have new sheet-mulching materials on hand.

MANAGING PESTS AND DISEASES

The urban farmer may have one pest advantage over our rural counterparts. Being isolated from the countryside, we may not be as susceptible to the catastrophic pest outbreaks that can happen in large areas of monoculture agriculture. Cities by their nature have a wide diversity of plants—though they may be mostly ornamental—and plant diversity is one of the greatest deterrents to pest and disease outbreaks that can be seriously damaging. A square mile of corn is much more likely to attract the critters that love to munch on corn than a square mile of mixed plants with a small corn patch.

On the other hand, many areas of cities lack much vegetation at all. Gardeners in these areas can find that they provide the only wildlife habitat around. Birds, insects, deer, gophers, moles, people's pets, and rodents—the dark side of urban farming—may flock to your oasis of food to feast and generally cavort in the greenery. Bird netting may be needed for plants not previously known to attract birds. Aphids may be all over your plants in spite of all your soil-building work. Your neighbor's cat may love to sun herself (and do other, uncute, things) in your garden beds. The point is, Don't take it personally. Anticipate and deal with it as best as you can.

HOW DO I KNOW IF I HAVE A PEST OR A DISEASE?

Is that insect beneficial or harmful to my plants? Are the leaves mottled because of a nutritional deficiency, watering problems, or a disease? Even the most seasoned farmer can have trouble identifying pests and diseases. There are many different soil-borne fungal or viral diseases that can look alike, and also look like nutrient deficiencies. Bacterial diseases can also be very confusing to diagnose. When in doubt we turn to our trusty U.S. Department of Agriculture–funded Cooperative Extension Service. Our tax dollars are at work in the agriculture sector, and now *you* are a part of the agriculture sector. Give 'em a call. They can be found on the Web. Locally owned plant nurseries can also be very helpful. It's best to bring in a sample rather than calling with a description.

To ID pests and diseases yourself, books and the Web can be very helpful. With insects, note down the number of legs and wings, a description, and the plant or plants it was on or near. Sort through pictures of insects until you find the one. Then you will know if it is damaging or beneficial. For possible diseases, bring a sample in from outside. Look through pictures of diseases that commonly attack the particular plant. Also check images showing the results of nutrient deficiencies.

Organic pest and disease management rests on the following principles:

‡ Healthy plants are less susceptible to pest and disease problems.
‡ Pest and disease management efforts should make building fertile soils with a high degree of biological activity a priority, to provide the necessary nutrients to plants; providing sufficient water is important as well.

✻ Natural predators should be encouraged; they will keep pest populations from exploding. Natural predators can be encouraged through planting their habitats in ornamental borders and among planting beds.

✻ Rotating crops and crop families can help break pest and disease cycles.

✻ Choose pest- and disease-resistant varieties suited to your climate (know your area's problem pests and diseases).

✻ Planting a diversity rather than a monoculture of plants will limit any one pest population and create a healthier ecosystem.

✻ The spread of pests will be minimized by good garden hygiene practices that allow proper airflow around crops, the containment of plant residue, and the disposal of diseased plant material.

You will notice that we have left one way to manage pests and diseases glaringly absent: spraying something to kill pests or disease. It's a state of mind of our modern times to think that some *thing* must be applied to kill pests and diseases, and this is a common attitude even among some organic gardeners. Just buy this natural stuff and spray it on there and the pests are goners. We think, for the most part, that sprays, even organic ones, aren't very effective and are a waste of money and resources.

When to Remove Plants and When to Live With the Problem

It's important as organic gardeners to learn to live with imperfection. Many pest and disease problems can and should be lived with. It's also important to learn to identify diseases that can spread to other plants. If you leave such plants in the ground, you may be stuck with a serious problem for years. If you rip out every plant with a spot, wilted leaf, or aphid infestation you won't have much, if anything, growing. Many plants can still produce delicious crops with quite a bit of pest or disease damage. However, some diseases and pest infestations can get so bad, your plant may fail to grow or the problem will spread. Identify and read about the specific pest or disease you have to help you make the call on whether to yank or live with it.

Bug Fear

Fear of bugs is real for some people. With a spray you don't have to get too near. In our culture we are generally very removed from any killing that is done on our behalf. In farming, it's done at a distance with chemicals. It's better to come face-to-face with the killing that needs to be done to grow our food and not harm people and wild animals with chemicals. We do feel regret when killing any living thing, so humane treatment, even of insects, is important to us. Handpicking and squishing requires close contact with death, but it ensures that the killing is done quickly, humanely, and effectively. For those who are afraid of bugs, it will take some getting used to. We recommend that you desensitize yourself to touching and squishing bugs slowly, through a measured process of exposing yourself to the experience. The thought is almost always worse than the reality.

Encouraging beneficial insects is also an important part of organic farming. Spiders, various species of wasps and flies, and of course, the beautiful red ladybird bugs all eat garden pests. In spite of new concepts of ecology as a web of life, in which all creatures serve a function, the vestiges of fifties' attitudes—there's a chemical to solve every problem—still linger. Many people have an instinctual fear of all bugs, especially spiders, and automatically reach for the spray can whenever they see flies (many beneficial wasps look like flies) or spiders. If you have this fear, try to remember that these creatures are helpful.

A good practice for all of us attempting organic farming and gardening is to become excellent observers when we first see a pest problem. Instead of following our first instinct to go grab that "natural" insecticidal soap spray, we might spend some time thinking about our soil management and watering practices of the last few months. Did we decide we didn't have time to add more compost when we planted our last crop of broccoli (broccoli *is* a heavy feeder)? We might think back to when we propagated our seedlings and remember that, Yes, we did go on vacation, and when we came back they were bone dry— neglected seedlings often grow up to be sickly, pest-attracting plants.

As with holistic doctors, organic, sustainable farmers use preventative measures to maintain plant health. The focus is on health, not disease. To be honest, we haven't used sprays in years. It's not that we don't have pests; it's just that we feel that sprays are too expensive, take too much to time to apply, and aren't very effective. While "organic" and "nontoxic" industrially produced pesticides may have some effect in the short term, factory production is heavily dependent on petrochemicals and not organic in the true sense of the word. We'd rather live with some pests and focus on the positive work of building healthy soil.

It would be a mistake, however, to always take pest and disease outbreaks as a black mark on your farming skills. Your garden is just one tiny outpost of a (more) balanced ecosystem amid a sea of neglected urban street trees, empty lots, and nonorganic ornamental landscapes. Just like insects, ecosystems don't respect fences. If you view your pest and disease problems from this larger perspective, you may come to understand why you can't achieve perfect balance within your small patch. In fact, you may begin to concern yourself with watering and mulching those scale-infested street trees as a preventative measure. Every act we do to encourage health and balance through growing plants in the city is a good act. Greater restoration of city ecosystems will take time.

Another caveat to the healthy soil/healthy plants theory is that there are some pests and diseases that if imported to your garden will be difficult to eradicate even if you are a paragon of well amended beds. Examples include snails and slugs, leaf miner, and soil-borne diseases. If your soil is sandy and you live in a temperate climate, ants basically have no limiting factor. Their nests can be myriad. Because of the symbiotic relationship between ants and honeydew-producing insects, during the warm season you will always be in a catch-up game with ants and aphids.

One more note of caution: Because one rule of organic farming says that a wider diversity of crops helps create a healthy ecosystem, some urban farmers plant a jumble of different plants in each bed. Let's go back to that square mile of corn. Monoculture on this scale is a recipe for disaster, leading farmers to believe that they must use pesticides. The suggestion that plant diversity will help with pest and disease management emerged in this context. Home gardeners have always had diversity because, guess what, we don't eat just corn. As long as you have a diversity of crops in your garden *at large,* there is no need to mix crops within beds. You already have diversity greater by a scale of magnitude compared to any large farm, and this diversity decreases pests. Mixing too many different crops in each bed can make your work much harder.

Primary pest and disease management techniques:

- Get over your fear of touching bugs.
- Accept that your plants may have pest damage and blemishes.
- Get used to eating moderately pest-damaged produce.

- Build healthy soil and provide adequate water.
- Rotate crops and crop families.
- Plant crops in the right season.
- Plant beneficial insect attractor plants.
- Learn about the common pest and disease problems in your area and good natural controls for them.
- Handpick pests off, and spray others off with water.
- Be careful not to import pests and diseases.
- Remove and dispose of disease- or pest-infested plant matter in the garbage or feed it to animals.

Secondary pest and disease management techniques:

- Use barriers such as floating row covers to keep out pests when plants are young.
- Make your own spray solutions (see recipes later in this chapter).
- Purchase and release beneficial insects. The publication "Suppliers of Beneficial Organisms in North America" is available at www .cdpr.ca.gov/docs/pestmgt/ipminov/bensuppl.htm.
- Suspend cultivation of varieties or related plants for a season.
- Solarize your soil to kill soil-borne diseases.
- Don't handle plants after handling tobacco, and don't allow butts in your garden. Tobacco mosaic virus affects Solanaceae family plants and can be present in tobacco products.

If you do get an infestation, below are some methods to get the pests in check.

Handpicking

Once a pest has become a problem, handpicking is the best defense. Many will argue that it is too time consuming. In the small-scale urban farm, however, handpicking is actually the most time-efficient method, because it is so effective. Other methods, such as spraying with water or soap mixtures, setting out traps, etc., are much less effective at killing enough pests to disrupt their life cycle, therefore requiring constant re-application. Sprays can also inhibit plant growth over time. With handpicking, you know you killed the insect. If you begin to observe the pest eggs and remove those, you will have even greater success. In terms of time, as you begin to observe where specific pests like to hang out, you will become very quick at spotting and removing them, and will be able to go through an entire bed within a few minutes.

Floating Row Covers

While established plants can tolerate some pest damage and still produce a good crop, direct-seeded and newly transplanted seedlings can be wiped out in a few days by snails, slugs, root maggots, and other soil-dwelling creatures. If your region or garden tends toward a particular pest that attacks young seedlings, it may be worth it to use a floating row cover—a lightweight synthetic fabric—which lets in water and light and also insulates seedlings from the cold. After reseeding your beets three times you will really come to appreciate it. It is sold in long rolls about the right width for intensive garden beds. To be effective the edges should be buried or staked down tightly, so that nothing can get in. Give a little slack so the seedlings can push up the lightweight material as they grow. Remove the row cover when the seedlings have a few sets of true leaves and can tolerate a bit of munching without being killed.

Homemade Sprays

While we don't see sprays as a first line of defense, they can prove useful for some. The following homemade sprays can help control mildew and many insects. The soap spray controls insects, while the pepper spray will work for almost any insect or animal pest. If you have to spray more than one plant either with nutrient or insecticide, you will definitely want something larger than a hand-pump spray bottle. Your hand just can't keep up the pumping action even for, say, five plants. Manually pressurized spray tanks for farmers can be purchased from farm supply companies.

Fungicidal baking-soda spray:

Mix 1 teaspoon baking soda to each quart of water.

Add 1 tablespoon of vegetable oil or soap (not detergent) to help the mixture adhere. Shake vigorously prior to each use.

Insecticidal soap spray:

Mix 2 tablespoons liquid or flake soap (not detergent) with 1 quart of water.

Shake vigorously prior to each use.

Insecticidal pepper spray:

Mix six or so fresh, minced chilies with ½ cup of vegetable oil and soak for a few days.

Strain the peppers out of the oil (wear gloves to protect your hands when handling peppers).

Combine the pepper oil and 4 tablespoons liquid or flake soap (not detergent) with 1 gallon of water.

Shake vigorously prior to each use.

Water

Surprise! Simply using a fairly hard spray of water works quite well to knock off aphids and other small, soft-bodied insects.

HOW TO AVOID IMPORTING PESTS AND DISEASES

The most important first step is to purchase certified disease-free perennial plants, bulbs, and seed potatoes. These may seem quite expensive. However, once you've purchased a guaranteed disease-free item, you can propagate more plants yourself by taking cuttings or saving tubers or bulbs for next year's seed. It's not worth the risk: Once you have a soil-borne disease you may never be rid of it. For instance, tomatoes are quite susceptible to verticillium wilt—it's undetectable in seedlings. Choose cheap plants at your own peril.

Cost can also be linked to the quality of annual seedlings. You may ask yourself, Just why *are* these starts so cheap? Generally speaking, it's more likely that the very large wholesale nurseries that sell to chain garden stores will have worse practices. Some are even known to neglect plants and then beef them up with chemical fertilizers before they go to the retail outlet. Local, small-scale, wholesale nurseries are often more accountable with regard to quality. If they receive complaints from their retail nursery customers or the end user, it can cost them business. If you're going to spend the money, don't create a monster by trying to save a few pennies. Poor quality plants can carry pests, soil-borne diseases, and generally fail to thrive.

Specific Pests and Diseases

It is not within the scope of this book to be an exhaustive source of information on pests and diseases. Following are some suggestions for the most common pests and diseases, and if you don't see what you've got, consult one of the resources in appendix 11.

APHIDS AND ANTS

Control the ants, control the aphids. Ants live a symbiotic life with aphids and other sap-sucking insects by carrying them to plants, protecting them, and then eating the honeydew produced by the bugs. In the greenhouse and with beehives, control ants by painting sticky bug paint (be careful; the stuff's like superglue) around the legs of all tables and the perimeter of the propagation area. Trees can be protected by painting sticky bug paint around the trunk about 5 inches from the ground. Table legs can also be set inside buckets of water (ants can't swim). Ants thrive in dry soil, so make sure your beds and plants are receiving adequate water. To destroy ant hives, find their trail and pour a stream of boiling water all along it until you reach the hive (the trail has scent that helps ants find where to go). Pour more boiling water down their hole.

Aphids are a little bigger than a poppy seed, can be green, black, gray, or white, and live on the underside and sometimes upper surfaces of plant leaves. The leaf edges will often roll under as sap is drained. They can be controlled by washing or spraying off with water or insecticidal soap (not detergent). One of the most effective methods is to wipe them off with a soapy water–soaked rag and rinse them into the bucket of water. This actually kills them instead of just knocking them off. Special spray nozzles can now be purchased that provide the correct water pressure to kill aphids on contact as well as the right angle to reach under leaves. They are useful when hand washing will take too much time. For best results, repeat the picking or spraying for three weeks. Introducing ladybird bugs into your garden or greenhouse can help too, as they love to eat aphids and other bugs. Introduce them at night or they'll fly away.

CABBAGE LOOPERS

Cabbage loopers attack all *Brassicas* (broccoli, kale, etc.), eating progressively larger holes in leaves or even gnawing off the entire plant.

These soft, squishy caterpillars are exactly the same shade of green as the plant, and eventually hatch into white moths. If you see these moths, go looper hunting by handpicking eggs and caterpillars. They live on the underside of leaves along the ribs, and sometimes migrate to the tops of leaves—they are very well camouflaged. They start out as almost microscopic lines and grow to an inch. Go over each and every leaf to find and destroy. You can squish 'em with your fingers or step on them. You'll get the hang of it after a while. For best results repeat the handpicking for three weeks.

WHITE FLIES

White flies aren't the worst thing in the world, but if they get out of hand they can cause a bit of damage. They look like very, very tiny white or gray gnats. Often they are present if the soil is constantly wet. If you have an overabundance, allow the soil to dry between waterings. To control them, put sticky bug paint on sheets of yellow-colored card stock and punch a hole in the top. Suspend these close to the ground with string or on wire stakes (clothes hangers work well) around your garden or propagation area. The flies will fly toward the attractive color and get stuck. That stuff is so sticky it even works if it gets wet.

LEAF MINER

Oh, the devastation! It is *so* hard to eliminate leaf miners. With diligence it is possible. Leaf miners are sapsuckers that live in-between the walls of tender leaves, most often on chard, beets, spinach, arugula, tomatoes, and various ornamentals. It is very difficult to find the actual larvae or flies. The most effective way to control them is to remove any leaf that shows sign of a tunnel, even if it means removing all the leaves down to the tiny emergent leaves in the center. Throw infested leaf parts in the garbage, not the compost. Luckily you can eat infested leaves; just tear off the damaged part. Another method is to suspend cultivation of susceptible beet family seedlings for a season.

SCALE

Scale are also farmed by ants. The ants tend to place them on perennial trees and shrubs. They damage plants and strengthen ant colonies, so they should be controlled. They look like hard black or brown bumps on stems and the undersides of leaves. Soft-bodied insects are harbored within these bumps. You will also see ants busily at work collecting honeydew. Scale can be very difficult to eradicate. First, make sure to water and amend adequately. Regular application of compost and mulch around your trees will be very beneficial. To control them, get a bucket of warm, soapy water and an old toothbrush. You may need a ladder. Plan to spend some time carefully scrubbing the scale off each branch, twig, and leaf, starting from the trunk and working your way up. Rinse the bugs off in your bucket. Repeat as needed a few weeks later.

SLUGS AND SNAILS

If you don't currently have slugs or snails, carefully inspect all seedlings you bring into your garden. Slugs and snails usually arrive on the underside of pots. They can eat entire leaves to a nub and will attack many types of plants, leaving a telltale trail of slime behind. To control them, you're going to have to get comfortable with killing. It's *not* okay to throw them into your neighbor's yard. Not only is it unkind to your neighbor; these mollusks are likely to come right back to you (with their spawn).

Snails and slugs are active and feed during the night, whereas they hide out in dark places during the day. Night picking is a wonderful way to control them in your garden, and a fun adventure. Get a headlamp and a sealable bag or bucket, and go forth into the dark, either a few hours after sunset or before sunrise. Look over every square foot of your garden. You may come away with pounds. If you'd rather hunt them during the day, look in shady places and along the soil line where it meets wood boxes or fences. Another good method is to make small rock piles in your garden and take them apart periodically during the sunniest, hottest part of the day. The little critters will have congregated there in the shade. To control in the greenhouse, take a half hour and lift each and every tray and seedling pot to find what's lurking.

To dispose of snails and slugs feed them to your hens, ducks, or geese or dump them out on the pavement and step on them. It's the most humane way. Do this every week for a month, and you'll make a dent.

CABBAGE MAGGOT

As with cabbage loopers, these insects do their damage in their larval stage. Unfortunately, because they live underground, you may not realize they're there until you've lost some plants. They attack *Brassica*s (broccoli, kale, etc.). If you are watering sufficiently and you see extremely wilted-looking plants, dig around the root zone and see if there are white grubs eating the roots. If so, remove the infected plants and try again the next season. Cabbage maggots are seasonal to spring, and removing their source of food until the following spring will break their life cycle. If you are not willing to stop growing *Brassica*s, wrap the seedling stems with a layer of brown paper at planting time as a preventative measure. This keeps maggots from getting through to the base of the stem and roots. The stem must be covered both above and below the soil line for this tactic to be effective.

BIRDS, CATS, DEER, GOPHERS, RACCOONS, SQUIRRELS, AND HUMANS

Some say reflective material such as special tape or old CDs scares birds. Unfortunately, in our experience this has not been very effective with city birds. Perhaps its all the reflective car and building windows they've grown used to. Surrounding beds, shrubs, and trees with bird netting or aviary wire works much better, but bird netting can be difficult to secure to planting beds. If it's not secured with some sort of frame it will simply flop onto the bed, and the birds will peck right through. A more effective method is to construct a 2-foot-high cage of aviary wire and place it over beds that have succulent lettuce or fruit. The wire itself will provide enough structure to keep the thing up. The same method will also keep cats, raccoons, and squirrels out. Dare we say BB gun?

If you have a gopher or mole problem, set traps. Find a fresh tunnel and dig it out, so you can see both branches of the tunnel. Put two traps back-to-back inside. Mound dirt back so no light gets in, and mark it with a stake. Check weekly so any kill doesn't get too old. To deter them, line the bottom of planter boxes or beds with hardware cloth or chicken wire before installing them.

Yes, deer do exist in large numbers in some cities, and they eat everything. If you notice nicely pruned plants—all roses and tender tomato tips neatly snipped off—it may be due to deer. Install 6½-foot-high deer netting around the entire perimeter of your garden, including the entry gate. Attaching thin vertical lath strips or grape stakes to your existing gate is an easy way to extend its height. They needn't have horizontal supports, as strength is not important. Deer simply won't jump over a high fence for fear they won't be able to escape. We've heard of many scent-based deterrents, but the need for constant reapplication makes us lean toward a more lasting solution.

For cats to use your fluffy garden beds as a bathroom, they must circle around; sticking closely spaced twigs standing up in the soil prevents this. A squirt of water square on the nose will make them run, but you're not around twenty-four hours a day to maintain this tactic.

In urban gardens, other people can sometimes be categorized as pests. Yes, we hate to inform you, but sometimes, just as your prize pumpkin is nearing maturity, it magically disappears. In our view this is a good kind of peskiness, since it may open the door for more people to begin urban farming. After the first burn of anger wears off, ask yourself why someone would steal garden produce. If this kind of benign larceny becomes a real problem (usually because people are picking unripe produce), post signs in your garden with notices such as FREE PUMPKIN PLANTS, INQUIRE INSIDE or WE FREELY SHARE VEGGIES, BUT PLEASE ASK FIRST SO WE CAN HELP YOU PICK THE RIPE ONES. Another approach we have tried, very successfully we might add, is to build planter boxes *outside* the fence with a sign saying HELP YOURSELF, with some harvesting tips. You may even be able to recruit this kind of pest to help you in your garden.

RODENTS

Rat and mice infestations *are* indeed the dirty little secret of urban farming. We may like to pretend it isn't so, but let's just get that off our chests now. Urban farmers aren't the only culprits. Check out the alley behind your favorite strip of restaurants at midnight, and you'll see a cavorting mass of fuzzy bodies. Keeping rats in check is important. Out-of-control populations can begin to eat large amounts of your produce right out of the ground and off the bush. They suck down chicken eggs too. In fact, initial rodent population growth is often the result of a poorly built chicken coop that allows the rats to gain access to an endless source of food—chicken feed. Then their population explodes. The other common culprit for urban farm rat infestations is open compost piles that have a lot of starch, fat, and meat kitchen scraps.

Once rodent overpopulation begins it can be extremely hard to check. The best control is prevention; see page 389 for directions on building a rodent-proof chicken coop. Don't leave other edibles, such as seeds and curing squash, out for long. Remove rat habitat, like piles of wood and lumber. Creating rodent-proof compost bins is a bit of a fantasy. Rodents have found ways to enter nearly every homemade or manufactured bin we have seen. A great way to ensure that your compost isn't too yummy is to handle as much of your kitchen scraps as possible using worms. Worm bins are easier to construct securely, and the food scraps are usually chopped up finely and digested by the worms before the rats can figure it out. Keeping rat-catching cats or dogs also can be quite effective—that is, as long as you have the right cat or dog. Barn owls, which find rodents an appealing meal, can be attracted to your neighborhood if provided with the right housing. Purchase or build an owl box and hoist it up high on a tall post (see appendix 11 for sources). If you also give owl boxes to your neighbors, you will attract these beautiful rodent predators more successfully.

If all else fails and you are finding too many rodents for your taste, use snap traps with bait, or purchase a number of poison boxes. With rodent-size entrance holes, the poison is kept away from other critters. Use rodent poison that kills the rodent but leaves no poisonous residue that could harm a creature that would eat a dead rat, like your dog. There have been brands on the market that cause rodents to die of dehydration or blood coagulation through a particular food formulation, such as Rode-trol, rather than a poison. If you set traps, be prepared to check and restock them weekly for six months to a year to get things under control. You may want to watch the movie *Caddyshack* for inspiration. We have found all other methods ineffective, however ingenious they may claim to be.

MOSQUITOES

While these insects won't harm your plants, they can really annoy humans. In cities it is important to avoid creating a breeding ground—standing water—for mosquitoes. Allowing water to pool anywhere on your property (including buckets, tires or other containers) creates such a breeding ground. If you plan to keep open barrels of water or make a pond, call your county mosquito-abatement district. They will provide you with small fish that eat mosquito larvae (appropriately called mosquito fish) free of charge.

THRIPS

Thrips are very tiny soft-bodied insects that attack a variety of crops, causing leaf, flower, and fruit damage as they suck juices from plant cells. The tops of leaves gain a silvery sheen. Onion leaves will show silvery streaks. Leaves can become curled at the edges. Because thrips are so tiny and tend to hide in hard to reach places, hand picking won't work. Use a soap or hot pepper spray if the problem gets out of hand. With mild infestations, crops should still produce fine yields.

DISEASES

Blossom End Rot

Blossom end rot is a syndrome that affects fruiting crops such as squash and tomatoes. It is caused by a calcium deficiency that weakens the fruit tissue, causing a sunken brown area. Mold or mildew can also grow on the damaged tissue. Blossom end rot is usually more of a problem at the beginning of the harvest season, and diminishes as the season progresses. To avoid it, ensure ample absorbable calcium by applying sufficient compost prior to planting. Consistent, even watering, especially with tomatoes, will ensure that the available calcium is absorbed. You may also want to lime your soil, but be sure to do so a bit in advance of planting. Blossom end rot on particular fruits can't be cured. To control it, religiously remove all fruits that show any sign of end rot and throw them in the garbage. It may feel like you are removing all your produce, but it's essential to eliminate anything with mold growth.

Damping Off Disease

Another fungal disease, damping off is common in greenhouses. It attacks seeds as they're sprouting, causing them to die, as well as the stems of tiny seedlings, severely constricting the flow of water and nutrients to the plant and causing them to keel over. Correct transplanting technique is important to avoid damping off. Do not pick up seedlings by their stems, as you will cause unseen damage to the tender sprout. Grasp them only by a leaf or by gently cradling the root ball when transplanting. Seedlings will easily grow more leaves but have only one stem. Damping off takes the opportunity to invade crushed stem tissue. Compost can also contain the fungus, so use high-quality compost in your potting mix. Cold, damp conditions can also be the cause, so if you have an outbreak, allow seeded trays to dry out a bit between waterings and ensure that the temperature of your growing area is optimum. Sterilize your trays between uses.

Powdery Mildew

This fungal problem shows up as white powder on plant leaves. Powdery mildew generally attacks peas, squash, and cucumbers but can show up on any leaf growth, especially young, succulent leaves. Spray the leaves with water or a baking soda solution during the morning. This may seem counterintuitive, as we associate fungus with wet conditions, but water spray actually helps disrupt the mildew. Remove leaves, plants, or fruit with heavy infestations and dispose of them in the garbage. Unfailingly remove every fruit that shows signs of mildew or mold until they start to appear clear. When you remove mildewed fruits, make clean cuts or breaks and put them in the garbage immediately. You can also remove squash leaves that are hopelessly covered. To reduce this risk, plant mildew-resistant varieties. Also remember that some mildew on leaves is okay. It's ubiquitous in some climates, and total eradication can be near to impossible.

W ◄►━ E

CHAPTER 14

SEED SAVING

Seed saving allows you to experience the complete life cycle of your farm and is incredibly satisfying. It also reduces your input costs and allows you to develop seeds especially tolerant of your growing conditions. It's interesting to note that we eat leaves, as with chard and lettuce, immature blossoms, as with broccoli, stalks, as with celery, tubers and roots, as with potatoes and carrots, and immature seed pods, as with green beans—so with some plants we may never actually have seen the process of blossoming and seed setting. It can be quite eye opening to allow these plants to grow to full maturity and seed-set.

When you save seed you get to see an amazing stage of the plant life cycle often missed. The dainty lettuce sends up a grand spike. The creeping parsley sends up umbrella-topped spires. The straight celery becomes a frothy mass of tiny branchlets and seeds. Obviously this can make your vegetable garden messier and take up a lot of space that could be used for growing crops. One solution is to move seed plants to a special propagation bed just before they begin the process of going to seed. Another option is to save seed from only a few crops each year, rotating through the list of crops you want to save seed from in about four or five years (the average period of viability of saved seed). Get together with a group of local urban farmers and each agree to save seed from a few varieties each season to share. Starting local

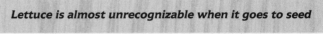

seed-saving groups and seed banks may turn out to be important to preserving community access to seeds in the future.

Only in recent times have farmers stopped saving their own seed and begun to rely on commercial sources. If you look back in your family tree, you will likely find avid farmers who saved and shared seed as a matter of course. While recently riffling through a box of old family letters from the 1800s I (Willow) found that this was true in my family tree as well. The letter begins (we've preserved the unique spelling and grammar):

> Hanover york Co Pa.
> *January the 7 th d 1888*
>
> Dear and beloved sister and friends I seated my selfe In the
> name of god to write A few lines unto you to let you know that
> we are well at present and All the friend as far as we know.

After a few paragraphs outlining the health and well-being of family and friends, the author moves on to food:

> How did your cabbage seed do we had very nice one we had
> all kinds plenty in the garden but apples was scarce in our
> neighborhood. They sell at $2.75 a barel they come from new
> yorke state we get 22 cts for butter 20 for eggs. . . .

Even more interesting was the following postscript:

> Sister, I sent you some sallet seed that's summer salet sow
> it thin then it gives heads like winter salet I must stop now or

the paper soon gets full no more for this time I sent my love to
you all Eliza Hohf

written by your sister
Maria Torry

Perhaps we will once again enter a time when the price of apples
is noteworthy, the profit from butter and eggs essential, and the salad
seed shared by letter. Hopefully, we will retain our standard spelling
and grammar.

Urban farmers can grow their own seed crops, starting with open-
pollinated seeds. Open-pollinated plant varieties grow true to type,
meaning that the offspring will be very like the parent as long as it's
pollinated by the same variety. It's easier to save seed with some crops
than others, so we recommend you start out with something easy and
keep going if the bug bites you.

BEGINNING URBAN SEED SAVING

For those of you who want to save some seeds but aren't interested in
getting into the technical bits, we're going to start with the easiest
crops. Later in the chapter, starting on page 312, you'll find more in-
depth information if you are ready to take a further step. Even the easiest crops may have some degree of crossbreeding, so if you are very
concerned with breed purity, follow the instructions in the next section starting on page 313. Be aware that viruses can be transmitted
from seed to plant to seed, so extra care must be taken to start with
clean seed stock (purchased seed is usually guaranteed disease-free).
If you're not that concerned with breed purity (though if you're sharing your saved seed, the others may be), many crops can be grown out
to seed without the precautions taken to ensure purity—just be ready
for some weird crops and decreased plant vigor over time. For growers in colder climates, crops that take two years to produce seed (bi-
ennials) can be tough, so check the table in appendix 10 to find out if
your chosen seed crops will mature in one or two years.

You can save seed from the following inbreeding self-pollinating
crops without having to isolate blossoms from other plants' pollen.
These crops lack common related weeds to crossbreed with, and there
is only a small degree of crossing between cultivars in the same species, so if you have a number of varieties of, say, lettuce, going to seed

at the same time they won't crossbreed. With these crops you won't need to worry much about whether other cultivars in the species are blooming at the same time. You can save seed from as few as one plant, but it's always better to save seed from a few to maintain genetic diversity. Note: see page 311 for a glossary of terms.

❁ beans
❁ lettuce
❁ peas
❁ malabar spinach
❁ New Zealand spinach
❁ tomato
❁ tomatillo

The following crops are self-pollinating and aren't in danger of crossbreeding with weeds. As long as only one variety in a species is allowed to flower and set seed at one time, there won't be a danger of crossbreeding. If your neighbor has flowering plants in these species there is a chance for some crossbreeding, but it's unlikely to be very significant unless you are concerned with absolute breed purity. You can save seed from as few as one plant.

❁ basil
❁ beets and chard (don't grow for seed at the same time, as they will cross with each other)
❁ celery and celeriac
❁ cilantro/coriander
❁ dill
❁ eggplant
❁ endive and escarole
❁ okra
❁ parsley
❁ peppers
❁ spinach

The following crops require a number of plants of the same variety flowering at the same time to pollinate correctly and set seed. As long as only one variety within a species is allowed to flower at a time they won't crossbreed. Be sure, however, that any wild varieties aren't growing nearby.

❁ arugula
❁ bunching onions

- chives
- garlic chives
- leeks
- onions, shallots, and green onions
- sunflower

The following crops can be saved as tubers, heads, or side sprouts to be used to seed the following year's crop.

- artichokes
- chayote squash (save whole fruits to use as seed)
- garlic
- potatoes
- sweet potatoes
- yams

Although *Brassica*s are notorious for crossbreeding with other cultivars within their species, as well as with wild plants, if you don't care much about crops remaining true to type, feel free to save seed without isolating. You might get a nice mustard with characteristics of nearby wild varieties. You could get a kale that seems a bit cabbagey. Since most *Brassica*s require more than one plant for sufficient pollination, be sure to allow a few to flower at once. An exception to this carte blanche for experimentation is radishes. Wild radish can cross easily with domesticated radish, resulting in a plant that may never form a bulb.

SEED HARVESTING

Seeds contained in pods and seed heads should be harvested when they are dry but before they burst open or fluff begins to carry seeds away on the wind. Once you see that a seed head is almost fully mature, you can tie a wax paper or brown paper bag onto it to ensure that the seeds aren't lost. If you live in an area where rain is a possibility close to harvest time, you may want to watch the weather closely and protect pods and heads as needed. As long as the seeds have fully matured, they can also be cut and dried in bags indoors in the event of rain. When fully dry, cut the seed head off the plant. You can thresh (separate the seeds from the plant matter) seeds from their pods or heads by placing the dried plants in a sack and jogging on them, or by simply removing them from the plant matter by hand. The sack method is great for seeds heads that tend to shatter easily, spilling your precious seeds everywhere. Separate them from dried plant bits by using a series of finer and finer screens, or simply toss them gently into the air from a bowl, blowing away the lighter chaff. You can also place a fan near your bowl of seeds to accomplish the same result. Air-dry the seeds in paper bags for about five days.

Seeds that are contained in fleshy fruits, such as eggplants and peppers, can be harvested by letting the fruit fully mature on the plant and then removing and air-drying the seeds on trays out of direct sun or extreme heat for about five days.

Those contained in fleshy fruits that are also each covered with a pulpy coat, such as tomatoes, must be fermented to break down the coat. Squeeze out the seeds in their pulp and wash them in a fine sieve or colander. Place them in a container with a few inches of water in it. Let them sit for a few days. Strain and wash in a sieve. Repeat the process until the fleshy coat has dissolved. Air-dry the seeds on trays out of direct sun or extreme heat for about five days.

For tubers and bulbs, save out as many of the mature tubers or bulbs when you harvest the crop as you will need for seed for the following year. Each clove of garlic will form a whole bulb next season. With potatoes, you can also use the ones that are too small to merit eating in addition to mature tubers. Each large one will provide enough eyes for two to six seed pieces. To expand your artichoke empire, simply separate side sprouts, ideally with a good bit of root attached, and root in a greenhouse or directly in the ground.

STORING SEED

High temperatures and moisture are anathema to seed longevity. Your typical garden seed packet is certainly not protective against moisture. And toolsheds, where many store their seed packets, can get pretty hot. The best way to store seeds is to put them, either loose or in paper packets, within sealable canning-type jars (jars with a rubber seal) and store them in a cool basement or closet. If you must store your jars in another area of your house, avoid the humid kitchen, any shelving with direct sun exposure, or any high-up shelves, since heat rises.

Crossbreed: a plant produced by breeding two plants of different varieties, cultivars, or species. This comes about when pollen from one variety fertilizes the flowers of a plant of another variety resulting in seed that is not true to type

Dioecious: a plant species in which the male and female parts are separated in different plants

Imperfect flower: one that has a stamen or pistil but not both

Inbreeding (autogamous) species or variety: one in which individual flowers or plants are fertilized with their own pollen. However, with more than one individual, some cross-pollination always occurs, and variability is maintained

Inbreeding depression: loss of vigor and variation due to pollination between two genetically similar plants

Outbreeding (allogamous) species or variety: one in which cross-pollination with other plants in the species or variety is normal and sometimes required

Perfect flower: one that contains both a stamen and a pistil

Pistil: the female portion of a flower, consisting of an ovary, style, and stigma

Pollen: the male spores of reproductive material produced by the anthers

Self-compatible: flowers or plants able to pollinate themselves

Self-pollinating: the transfer of pollen from an anther to the stigma in the same flower, or to the stigma in another flower on the same plant, and the subsequent fertilization of the pistil

Self-sterile: a trait associated with some perfect flowers whose pollen cannot fertilize a flower on the same plant but fertilizes normally when transferred to a flower on another plant of the same variety

Sibbing: the hand transfer of pollen between flowers on different plants of the same variety

Stamen: the male portion of a flower that produces pollen grains, which consist of filaments and anthers

Stigma: the portion of the pistil that receives the pollen grains during fertilization

Style: the elongated portion of the pistil that connects the stigma and the ovary

True-to-type: a plant that conforms exactly to the known characteristics of that particular variety

Umbel: a flower/seed head

ADVANCED URBAN SEED SAVING

Advanced seed saving starts with sex—plant sex, that is. We're sure you've been wondering, When are they going to get to the sex part already? While we do love to garden in the nude, we're not talking about humans here. Knowing how to save seeds begins with understanding how plants reproduce. The sexual organs of plants are inside their flowers, and since plants cannot move about, pollination—when sperm (pollen granule) meets egg (ovule)—is achieved with help. Plants have sexual relations either within their own flower or with that of another plant of the same species with the aid of pollen-spreading insects, birds, wind, and gravity.

Vegetable flowers are of two types, perfect and imperfect. Perfect means that each blossom contains both male and female parts. Plants with imperfect flowers have two types of flowers on the same plant, one containing the male part and one containing that of the female. Sometimes perfect flowers self-pollinate, with pollen falling from the stamen to the pistil within a blossom of its own accord or when jostled by the wind, but this isn't always what happens. Some perfect flowers are outbreeders: Insects bring pollen from nearby plants of the same species to pollinate the female part and take pollen from that flower to others, on the same or other plants. Other perfect flowers are pollinated by the wind, resulting in both self-pollination and out-pollination.

Corn presents a unique and astounding example of a wind-pollinated plant. We have always felt watching corn mature to be nearly miraculous, knowing that each and every kernel is actually formed from an

PERFECT FLOWER (BEAN)

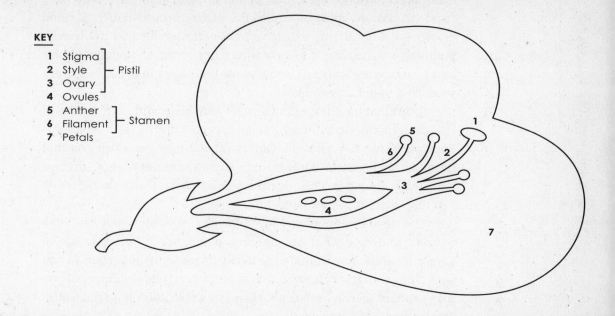

KEY

1 Stigma
2 Style ⎫ Pistil
3 Ovary
4 Ovules
5 Anther
6 Filament ⎫ Stamen
7 Petals

Perfect flowers have both male and female parts within the same flower. This bean flower is self-pollinating.

individual flower that must be pollinated within the husk. (Sunflowers are similarly multiflowered.) Corn pollination is achieved with three plant parts: the tassel at the top that produces the pollen; the silk emerging from the husk; and the ovules at the end of the silk that will become the kernels. The silk itself is the flower, and each and every silk strand must receive pollen to form a kernel. This is why corn ears can have gaps, and why it is recommended to plant corn in at least a ten foot by ten foot block. Since it is a wind-pollinated plant, there must be enough pollen flowing through the air to fertilize all those tassels.

PRODUCING "TRUE TO TYPE" SEED

When growing a plant to save seed, most often purity of variety or cultivar is what you're after. This means that unless the variety is self-pollinating and can't be contaminated by weeds or nearby flowering vegetables, the particular plants you intend to save seed from must be isolated from others of the same species that they could crossbreed with. Plants that have perfect flowers *and* naturally self-pollinate are the easiest to save seed from because little or no isolation is necessary. You can simply allow the plants to produce mature seeds and then harvest them.

Crops that have imperfect flowers and those with perfect flowers that cross-pollinate within the same species must be isolated to produce pure true to type seed. Otherwise you may get a broccoli that never forms heads or squash that's somewhere between a zucchini and an acorn squash. This isn't always bad—in fact, plant breeders do just that to produce new cultivars.

The table in appendix 10 lists the characteristics you will need to know to decide what isolation and pollination methods to use to save pure seed. You will need to decide if purity is important to you with crops with few cultivars, such as arugula, parsley, or chives. If you plan to share seed, however, people may want to know if it's not a pure variety.

GROWING OUT CROPS FOR SEED

Tend your seed crops well: Although seed from stunted or untended crops isn't necessarily affected in subsequent generations, good-looking crops will make it easier to select for traits and will set larger quantities of seed.

You can save seeds from some root vegetables, *Brassica*s, umbelliferae, and all the summer fruiting crops and beans at the end of the growing season in the late summer or fall. If you live in a place with summer rains, carefully monitor crops that produce dry seedpods or heads. Try not to let them get wet, but if they do, make sure they dry as fully as possible again on the plant before you harvest them.

Most of the root vegetables, onions, *Brassica*s, and some umbelliferae are biennial when grown out for seed. This means that they will take two warm seasons to produce seed. Most of these crops require a cool period and/or a period of dormancy to begin setting seed. The seed saving table lists specific needs, but in general, depending on your climate, you should either let the plants overwinter or dig them out and store them in a sheltered, dark place for the winter. In mild climates, the root vegetables and *Brassica*s may be started in the fall and overwintered in the ground. In areas with some freezing, plants should be started in the spring, thickly mulched in the fall, and over-wintered in the ground. In far northern areas, plants must be dug out in the fall and replanted in the spring.

To select specific traits in root vegetables you will need to dig them up to examine them regardless of where you live. In mild climates the selected roots can be replanted in the ground in the fall after examination and will flower the following season.

Following are some guidelines for cold climates.

Root vegetables:
- Plant in the spring.
- Dig up the roots in the fall and save the best out, trimming the tops to two inches.
- Store in sawdust, moss, or dry leaves in a dark place at 32 degrees to 40 degrees (basement, garage, etc.).
- Moisten periodically to maintain 90 percent humidity.
- Replant in the spring.

Other biennials:
- Plant in the spring.
- In mid to late fall gently dig up the plants.
- Trim the roots to 8 inches to 12 inches.
- Pot in sand.
- Moisten periodically.
- Store in a dark place (basement, garage, etc.).
- Replant in the spring.

Onions:
- Plant in the spring.
- Harvest in the fall when the tops are two-thirds to three-quarters brown.
- Trim leaves and cure outdoors in a garage or a shed for a few weeks.
- Store bulbs in a dry, cool, dark place.
- Replant in the spring.

MAINTAINING GENETIC DIVERSITY

Genetic diversity is important to open-pollinated crops because it ensures ongoing vigor through the generations and a wide range of resistance to diseases, insects, and weather extremes. To maintain genetic diversity within a cultivar, save and mix seed from twenty inbreeding plants and one hundred outbreeding plants. This can be quite difficult for urban growers. With some plants, such as green beans or peas, harvesting from ten or twenty plants can be easily achieved by simply letting the last pods on a row of vines ripen. It's pretty hard, however, to devote an entire bed for a year to one lettuce cultivar.

The options for urban farmers are many. If you simply save seed from as many plants as you feel you can sacrifice, the room for it will probably be fine. If your strain loses vigor, start over with purchased seed after a few years. Another solution is to alternate purchasing seed and saving seed, saving seed only from purchased seeds and not your own. With some crops you can increase the genetic diversity while still only growing one or two plants for seed by "sibbing" the plants. To sib, spread pollen by hand to each flower of the plant or plants for seed from flowers on other plants you plan to eat.

If you really want to develop your own lasting varieties by saving successive generations of seed, you will need either a large garden or to band together with other urban farmers, each saving seed from a few plants of a specific cultivar and mixing the resulting seeds together. Or, on a rotating basis, each urban farmer could devote their entire garden to seed saving for a year or two and share the results.

SELECTING INDIVIDUAL PLANTS FOR SEED SAVING

Once you have decided to save seed from a crop, select the most vigorous, true to type plants as your seed plants and mark them in some way, such as with ribbon or a stake. You can eat the rest. Sometimes you may need to grow a crop out to maturity to tell which plants are most true to type. With crops eaten for their leaves or other plant parts, be sure *not* to save seed from the first plants to flower, as this will select for early bolting, which is a bad thing. Early flowering on fruiting crops can be a good trait, resulting in earlier crops.

Obviously, many traits you might want to select for can be influenced by the growing conditions. If you don't provide good conditions you may be fooled. It can be difficult to select for each and every beneficial trait, but if you save seed successively over many generations, you will become adept at the observation skills needed for quality selection.

Traits to select for:

- earliness
- disease and pest resistance
- high yields
- color
- cold hardiness
- heat tolerance
- germination speed
- plant stockiness
- late bolting
- drought tolerance
- uniformity
- trueness to type
- color, size, thickness of flesh, productivity, storability, and the flavor of the fruits

ISOLATION METHODS

The table in appendix 10 lists the necessary information to determine which isolation method to use for each crop.

There are three main isolation methods:

- distance
- caging, bagging, or taping
- time

The distance method is used if you are growing two or more crops within the same species that are flowering at the same time. For instance, if you wanted to save seed from curly parsley and flat-leaf parsley at the same time, you would need to separate them by one mile. This is obviously impossible for urban farmers. The necessary distances between flowering plants of the same species vary, but regardless, it's still nearly impossible to achieve these distances in the city. In addition, many nearby wild plants, especially in the *Brassica* family, can easily spoil your seed crop with their pollen. For these reasons we don't recommend the distance method for urban farmers.

Caging, bagging, and taping are all variations of the same method. Depending on the crop, one is more suitable than another. They all keep pollen from other varieties within the same species, including wild plants, out of the blossoms of the plants intended for seed saving. Because of the potential for contamination from wild plants for some species, these techniques can be essential, even if you are only growing one cultivar from a particular species at a time. Caging is also essential when saving seed from outbreeding plants that require a number of individuals to pollinate and set seed correctly. This is because you can cover a number of plants with the cage.

The hands-down easiest isolation technique is the time method. Large-scale seed-saving operations can't do this, because they need to grow many varieties at a time, but it is very useful for small-scale urban seed savers. Time isolation consists of allowing blossoms from only one variety within a species to mature in a given time period. This easily ensures that there will be no cross-pollination. If you plan to save seed from two vegetables within a species, such as beets and chard, or from two cultivars of a vegetable, stagger the plantings and carefully monitor the maturation of the flowers to ensure that they are not blooming at the same time.

In spite of the fact that time isolation is the easiest method, due to the vast geographic range of pollinating insects, wind-borne pollen, and the presence of wild relatives, caging, bagging, and taping are necessary with many crops if you wish to ensure purity. If another gardener has let beets go to seed in their garden, it can easily contaminate your chard seed crop.

On the other hand, since some of these species contain only one or two garden vegetables, unless you want an absolutely pure strain, you may opt not to bag or cage with good enough results. In your own garden you can isolate these crops with time and hope that contamination from neighboring gardens will be slight. The vegetables of the umbelliferae family—celery, dill, cilantro/coriander, and parsley—while outbreeders, are seldom grown in a very large number of varieties. You may not care much if your leaf dill variety is a bit crossed with a heading variety. Similarly, you may not mind an amalgam of flat-leaf and savoy spinach.

Caging

Lightweight cages can be built of wood, PVC pipe, or other materials. Keep in mind that many crops grow to twice or three times their normal height and width when they go to seed. The cage should be encased in floating row cover material (it lets in light and moisture but keeps out insects and wind-borne pollen), with 4 inches or 5 inches of extra material at the bottom. Stake the cages down (wind can blow them away) and pile dirt onto the bottom material to ensure that no insects can crawl in. There should be a tight-fitting lid that can be removed for hand pollinating if necessary. You can use packing or duct tape to seal the edges.

Plants should be staked in place before caging, and they should be caged before the blooms open. Then you can use a brush or shake the plants to transfer pollen between the flowers of all the plants in the cage. The ideal time for pollination is 10:00 A.M. to 2:00 P.M.

A number of crops require insect pollination. Hand pollination just won't work, and if you open the cages, insects laden with pollen from miles off will come prancing in. What to do? Well, perhaps we just accept that some crops are best grown by professionals who can either raise and introduce baby insects or maintain the required vast isolation distances to grow the crops free of cages.

Depending on their makeup, crops can be pollinated by either honeybees or flies. One effective method for urban seed savers is to introduce newly hatched honeybees from your own hives or the hive of a friend. Although it is possible to attract mature bees with honey, they will not be used to confinement and won't do much work. To introduce baby bees, set a nuc (queenless hive) with a ready-to-hatch brood in the cage. A disgusting method of catching flies that works is to set out rotting meat, and then place the dish with the feasting beasts into the cage. Consult appendix 11 for information on purchasing insects.

Alternate day caging is a method you can use for crops that require insect pollination. You simply remove the lid from the cage of each different cultivar on successive days to let insects in, but ensure they are not also collecting pollen from your other varieties. Because it is nearly impossible in urban settings to ensure that no similar crops are being grown by your neighbors, we do not recommend alternate day caging for urban farmers.

Bagging

Bagging is the easiest method for varieties that are self-pollinating. However, if a few plants are needed, a large bag encircling a few plants' flower heads can be used. If bagging a variety that needs pollen from other plants of the same variety to set seed, simply unbag and use a brush to spread pollen from as many other plants as possible. You may decide to save seed from only one plant and eat or remove others that supplied pollen.

Except for spinach and chard, which have tiny pollen that can get through the mesh, floating row cover material makes the best bags. Simply cut a large circle and tie it securely around the plant stalk below the flower buds. Be sure to stake the plant well. You will likely need to wrap the edges of the bags in rags or shove cotton balls into the gaps to keep out insects. To spread pollen, simply shake the bag each day, or remove the bag and use a brush to spread pollen from other same-variety plants nearby.

There are some plants where the bagging method might be useful but the form of the plant doesn't lend itself well to the technique. For instance, because of their low-branching habit, pepper plants would be nearly impossible to bag well without large amounts of cotton wool stuffed between each branch. As long as the blossoms are large enough, as in the case of peppers, eggplant, and okra, the blossoms themselves can be bagged. This can be a tedious process, especially with pepper plants, which also require mixing the seeds of five to ten plants to maintain diversity.

Bagging corn

If you wish to save a pure strain of corn, care and time is required. You will need a sharp knife, a stapler, and both ear and tassle bags. Because corn is susceptible to inbreeding depression, you will need to save seed from as many plants as you have space for, ideally two hundred. The process of hand pollinating and bagging can last from a few days to a week, depending on the level of development of the tassels and ears. You will begin the process just before the silks begin to emerge from the ears. You may need to stretch the bagging of the ears over a number of days if they aren't all at the same stage of maturity, but you must bag them before the silks emerge from the ears and the tassels start shedding pollen.

Tear off the leaves covering the corn ears that are ready for bagging.

With a sharp pocketknife cut the tips of the husk leaves covering the ears, so you can see the silks inside.

Put ear bags over all the ears, and staple at the bottom.

Plan to bag the tassels at the top of each plant in the evening of the day before you will hand pollinate. The tassels are mature when the anthers have begun to emerge along the tips of the horizontal and lateral branches of the tassel.

Give the tassel a shake to dislodge any foreign pollen, then hold all the tassel branches upward to put the bag on.

Fold the bag close to the stalk and staple.

Make sure you staple above any pollen that may have collected in the leaves just below the tassel.

Remove those leaves.

Harvest pollen from the bags the next day in late morning or early afternoon.

Bend each tassel down without breaking it and shake the tassel.

Remove the staple and bag while holding the tassel downward.

Mix pollen from as many bags as possible and pour it into an ear bag.

Going from ear to ear, remove each bag, sprinkle about ½ teaspoon onto the tassels, and rebag the ear with a used or new brown tassel bag.

Pull the bag down between the ear and stalk and staple the bag onto the ear a bit loose, so it can grow.

The ears can remain bagged until harvest.

Taping

Taping is used with plants from the Cucurbitaceae family, such as squash and melon. These species produce both male and female flowers and are pollinated by insects. Since your neighbors are very likely growing Cucurbits, taping is a must to save pure seed. This isn't just a problem for home seed savers. We recently grew the heirloom melon Prescott Fond Blanc in a special hot box (we have cool summers here). The plant set fruit, and we watched as it ballooned in size. We decided to look it up on the Web to find out what it should look like when ripe, only to see that what we had grown looked nothing like what we saw. Since various "cucumbers" are actually in a melon species, we really weren't sure what we were getting.

The technique for ensuring breed purity with Cucurbits is to keep insects out of female flowers and hand pollinate. You can fertilize the female flower with pollen from a male flower on the same plant or from a different plant as long as it is the same variety. The first step in the taping method is learning to distinguish male from female flowers. The male flowers have straight stems, while the stems of the females have a slight bulge at the base. Next you will want to get a feel for what flowers look like when they are mature and one day away from opening. This will simply take a few days of observation, either of the same plant or another similar one. If female blossoms open on the plant you plan to save seed from before you have gotten to taping them, remove them.

To use the taping method with Cucurbits follow these steps:

In the evening, use masking tape to shut all the flowers that will open the next day.

The next morning, pick (remove) one or a few male flowers from the same plant or variety along with their stems.

Remove the tape on a male flower and gently tear off all the petals.

Remove the tape from a female flower and gently rub the pollen from the male flower into each part of the stigma of the female (using more than one male flower will improve success). If an insect flies in before you can do your pollination, remove the female flower.

Retape the female flower and tie a string around the stem to mark it.

Repeat with any other female flowers on the plant.

Once the fruits with markers are fully mature, harvest and allow them to sit for two or three weeks. Then open and remove and dry the seeds (melon and cucumber seeds should be processed using the fermentation method).

TREATING SAVED SEED FOR DISEASE

If you acquire a disease in your seed crop, or to ensure that you don't acquire one, you can treat your harvested seeds to kill pathogens. Common diseases that can be spread through seeds include black rot, black leg, and black leaf spot, all of which attack *Brassicas*; bacterial canker and target spot, which affect tomatoes; septoria spot on celery; and downy mildew, which attacks spinach among other crops.

A simple hot water treatment is effective in killing diseases that might be present in saved-seed. The seeds are consistently heated at a temperature high enough to kill pathogens but not so high as to kill the germ of the seed. The temperature must be kept constant, so some care and attention is in order. You will need to acquire a thermometer intended for use with liquids and an electric frying pan. A practice run will help, but other than that it's as easy as pie. Follow the chart for treatment times.

Heat water in a saucepan to 122 degrees.

Pour one third of the water into a warmed electric frying pan.

Set the saucepan into the frying pan.

Regulate the heat by removing the saucepan or turning up the frying pan.

Prewarm the seeds in 100 degrees to 105 degrees water for five minutes.

Once the temperature is stable, pour in the seeds and stir.

Pour the water and seeds through a fine sieve and spread the seeds on trays to dry.

CROP	AMOUNT OF TIME	TEMPERATURE
Broccoli	20 minutes	122 degrees
Brussels sprouts	25 minutes	122 degrees
Cabbage		
Carrots		
Cauliflower		
Celery	30 minutes	118 degrees
Chinese cabbage		
Collards		
Eggplant		
Kale		
Kohlrabi		
Lettuce		
Mustard greens	15 minutes	122 degrees
Pepper	30 minutes	125 degrees
Radish		
Rutabaga		
Turnip		
Spinach		
Tomato		

CHAPTER 15

FRUIT TREES

Fruit trees, because they are perennial crops (e.g., you don't have to plant them year after year), can be one of the easiest— and most productive—things you will grow on your urban farm. Just provide your trees enough sun, water, and nutrients, and you'll have tons of delicious homegrown fruit. After a tree is a number of years old, you'll probably even have enough fruit to share with friends and neighbors. This is great news for urban growers.

Although this is changing, rural fruit growing has tended to feature tall standard-size fruit trees pruned in a round canopy form, planted in a traditional orchard row format to allow for easy tractor access. This takes up a lot of space. In terms of fruit varieties, commercial fruit production focuses on types suited to easy harvest, shipping, and long shelf life rather than good flavor. Urban fruit forestry is the total opposite: We favor dwarf trees planted closely together, with flavor as our top priority. A dwarf apple tree only takes up a few horizontal feet in an urban farm, and it can be maintained at 6 feet tall. Urban growers also want a wider variety of types of fruits and trees that yield over a longer period of time. As an ideal, we suggest the following plan:

❖ Grow up to two to five trees or shrubs per person (up to 100 to 150 pounds of fruit per year per person).

❖ Choose a wide variety of types of fruits that mature throughout the year.

❖ Select different varieties of each fruit that will mature at different times, for instance an early and a late apple.

❖ Grow fruit trees that yield over a longer period of time.

❖ Plan to plant trees and shrubs so they don't shade annual vegetable beds.

❖ Prune your trees to be small so that they can be cared for and harvested without the use of dangerous ladders.

❖ Grow fruit with the most delicious flavor.

In this chapter you will learn effective techniques for pruning and training fruit trees to remain short and easy to work with while providing you with a staggered yield of delicious fruits throughout the growing season. Home fruit growing is intimidating to many, because tree pruning has been made out to be complicated and risky: "If you make the wrong cut at the wrong time, you'll permanently damage your tree." "You must prune when the tree is dormant in the winter." These "rules" are completely unnecessary and often contrary to the interests of home growers. We're here to demystify the techniques, so you can become a confident pruner and fruit grower. Grafting—making your own fruit trees—is similarly portrayed as quite difficult. Not true. We will also outline money-saving methods for starting your own trees.

WHERE TO BUY TREES

Start with unpruned trees so that you can make the initial pruning cuts that will shape them as you want them to be shaped (unpruned trees have a growing tip on the trunk). Because we want you to start out with unpruned trees, we recommend you purchase bare-root fruit trees from winter to spring from a locally owned nursery or by mail order from a regional nursery (see appendix 11 for sources). "Bare root" means the tree has arrived fresh from the growing grounds, and hasn't been potted up. Their roots aren't fully established in a rootball and will transplant easily.

TYPES OF FRUIT TREES, SHRUBS, AND VINES

Although there are many types and species of fruit trees and shrubs, we will limit ourselves here to the most common. While we will touch on citrus, vines, nuts, and other important species, most of our attention will be devoted to the Pome and Stone fruits.

Pome Fruits	Stone Fruits	Citrus	Other Subtropical	Other Tropical	Vines
Apple	Peach	Lemon	Mulberry	Avocado	Kiwi
Pear	Asian plum	Orange	Fig		Grape
Asian pear	European plum	Tangerine	Pomegranate	Banana	Passion fruit
Quince	Nectarine	Grapefruit	Persimmon		
	Cherry	Lime	Olive	Guava	
	Apricot	Mandarin	Loquat	Mango	
	Mixes (e.g., Pluot)	Kumquat	Pineapple guava	Papaya	
	Almond				
	Sour cherry				

Pome and Stone Fruits

These fruits are the bread and butter of home fruit production. Varieties have been developed for all climate regions. They are deciduous (they drop their leaves in winter and go dormant), bloom in the late winter through spring, and produce crops from summer through fall.

PARTS OF A TREE

Pome and Stone fruit trees consist of a rootstock, a graft union, the central leader trunk, scaffold or main branches, and additional branches off the scaffolds called lateral branches. They can also have suckers that grow vertically up from the roots and water sprouts, or weak sapling sprouts that grow vertically up from the scaffold branches.

Buds and fruits form in different ways on Pome and Stone fruit trees. Pome fruits set buds on short spurs in tight clusters that can mature into many fruits on the same spur. Stone fruit buds swell from within a branch and, in the case of peaches, create two blooms at each spot, or in the case of other Stone fruits, create a number of blooms at each spot.

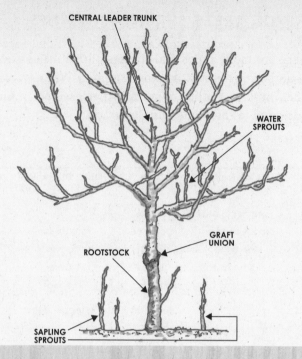

CENTRAL LEADER TRUNK

WATER
SPROUTS

GRAFT
UNION

ROOTSTOCK

SAPLING
SPROUTS

The parts of a tree (Pome and Stone fruits)

Rootstock

Traditionally Pome and Stone fruits are propagated vegetatively. This means that trees are produced from cuttings from a particularly good tree rather than grown from seeds. Fruit from seed-grown trees can be extremely variable—the offspring won't necessarily have the good qualities of the parent. Trees grown from cuttings are genetic clones of the parent—the same in every way. But because soil and climate vary greatly from place to place, cuttings of fruit trees are grafted, or joined, to tree root systems that are well adapted to the soil and climate conditions of the particular region. These well-adapted rootstocks make for healthy, vigorous trees that produce delicious fruit. Where the cutting and rootstock were originally joined there is often a ridge or ring in the bark. When you buy trees you will be able to select the appropriate rootstock for your garden.

Both rootstock type and pruning techniques can determine the size of fruit trees. The three main categories of rootstocks are standard, semidwarf, and dwarf. In an effort to create trees that stay small without being pruned for that purpose, breeders have created dwarf and superdwarf rootstocks. In many cases these last two rootstock types make for less vigorous and healthy trees. We don't recommend them. In reality, any deciduous fruit tree grafted onto standard and semidwarf rootstock can be pruned to any size. **In this book we use these names—standard, semidwarf, and dwarf—to refer to the general height categories of trees as achieved through pruning, not the type of rootstock.**

POLLINATION AND CLIMATE REQUIREMENTS

Pome and Stone fruits are pollinated by insects, mostly bees. Depending on the variety, trees can be self-fruitful or self-sterile. Self-fruitful trees can be planted singly and will yield fruit; self-sterile trees must be planted along with a tree of another similar variety that will pollinate it. Self-sterile trees' flowers don't produce mature male and female parts at the same time, so for the female part to receive pollen, it must have another tree within the same species producing pollen at the right time. Plant catalogs and nurseries have lists of pollinators for the different self-sterile varieties they offer. If you know your neighbors have a tree that will pollinate yours, you may be able to get away with planting a self-sterile variety without a pollinator.

Pome and Stone fruits must receive the correct number of chill hours to enable them to set abundant fruit crops. These are the number of hours the temperature drops below 45 degrees each year. In colder regions chill hours are ample, but gardeners with warmer winters must pay close attention to chill-hour requirements and plant appropriate varieties. On the other hand, some maritime and colder regions lack sufficient warmth during the spring and summer fruiting season to grow such heat-loving Stone fruits as peaches, cherries, and apricots. Special cultivars have been developed for these regions.

The most successful fruit growers select varieties that do well in their climate region. We would love to grow cherries and peaches, but in our maritime climate we never get the required heat. Although we have had limited success with special cultivars, in our experience, the heat lovers are forever unhappy and bug-infested, while our pears, plums, and apples thrive. Another example: While northern dwellers *can* grow tropical crops such as citrus, the required care (moving them to overwinter indoors) is extremely inconvenient and time consuming.

PLANTING YOUR FRUIT TREE

Materials:

- bareroot or potted fruit tree
- shovel
- hose
- one bucket of compost or rotted manure

Planting trees is fun and easy. First, find the perfect spot for it. Bear in mind that trees need full sun to thrive. Before you plant, envision the tree at its full height and width—is it going to block out too much of your garden's sun? Is it okay if it blocks a window or fence? Think before you plant! Dig a hole a bit bigger than the root ball or the pot. Toss some composted manure or compost into the hole. Place the tree in, and rotate it around until the branches are oriented to your liking. Making sure to hold the tree straight, start filling in the soil. As you fill, turn on the hose and water it, so it compacts. The goal is to plant your tree so that once it is in the ground the soil level is the same as it was in the pot. If you don't water as you fill, your tree may end up sinking down too low. If you're planting bare-root trees, the soil should cover the roots, but not the graft union (which looks like a ring of puckered bark on the trunk). You may have to pull the tree out and start over again to achieve this level. Compact the soil with your heel and water the tree deeply for at least fifteen minutes with the garden hose. Finally, you can make a doughnut-shaped ring about 1½ feet out from the tree to hold water. Mulch the tree with wood chips right up to the base.

Our advice—stick with what thrives. To find out the best species and specific cultivars for your area, consult with your local cooperative extension service and plant nurseries.

Following are some recommendations for tree forms that you can create through pruning (see pages 338 through 344 for instructions). Once you have decided the height and form your trees will take, you will be able to decide planting distances and layout.

Tree Height and Form	Height Range (in feet)	Minimum Planting Distance (in feet)	Notes
Standard open center and central leader	15–20	25	Not recommended for urban farmers—too tall
Semidwarf open center and central leader	8–15	15	Not recommended for urban farmers—too tall
Dwarf open center and central leader	6–8	10	Can be achieved through pruning on any root stock
Superdwarf open center or modified central leader	5–6	6	Can be achieved through pruning on any rootstock
Fruiting hedge	4 and higher	2	Can be achieved through pruning on any rootstock
Espalier: horizontal, candelabra, Palmette Verrier, double or triple-u, Belgian arch, step-over, and fan	4–10	6–15	Can be achieved through pruning on any rootstock
Espalier: Belgian fence, Belgian doublet, column	4–10	1–2	Can be achieved through pruning on any rootstock

PRUNING

Pruning Tools and Supplies

Useful pruning tools and supplies include:

- hand pruning shears
- pruning loppers
- pruning saw
- pole pruner
- ladder
- sharpening stone and oil
- "spreaders" used to prop branches apart to spread them. (Available by mail order.)
- natural fiber twine for staking or tying branches to trellises
- metal irrigation stakes for training branches downward
- weather-proof labels or cut-up pieces of tin cans. Write the name of your trees on the tin can labels and you will be able to read the marks

As much as possible, keep your tools cleaned, oiled, and sharpened. Poor cuts can lead to damage and disease.

Pruning Cuts

We can't see the life force inside trees (or at least, most of us can't), but if we could, pruning would be very easy. Pruning works with the life force of the tree to direct it to grow in the direction we wish. If you develop the ability to imagine water flowing from the roots of the tree out to the unpruned tips of the branches, pruning will be much easier for you. If you cut off, or shorten, a branch, the "water" must find another route to go. It will do so by pushing out the latent bud closest to the cut or going up the branch or trunk from which a side branch was removed.

There are two types of pruning cuts: heading and thinning. They can happen anywhere, from the trunk to the lateral branches. Heading cuts remove only part of an existing branch or twig, and generally encourage widening and new growth from latent buds. Thinning cuts remove single branches or twigs entirely, and generally cause energy to go towards strengthening the trunk and remaining branches of the tree.

The goal of a heading cut is to redirect the growth of the tree from the tip you cut to the branch or bud below the cut. This is also known as heading back. When doing heading cuts, you want to know exactly which branches or buds are next in line below, and make sure they are

HEADING CUT

Make heading cuts just above buds or existing branches. This cut redirects the growth to buds and branches below the cut and is used to limit the size of the tree.

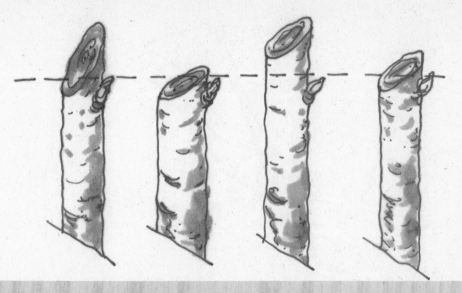

The first three heading cuts have been done improperly. The fourth is the ideal, at a 45-degree angle, ⅛–½ inch above the bud or branch.

pointing in a direction in which you want the tree to grow. If you are doing one on a small branch without side branches, make sure that the bud below your cut is pointing away from the center of the tree.

Heading cuts should be about ⅛ inch to ½ inch above the bud or branch below the cut, depending on the diameter of the branch you're cutting. The cut should be at a 45 degree angle, with the high side of the cut pointing in the same direction as the bud or branch below the cut.

The goal of thinning cuts is to clear out branches and twigs that inhibit airflow, are touching other branches, or are pointing into the center of the tree. They redirect the "water," or life force, to the remaining branches and twigs, making them grow faster. They limit new growth from buds, because the water can still flow down to the tips of the trunk or branch. Thinning cuts can happen on the trunk or main branches; they remove the entire branch or twig, down to the next branch or the trunk. The cut should be close to the branch or trunk at a 90 degree angle (flat along the bark). The cutting blade of your hand pruners should be against the branch you're leaving so that the cut is as close as possible.

THINNING CUT

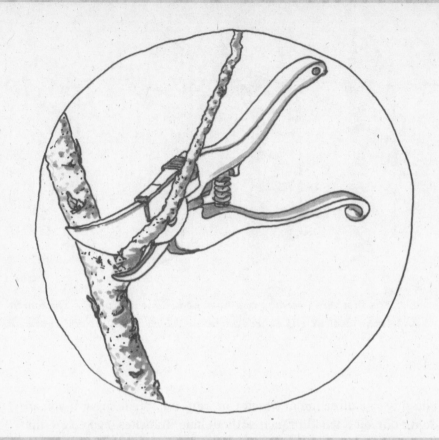

Make thinning cuts to remove any unwanted twigs or branches such as crossing branches or watersprouts. Note that the blade side of the pruning shears is flush with the branch.

If you need to thin out a branch that is very large, doing the cut in one pass can rip the bark of the tree as the branch falls. To avoid this, cut the branch off about 6 inches to 1 foot above the final cut to reduce the weight. This should be done in two steps. First make a cut going halfway through the branch, coming up from below to relieve pressure. Next cut the branch off about an inch or two from the notch. Then make your final cut ½ inch to 1 inch from the bark of the remaining branch or trunk.

Heading and thinning cuts can be used as you wish to create trees, shrubs, and hedges in whatever form you like. As long as you

REMOVING A LARGE LIMB

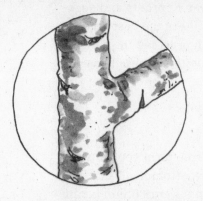 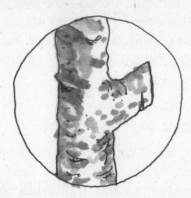 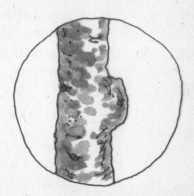

Step 1: Make a notch the branch about four inches from the trunk. Step 2: Saw the branch off a few inches from the notch. Step 3: Saw the remaining stump off one to two inches from the remaining branch or trunk. It will heal over.

don't injure the tree by making bad cuts, there really aren't too many rules. One is: don't prune a fruit tree like a hedge. This will create a dense thicket of crossing branches with a lack of airflow, increasing diseases and pest infestations. Pruning shears, like those you may use on your boxwood hedge, make a mess of fruit trees.

Timing of Pruning

Dogma says pruning should only be done during the dormant period in the winter. For urban growers, pruning in winter has one thing going for it—it's easier to see the branches when they don't have any leaves on them—but this is not a reason with any merit in itself. Pruning during dormancy in winter stimulates rapid growth, which is the last thing most urban fruit growers want. However, the dormant period is a great time to make a few important thinning cuts and work on training (explained on page 333). It is also a good method for older trees if you want to stimulate them to grow.

Pruning in summer limits rapid growth. The tree feels that it has just grown when it needed to grow (the spring and summer) and is heading toward its dormant period—not a time to put on new, tender growth. Summer pruning sends the tree a message: Don't grow! Summer pruning is the best strategy for urban growers with younger trees, since it helps you limit their size and avoid water sprouts and suckers (see the illustration on page 328). May to June is the best time for summer pruning. The challenge will be learning to see the form of the branches and the energy flows of the tree with all those leaves in the way.

The First Pruning Cut

The first pruning will depend on the type of tree form you want. To control tree size to around 6 feet to 8 feet, we recommend that you make an initial heading back cut 12 inches from the graft union (a ring of puckered bark where the fruit wood was grafted to the root stock). This will feel drastic, especially if there are a lot of nice branches on the tree, but don't worry. This cut will activate buds below to form new branches. The 12 inch initial pruning cut is appropriate for open center, central leader, and most espalier forms (see pages 340, 341, and 343 for illustrations of these forms). For a modified central leader, the initial heading cut should be between 18 inches and 24 inches.

The first pruning cut can be very shocking for novices. It feels as if you're cutting away most of the tree. This cut actually stimulates good root formation and allows you to maintain a healthy tree easily in the future. By limiting the tree size you won't have to drag out cumbersome ladders and are more likely to keep up with pruning. Because they are so tall, many backyard trees haven't been pruned in years and harbor insects and diseases that not only affect the tree, but spread to other plants in the garden. Stay strong and remember this when you do that first drastic cut!

SHAPING, TRAINING, AND MAINTENANCE PRUNING

Note: Please refer to the illustration of the parts of a tree on page 328. Achieving the final form will take a number of years. Forming of the shape includes pruning and training. Pruning limits the height, spread, and width of the tree and directs growth where you want it. Training moves selected branches to where you want them. This can

THE FIRST PRUNING CUT

12–24″

Though it seems harsh, cutting off a young tree at 12 to 24
inches will allow you to train your tree to a manageable
size and will not harm your tree.

be achieved through the following techniques: tying branches to trellis
wires or stakes; spreading branches with spreaders; training branches
towards the ground with string and stakes; or hanging weights off
branches.

After the final form is achieved, you will want to do maintenance
pruning to maintain the shape and allow the amount of fruiting you
want.

Maintenance pruning goals:

- maintain the tree at the shape and size you choose
- ensure good airflow and sunlight to the scaffold (main) and lateral
  (secondary) branches by removing extra lateral branches
- avoid the crossing of lateral branches by removing one
- stop the spread of disease by removing diseased branches

To maintain your chosen size you will need to head back scaffold and
secondary branches. Remember, if you're uncertain, you can do it in

stages, removing more length later. You can't add back what you've already cut.

Peaches, nectarines, plums, and pluots produce fruit on last year's growth, and some varieties also produce fruit on spurs. You will need to allow sprouts to grow on the scaffold and lateral branches for fruiting to occur. Since it's often important to reduce the amount of fruit, head back the previous year's branches you plan to keep, and thin the rest. To be clear: Allow new sprouts to grow each year on the scaffold and lateral branches, heading about half of them back to half their length in the summer. Thin the rest out. Prune the remaining sprouts back to one or two buds the winter after fruiting. In the meantime, new sprouts will be growing.

Apricots, apples, cherries, quince, and pears fruit on spurs, on scaffold, and on secondary branches, some of which produce fruit one year and leaves the next, so these don't need special pruning to encourage fruit production.

RECOMMENDED FORMS FOR POME AND STONE FRUITS

With work, you can coax almost any tree into almost any form. The forms we have suggested work best with the natural growth pattern of the tree, but feel free to experiment.

The open center form is preferred for urban farmers, though the modified central leader can also work very well. Trees that have a very upright growth form, such as pears and cherries, have a naturally small angle between the central leader and branches. It will take more work to spread the branches to be perpendicular to the trunk for the recommended forms.

Most Stone fruits (with the exception of European plums) have weaker wood than apples and pears, so be gradual as you spread branches or train them into shapes. Most Stone fruits also need periodic renewal of fruiting branches, so if you'd like to use an espalier form, they do best in the fan shape rather than the forms with permanent branch structures such as the candelabra form.

Feel free to work your own artistry with espalier shapes. The instructions for shaping trees that follow do not include information on pruning for fruit production. See the previous section for that.

RECOMMENDED FORMS FOR POME AND STONE FRUITS

Fruit	Trained Forms	Canopy Forms
Almond	Fan espalier	Open center; modified central leader
Apple	Fruiting hedge or nonfan espalier forms	Open center
Apricot	Fan espalier	Open center; modified central leader
Asian pear	Fruiting hedge or nonfan espalier forms	Open center; modified central leader
Asian plum	Fruiting hedge or fan espalier	Open center; modified central leader
Cherry	Fan espalier	Open center; modified central leader
European plum	Fruiting hedge or fan espalier	Open center; modified central leader
Mixes (e.g., Pluot)	Fan or column espalier	Open center; modified central leader
Nectarine	Fan or column espalier	Open center
Peach	Fan or column espalier	Open center
Pear	Fruiting hedge or nonfan espalier forms	Open center; modified central leader
Quince	Fruiting hedge or nonfan espalier forms	Open center

Open Center:

- When you plant your tree, prune back the central leader to 12 inches.
- Once new branches begin to form, select three well-spaced branches or two opposing sets of branches (if viewed from above they form a cross). Where these branches originate may be up to a foot or so apart from each other on the trunk vertically.
- Select branches that look healthy and have the widest angle between the trunk and branch.
- Remove all the other branches from the trunk using thinning cuts.
- Spread the branches as necessary in a gradual manner. Your goal is for the scaffold, or main branches, to grow at a 45 degree angle to the trunk. Begin with small spreaders or sticks wedged into the crotches of the branches, small weights tied to the branches about two thirds of the way down, or strings attached to the ground with stakes. Be gentle. Do not overdo it!
- Every three months or so, increase the spread again.
- At this stage you can begin to select secondary lateral branches that branch off the scaffold branches.
- Select branches that grow upward and/or away from the center of the tree and do not cross other branches.

❖ Remove any other newly forming branches from the trunk and the scaffold branches with thinning cuts. Leave fruiting spurs.

❖ When your scaffold branches have reached half the length you desire, head them back to an upward or sideways pointing bud.

❖ To maintain the shape of your tree, remove suckers, water sprouts, and unwanted branches each year.

❖ Head back any secondary branches that grow beyond the length you desire.

MODIFIED CENTRAL LEADER FORM

Year one

Year three

Modified Central Leader

Modified central leader trees can also be kept quite short.

❖ When you plant your tree, prune back the central leader to 18 inches to 24 inches.

❖ Once new branches begin to form, select up to six branches spiraling up the trunk at regular intervals up the central leader.

❖ Select branches that look healthy and have the widest angle between the trunk and branch, if possible.

❖ Remove all the other branches from the trunk with thinning cuts.

MODIFIED CENTRAL LEADER FINAL FORM

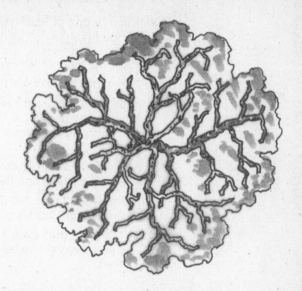

Bird's-eye view

‡ Spread the branches as necessary in a gradual manner. Your goal is to have the main branches grow at a 45 degree angle to the trunk. Begin with small spreaders or sticks wedged into the crotches of the branches, small weights tied to the branches about two thirds of the way down, or strings attached to the ground with stakes. Be gentle. Do not overdo it!

‡ Every three months or so, increase the spread again.

‡ At this stage you can begin to select secondary lateral branches that grow off the main branches.

‡ Select branches that grow upward and/or away from the center of the tree and do not cross other branches.

‡ Remove any other newly forming branches from the trunk and the main branches with thinning cuts. Leave fruiting spurs.

‡ When your scaffold branches have reached half of the length you desire, head them back to an upward or sideways pointing bud.

‡ Remember that extremely long main branches will require support boards, so it's best to maintain the width of the tree at no more than a 6-foot spread unless training to a special form.

‡ To maintain the shape of your tree, remove suckers, water sprouts, and unwanted branches each year.

‡ Head back scaffold branches and secondary branches that grow beyond the length you desire.

Fruit Hedge:

Creating a hedge of fruit trees is especially easy with the Pome fruits and can be done with some Stone fruits, such as plums.

- Plant the hedge with one or more varieties 2 feet apart.
- Allow the central leaders to remain unpruned at first.
- Select scaffold branches at 6 inch to 12 inch intervals that are growing parallel to the line of the hedge and train with spreaders till they are between a 45 and 90 degree angle to the trunk.
- Allow the scaffold branches to grow 6 inches to 9 inches on either side of the trunk, and then head back to a fruiting spur.
- In the second year, head back the central leaders of the trees at the desired height, traditionally about 6 feet high.
- To maintain the hedge, remove sprouts from the scaffold branches and keep the central leader and scaffold branches headed back as needed.

Horizontal espalier (Cordon):

- Allow the central leaders to remain unpruned at first.
- Select sets of opposite branches in the same plane at regular intervals, from 1 to 2 feet apart.
- Remove all other branches from the central leader.
- Gradually train the branch sets to grow horizontally in a flat plane to arrive at a 90 degree angle from the central leader.
- Head back the central leader just above the top set of branches.
- Head back the horizontal branches to maintain the tree width you desire.
- To maintain the form, remove branches that grow off the horizontal scaffold branches and the trunk, leaving the fruiting spurs.

Candelabra espalier:

- Head back the central leader to 18 inches to 24 inches.
- Select three sets of opposite branches in a flat plane below the initial heading cut 1 to 2 feet apart. This may take a few years. Remove all other branches from the trunk.
- Gradually train the sets of branches to grow horizontally in a flat plane.
- When the lowest set of branches is about 6 inches longer than the tree width you desire, begin to train them upwards to form an L shape.

- The arms of the candelabra will nest within each other. The lowest set should be about a foot or more wider than the set above it, and so forth.
- Gradually train each set of branches into the L shape.
- Head back the vertical branches at the desired height.
- To maintain the form, remove branches that grow off the vertical and horizontal branches, leaving the fruiting spurs.

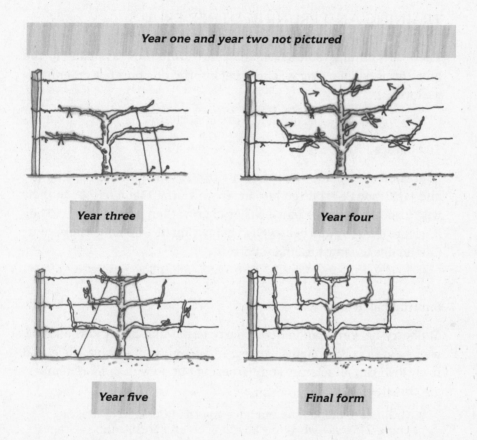

Year one and year two not pictured

Year three

Year four

Year five

Final form

Column Espalier:

These can be grown singly or in rows, and work well with Pome and Stone fruits. If planting a row, plant trees 2 feet apart. Provide a support stake, a taut wire, or a wall.

- Allow the central leader to remain unpruned.
- Allow lateral side branches to grow every 6 inches or so, and then head them back to a fruiting spur a few inches from the central leader.

- Remove all other side shoots.
- When the column has reached your desired height, head it back.
- To maintain the shape, remove lateral branches and head back the central leader as desired.

TRAINING AND PRUNING OTHER FRUITS

There are innumerable fascinating rare fruits now available to the home grower. Following is some brief information on a few other fruits we love.

Temperate Fruits

Sour cherries are one of our favorite fruits—the gold standard for pie and jam, though a bit too tart for fresh eating. Being closer to their wild forbears, they are much easier to grow than other cherries. They are more shrubby and bear well in many climates, notably where summer heat is too low for other cherries.

Subtropical Fruits

Mulberry, fig, persimmon, sour cherry, grape, and hardy kiwi all grow well even in cold climates. Pomegranate, pineapple guava, olive, and fuzzy kiwi can all tolerate some freezing but do well in more temperate climates.

- Grapes are ubiquitous, but have you tried Concord or Muscat grapes? We say, why grow anything else for fresh eating? Grapes need a strong trellis, preferring the horizontal. Prune hard or they will take over your space.
- Prune and train mulberry and fig quite hard. If left to their own devices they will both get very large, and since they produce soft fruits, the birds will get 'em before you do. Plant white or yellow mulberry cultivars to avoid staining and birds eating your crop. A great way to harvest mulberries is to spread a clean cloth under the tree and shake. Otherwise it is very time consuming. The best tasting, though messy, varietal of mulberry is the Persian, which bears a dark juicy fruit that looks like a blackberry but tastes like spicy wine. In the case of figs, be sure to select cultivars well adapted to your area. Otherwise the fruit may still be green when cold weather comes.

- Pomegranate and pineapple guava are delicious fruits that grow on beautiful shrubs. The flavor of both is divine. Pineapple guava blossoms are delicious too. Both can be grown in a variety of climates, as long as the right cultivar is selected. Pomegranates require summer heat.

- Olives are the staff of life. Olive trees are a symbol of peace, and brining your own olives will be a religious experience. Olives grow extremely slowly and are prone to attack by the olive fruit fly. The fly lays its eggs just under the skin, producing dimpled spots and tiny holes on the fruit and rendering it useless. The fly can be managed, so if you think you might have olive fruit flies, research eradication methods.

- Kiwi, both fuzzy and hardy, are beautiful if huge deciduous vines that prefer to grow on a horizontal trellis. They're great if you have a trellis shade structure over a deck or outdoor area, or they can be trained on a wire structure.

Tropical Fruits

Citrus, banana, loquat, guava, passion fruit, and avocado only grow well in the warmer climate zones, or indoors or in pots that can be moved indoors in cooler areas. If you can grow citrus where you live, be sure to select varieties that produce flavorful fruit in your climate. Well-adapted varieties of tropical fruits are growing in immigrant communities in many U.S. cities right now. But language and cultural barriers keep these varieties from spreading. Fruiting bananas, passion fruits, guavas, etc., are commonly brought from home countries and adapted to the new region. With some impromptu sign language and tact you may be able to get some cuttings or seeds.

- Citrus can be pruned however you want. They are totally malleable. Don't let them become thicketlike, since they often have horrid thorns. Thin branches growing into the center of the tree.

- There are now banana cultivars that produce fruits in temperate climates. Pretty exciting. Bananas get huge and require sun. They're great for front yards and median strips.

- Loquat trees grow in temperate zones with little freezing. They are extremely prone to fire blight, so maintain sanitation when pruning. They produce a dense shade if allowed to grow in a canopy form.

- Guava, avocado, mango, and papaya also prefer warmer environments and will get huge if allowed to do so. We have an avocado tree in our neighborhood loaded with fruit each year. Even with the tallest basket picker on top of the tallest ladder, we can't reach much fruit. We keep thinking we could rope the Pacific Gas & Electric guys into loaning us their lift. Point

Fruit	Good Forms	Notes
Avocado	Modified central leader	Prune for size, naturally grow very large
Banana	Don't prune	Remove sections if it gets too big
Citrus	Any espalier form, open center	Prune in warm weather
Fig	Fan form, modified central leader	Prune for size, naturally grow very large
Grapes	Horizontal or vertical trellis	Like to grow flat on a horizontal surface, fruit on last year's wood
Guava	Modified central leader	Prune for size, naturally grow very large
Kiwi	Horizontal or vertical trellis	Likes to grow flat on a horizontal surface
Loquat	Modified central leader	Prune for size, naturally grow very large
Mango	Fan espalier form, modified central leader	Prune for size, naturally grow very large
Mulberry	Any espalier form, modified central leader	Prune for size, naturally grow very large
Olive	Any espalier form, modified central leader	Grow extremely slowly, lightly prune, if at all, until larger
Papaya	Prune shoots to limit size	Prune for size, naturally grow very large
Passion fruit	Horizontal or vertical trellis	Like to grow flat on a horizontal surface
Persimmon	Open center	Let grow naturally; light pruning for size and shape
Pineapple guava	Shrub	Light pruning for size
Pomegranate	Shrub	Light pruning for size
Sour cherry	Modified central leader, shrub	Don't head back the central leader until scaffold branches are well established or growth will be stunted

being, prune hard. Avocados require a male and a female plant for pollination.

✤ Passion fruit creates gorgeous flowers and ambrosial fruit. Slurp the pulp right out of the shell, and you'll wonder if you've been transported to the tropics. In our subtropical climates, fruit can be on the tart side, but mixed with sweetener the pulp makes a delicious juice or canned syrup.

BUILDING TRAINING STRUCTURES

Vertical Structures

Espaliered trees and shrubs and some vines do well trained to a flat vertical structure, and fences will often do very nicely. Vertical woven or welded-wire trellises can also serve well, but it is important to shield tree wood from rusting metal. The downside to using fences and trellises is that branches can easily get away from you and intertwine with the wires of the fence. As branches grow in girth, the wire can begin to cut into the wood, causing damage to the plant and allowing easy entry for diseases. To avoid these problems, train trees and shrubs 6 inches to 12 inches away from fences using wood prunings or short pieces of PVC pipe as spacers.

Traditionally, espaliers are trained along horizontal cables set at 1-foot intervals between posts. Branches are tied or clipped loosely to the cables as they're trained. If you plan to build such a structure, set the posts in concrete. Use cable, cable clamps, a tensioner, and a swage-it tool for crimping the clamps. Wooden or bamboo stakes and fences can also be used without damaging the tree.

Horizontal Structures

Many vines, such as kiwis, passion fruit, and grapes (as well as some annual squash), prefer to grow on a flat, horizontal trellis suspended above the ground. These double perfectly as shade structures for lounging or as carports. They can be built solidly of wood or a combination of wood and cable. In addition, lower-row trellises can be constructed with solid wood T braces set in cement every 5 feet to 10 feet with cable stretched between them. The height of the T can be determined by your height for ease of harvesting and pruning, and the width of the T can be from 3 feet to 5 feet.

WATERING AND PROVIDING NUTRIENTS

In the first five years, water your trees periodically in the spring and summer if your area lacks rain. If your trees develop fungal diseases, such as powdery mildew or apple scab, you will know you are watering too much (if you have overly wet land, plant pears; they can tolerate it). Maintain a nice doughnut of soil around the tree, and fill the hole with water to the brim a number of times to give a good, deep soak. Once a month is plenty. After five years follow common sense and water your trees if they look parched or if there's a serious drought.

Trees love mulch. Mulch attracts the organisms that break mulch down, so it will eventually become compost and feed the tree. Wood chips are best; mulch your trees twice a year. You can also add a top dressing of finished compost under the mulch to provide extra nutrients. Cover cropping between and around your trees, and then chopping the green matter down to decompose right there, is also a great way of getting nutrients to trees. Layering kitchen scraps and wood chips between and around your trees is another great method of fertilizing trees (be sure to stay away from the trunk). Red wigglers will abound and provide nutrient-rich worm castings for your trees.

FRUIT THINNING

Thinning, or culling, fruit is an important part of maintaining tree health and fruit quality for apples, pears, peaches, and plums. You will need to cull to maintain tasty fruit and a healthy tree. Remember that if you don't cull, your fruit may be pest infested, small, and less tasty, and you may lose important branches to the excess weight. As the fruit

Once thinned, this cluster of apples will only have only one fruit

begins to reach the size of a quarter and larger, remove immature fruit from fruit clusters. Start by taking off just one of the clusters if they are closer than 2 inches together, and half of the fruit from each cluster. Your goal the final time you cull is to have one fruit per cluster, with the clusters at least 5 inches apart.

FRUIT HARVESTING

Because apricots, pears, apples, and plums are formed on spurs that fruit for years, picking must be done carefully, with an upward twisting motion, so as not to break the spur. You will notice a node between the spur and the stem. If the fruit is near or at ripeness, this node will break easily. Apricots, peaches, and nectarines will release from the stem when ripe, and cherries should break off easily in clusters from the spur. Getting the knack of just the right angle at which to lift up will minimize any spur breakage.

When harvesting fruit you want to pick it when it is ripe but not overripe. If you plan to eat it right away, or can it, you want it to be perfectly ripe. If you plan to store it, you want it to be mature but underripe. Many factors will let you know when your fruit is ripe. Size, changes in color, the ability to give a bit under thumb pressure, and the presence of ripe fruit on the ground are all indicators of ripeness. Once you have been through a harvest season with your trees, you will come to know the signs.

Because of the extremely early picking of store-bought fruits, many home growers overdo it in the opposite direction. Grocery store fruit is often picked totally green, before starches and sugars have had a chance to mature, and then matured artificially with the use of methane gas. This doesn't produce good flavor. On the other hand, overripe fruit has a zero shelf life in your kitchen. There is a point at which the carbohydrates are fully developed in your fruit, yet it is still not fully ripe. At this point it can finish ripening off the tree without harm—it produces its own ethylene hormone naturally. It's a good point at which to pick your fruit, so it can sit around a few days and be eaten over time as it ripens indoors.

To extend the shelf life of your fruit further, wrap each piece as you harvest it in paper (this keeps the ripening hormone from spreading and causing a chain reaction), and refrigerate or store in a cool basement or root cellar. Depending on the variety, apples and pears will last two to six months, while stone fruits will last, at most, one month.

Our favorite harvest method is to wear a backpack backward on the chest and pick into that, then carefully packing the fruit in shallow boxes so as not to bruise. Ground-fall fruit that has fallen to the ground on its own, while fine for eating if you don't mind an occasional mealy, buggy, or rotten bite, must never be used for storage or canning, since it is full of microbes.

PEST AND DISEASE MANAGEMENT

Fruit trees can be attacked by a variety of animals, insects, bacterial and fungal diseases, and viruses. Giving the right amount of water and nutrients will minimize outbreaks, but some are unavoidable. Ants are special enemies of trees. Not pests themselves, they farm a number of especially damaging pests. Anything you can do to control ants will help, see page 297 for details on ant management.

You can live with minor outbreaks of many pests. But some of these, and blights, can cause severe damage. If you are headed in this direction, call your local cooperative extension service for help with diagnosis and treatment, but be sure to ask for organic control recommendations. Your local nursery will also be a source of information on applying superior oil, dormant oil, or other sprays. Cutting away the infected parts controls some bacterial blights, such as fire blight. There is a debate about its effectiveness, but many believe that if you sterilize your pruners between each cut, with chlorine or bleach solution (nine parts water to one part), the spread of the blight to other parts of the tree or other susceptible plants will be limited. We lack the space to cover tree pest and disease management in detail, see appendix 9 for a table of some common fruit tree pests and diseases and organic controls.

GRAFTING MULTIFRUIT TREES

We have an apple tree that makes six different kinds of apples—all bearing at different times of year—so that in a small area against a fence we get apples July through November. This bounty is made possible through a method called "grafting." It is an ancient technique of taking a scion (a cutting from the desired tree), and connecting it to the mother tree during winter dormancy. If properly placed, the mother tree will accept the scion and, come spring, her sap will run into the new branch, nourishing that new variety. It is also possible to graft one kind of fruit onto a different mother, e.g., a peach onto an apricot, or a pear onto an apple. These so-called fruit salad grafts are rather difficult and aren't long-lasting, so we don't recommend doing them.

Most people will want to add a new variety of the same type of fruit to their existing tree; this is technically called topworking. Traditionally this was used by orchardists to test out new varieties without making a big commitment of buying an entire tree. For urban farmers, grafting fruit trees ensures an extended fruit season and increased variety (how many Granny Smiths do you really want to eat?). After two years, fruit will develop on your grafted branch. Below is one way to do it (there are many methods; see page 540 for grafting resources).

During the dormant season, a few months before your trees are going to "break bud"—leaf out—you should start collecting scion wood. This will involve either attending a scion exchange (a place where fruit geeks gather to share cuttings) or taking cuttings from admired trees belonging to your fellow farmers. Either way, make sure the scion is the diameter of a pencil and has at least four buds on the stick. They should be labeled, then wrapped into moist (but not wet) paper towels, placed in Ziploc bags, and stored in the refrigerator until grafting day (up to two months later). On that day—usually in early spring, when the tree is still dormant—gather your grafting tools together.

Grafting supplies:

- grafting knife (or just a sharp, flat-edged knife)
- pruning shears
- grafting wax or plumber's putty
- grafting tape or paraffin tape
- tree labels (copper or homemade labels made from aluminum cans that you emboss the name on with a sharp pencil)
- notebook for writing down what was grafted onto what
- scions (cuttings) of the fruit tree you want to grow
- rubbing alcohol

Top working a fruit tree by using a modified cleft is easy. First make a straight cut into a wide branch on the tree you wish to topwork.

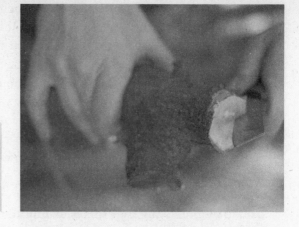

Next, take your scion wood and shape the tip of it into a V. Using your grafting knife for leverage, insert the V-shaped scion wood into the outside ring of the fruit tree. Make sure the cambium layers are touching. If you want to increase your chances of the graft "taking," place another piece of scion wood on the other side of the branch, thus doubling your chances of success.

Seal with grafting putty or wrap with grafting tape.

The easiest graft to do, in our opinion, is called the modified cleft.

1. First take your scion wood out of the bag and unwrap it. Orient it so the branch is right side up, as it will grow (hint: the buds are larger on bottom, then taper up). On the bottom of the scion, using your knife and cutting away from your body, make two cuts on each side of the scion, shaving off wood from the outside so it makes a V-shape. The cuts should expose 1 inch or 1½ inches of the branch. Look closely: You should see a green layer of the scion. This is called the "cambium." The cambium is the living tissue of a dormant tree, and your goal is to match it with the mother tree's cambium.

2. Next, find an end branch on the mother tree that is 1 inch to 2 inches in diameter. With your pruners, make a straight cut off the end of the mother branch. Then, being careful not to cut off your thumb, hold the branch, and slowly work the edge of your knife into the flat part of the branch, effectively splitting it in half, but only so it goes down about 1½ inches—do not halve the entire branch.

3. Take up your V-shaped scion wood and gently pry it into the cut created in the mother branch. Your goal is to join at least one part of the cambium of the scion with the cambium of the mother branch. With a branch that is 2 inches in diameter, you can usually insert two scion sticks onto each side of the mother branch, which increases your chance of success.

4. Next, smear the plumber's putty or grafting wax or wrap with grafting tape so it is fully coated. The idea is to keep the scion from drying out. Dab a bit of putty or wax or wrap with paraffin on the top exposed end of the scion too.

5. Now wrap the bonded area with paraffin. Be gentle while you do this—you don't want the scion to shift.

6. Label the scion immediately (you will totally forget in a few hours what was grafted where).

7. Wipe off your knife with the rubbing alcohol to sterilize the blade from any disease, and continue grafting.

Come spring, if you did it right, conditions were perfect, and your scion wood was sound, the mother tree flows her nutrients into the cambium of the scion, which feeds it. A callus forms around the graft union, and you'll have a grafted tree. Success rates are rarely

100 percent, which is why we recommend grafting multiple scions on multiple branches, to ensure that at least one will take. Keep practicing, or take a class from a wise elder, and soon you'll be an expert grafter!

CREATING YOUR OWN FRUIT TREES

Once you master topworking, you might be tempted to raise your own fruit trees by grafting onto rootstock. To create a fruit tree you need two things: rootstock and scion wood. The rootstock forms the roots of the tree and is selected to be vigorous and disease-resistant in your climate and soil. The scion forms the fruiting part of the tree and is selected for delicious fruit and vigor and disease-resistance in your climate.

What You Need to Make Your Own Fruit Tree:

* rootstock appropriate for your area
* scions of the variety you want the tree to become
* pruners
* grafting knife or straight-edge knife
* plumber's putty
* grafting tape
* labels
* wax pencil
* rubbing alcohol

As explained in the beginning of this chapter, most fruit trees grow on rootstock. That is, commonly known varieties of fruit trees—like Gala apples or Bartlett pears—are actually grafted onto the roots of a vigorous, disease-resistant apple or pear tree. Different rootstocks are used depending on your area's climate and disease profile. While very cheap, rootstock can be difficult to come by in small enough quantities for small-scale growers (typically they are sold to commercial growers in bundles of one hundred). There are a few mail-order nurseries that sell rootstocks in bundles of ten. They are listed in appendix 11. These mail-order catalogs will tell you which rootstock will match with the kind of fruit trees you're hoping to create.

After you get your rootstock, which usually arrives in the mail in late winter or early spring, pot it up in potting soil. Then start scouting for scions. Did you love an apple that a neighbor grew last year? Did you hear about a scion-exchange? Do you know a rural farmer who has

To make your own fruit tree, gather supplies (left to right): grafting tape, grafting seal, scion wood, simple knife, pruners, and rootstock.

First, make an angled cut on the rootstock.

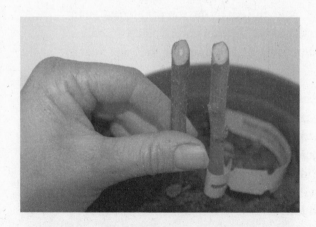

Second, cut a piece of scion wood at the same angle and the same diameter as the rootstock. Note the cambium layer just next to the bark. These should be bright green, and your job is to get them aligned with each other.

Next, firmly wrap the scion wood and rootstock with grafting tape so the cambiums are touching and are held snugly together. Finally (not pictured) dab the top of the scion wood with grafting seal so it doesn't dry out and label the fruit tree.

a terrific heirloom variety of pear that you'd like to have on your urban farm? In the winter, start collecting. Be sure to take pencil-size cuttings, with at least four buds on the cutting. Wrap them in a damp paper towel, label with masking tape and a Sharpie, place in a Ziploc bag, and store in the fridge for up to a month or two, until bench grafting time.

Watch your potted rootstock in early spring: Are the buds starting to swell? If so, it's time to graft. There are a variety of different kinds of grafts: cleft grafting, whip and tongue, wedge grafting, bud grafting. For potted up rootstock, we suggest whip and tongue.

1. Place the potted rootstock on a table (correct ergonomics is a nice benefit of bench grafting), and place each scion bundle with its matching rootstock. Using the pruners, make a cut so that the rootstock is trimmed down to 12 inches from the dirt. Now select scions that are of the same diameter for each rootstock. The better they match the rootstock in size, the more likely the graft will be successful.

2. Being very careful not to cut yourself, and starting about 2 inches from the bottom of the scion, make a smooth sloping cut away from your body. Your aim is to take out a chunk so it looks like a one-sided stake.

3. Do the same thing with the rootstock, but hold the tree at the base, face it away from you, and make the cut at the top of the tree. Hold the end of the scion up to the top of the rootstock—their cambiums should match up. If they don't, whittle a bit more so they will match up completely.

4. At this point, some grafters choose to wax the edges of the union, wrap the ends up with paraffin so they are held in place, and call it a day. This can and does work—as long as the cambium layers are touching and they are wrapped so tightly they can't move. Advanced grafters will make a notch at the thickest edge of each of the sticks so that the pointed end of each can actually enter into the notch; hence the name whip and tongue. This union takes some fiddling—and swearing—but it holds the scion in place so it can't fall off easily. It also firmly binds the two cambiums.

5. Apply wax or putty or paraffin at the union and the top of the scion, and then wrap the whole thing with grafting tape.

6. Be sure to label your new tree! For double protection, use the wax pencil to mark the pot and make a label that attaches to the base of the tree.

7. Wipe your knife with alcohol before moving on to the next bench graft.

8. Keep the new rootstock and scion in a shady place outside, where birds won't land on them; by the time of bud break, you'll know whether your graft was successful, as the scion wood will sprout leaves.

PART III

RAISING CITY ANIMALS

In our experience, suitable livestock for urban areas include honeybees, chickens, turkeys, ducks, rabbits, and goats. We think the addition of livestock on the urban farm is invaluable. "Pets with benefits" will reward you with farm products for the table such as eggs, honey, pollen, meat, and milk. By-products may include manures that can be composted and then added back into the vegetable garden to create a closed nutrient cycle. Don't overlook entertainment value—watching the antics of various farm animals brings real joy to the farm.

There are a few farm animals that we don't recommend for urban areas: geese or guinea fowl, as they are too loud; pigs, as they are too smelly; cows and horses, as they are too large for most urban areas. But just because we don't devote a chapter to them, they might work for your farm. If you have a larger piece of property in an urban area, pigs or a cow might just work. And don't overlook the so-called exotic farm animals, such as quail or guinea pigs. Just because they aren't listed here doesn't mean it can't be done!

Keeping animals on the farm is an advanced topic and something that should not be entered into lightly. It's very important to start slowly and add a new type of animal on the farm only after you've mastered the ones you have before. Most people start with either chickens or honeybees because these creatures need minimal setup and are relatively easy to care for. After a year or more with chickens or bees, some people feel the need to start raising meat birds, such as turkeys or ducks, or rabbits. Others become enamored with the idea of having milk goats. Yet others are perfectly happy to stop where they are. It's your call.

Note that we do outline methods for slaughter of various farm animals like poultry and rabbits. The best way to learn how to do this intense

endeavor is to find a mentor who will demonstrate how they process their animals. By reading the sections on slaughter here, you will enter the process with a deeper knowledge and can return to the book after the hands-on demonstration, to refresh your memory. Of course, you don't have to eat your animals if you don't want to. This is your farm, and we admire vegetarians.

Always consult your local city ordinances about which farm animals are legal in your city. Every city has a code about which animals are allowed. If your favorite farm animal isn't permitted, you can get the law changed. In a much publicized battle, chicken lovers in Madison, Wisconsin (Mad City Chickens), successfully fought to change the law that banned backyard hens in their city by pressuring the city council and educating the general public about the benefits of chickens in the city. Already you can see how being an urban farmer will get you engaged in the civic community.

Each chapter in this section is enough to get you started raising each of our top six urban farm animals—it will cover what they'll need in terms of housing, food, and space. Of course there are whole books devoted to each of these animals, and we recommend the best ones in appendix 11 for deeper knowledge and urge you to find a mentor who will help you develop your animal husbandry skills further.

CHAPTER 16

BEES

Honeybees are what we call a gateway urban farm animal. Once you get hooked on bees you might have so much fun and success, you'll want to try experimenting with another farm animal. Bees are fairly easy to take care of, once you get over being scared of being stung. They require only a small input of work but give a big output in the form of honey, wax, pollen, and improved crop pollination. Another positive for beekeeping: You can go on a two-week vacation, and the bees won't notice at all. This is not the case with chickens or rabbits, which require daily feeding, water, and attention.

There's also something satisfyingly complex about bees: their habits, their society, and their biology. By hosting a colony of bees in your garden or on your rooftop, you invite nature in to teach you some lessons. Some of which are hard lessons, like when your bees swarm because you haven't been managing them properly, or when they die mysteriously from some contagion or another. Despite these pitfalls, beekeepers return again and again to the hobby of keeping bees—because the honey tastes so good, the apple yield on your tree doubles from the bees' pollination efforts, or just because your heart is glad to see the bees flying off to work on a sunny day, without complaint.

The methods and equipment outlined below are for a type of

modern beekeeping known as Langstroth. There are other methods—one of the most popular is called Kenyan Top Bar, which doesn't require as much equipment as using Langstroth hives. They don't use frames, and instead allow the bees to build a more natural, free-standing, wax honeycomb. Although this is an appealing idea, it requires a very hands-on beekeeper and can make for a difficult honey harvest. For these reasons, this book will not cover this method; for more information on the Kenyan Top Bar method, see Resources, page 535.

Requirements:

- a complete beehive (detailed information below)
- extra supers (boxes that are added to the main hive during honey flow) with frames
- hive tool
- sturdy beekeeper gloves
- veil
- smoker
- bees with a resident queen
- water source (could be just a pan of water that you fill regularly)
- extractor
- a legal place to keep the bees that gets at least half a day of sun; east-facing is ideal

If you're considering getting bees, the first question to ask is: Are you able to keep bees in your city? Check your local city ordinances to see if it is legal, and whether they have guidelines for where to place the hive—some cities require them to be sited 100 feet away from a neighboring structure, for example. Besides worrying about the law, consider the needs of your neighbors and community. If you plan to place the hive in your backyard, you'll need to find a place where their flight path won't disturb the neighbors, or go into their houses. We place our hives behind a fence or wall, because the bees are forced to fly up and away, toward unimpeded light. If you have a community garden, are all members keen to have bees buzzing around? Anyone allergic? If someone in your community is deathly allergic to bee stings, it doesn't make sense for you to keep bees.

After you've dealt with the possible human complications, you then should move on to whether they will thrive at your location.

A BEEKEEPER'S YEAR

Winter: Order packages. Stock up on equipment. Repaint/repair worn boxes. Feed the bees sugar water, if you think they might be out of honey stores.

Early spring: On the first warm day, open up your hive and do an inspection. Look for eggs and larvae—a sign the queen is alive and well. Feed sugar water to them if there's no forage. Monitor bees and nectar flow.

Late spring: If nectar flow was good and the boxes are filling up, add supers. Harvest in late spring/early summer, depending on your climate.

Early summer: Harvest honey if flows are good. Check on hives periodically to make sure they aren't building queens. Destroy swarm-queen cells, and add supers to prevent swarming.

Summer: Harvest honey. Make sure bees have access to water; it can be just a bucket filled with a stick floating in it. If it's extremely hot, provide shade for the bees. Monitor for ants.

Late summer: Harvest again if flows are good. Treat for mites by opening the top of the hive and sprinkling it with powdered sugar. If the bees had mites, they'll fall off the bees and to the bottom of the hive. A good reason to have screen-bottom boards.

Fall: Final harvest.

Late fall: Remove supers from the hive. Place hive entrance reducers on hives in areas with cold climates. Surround hives or apiary with straw bales as a wind block for cold or windy areas.

Early winter: Feed sugar water if necessary; infuse with spearmint oil to prevent mites.

IDEAL BEEHIVE LAYOUT

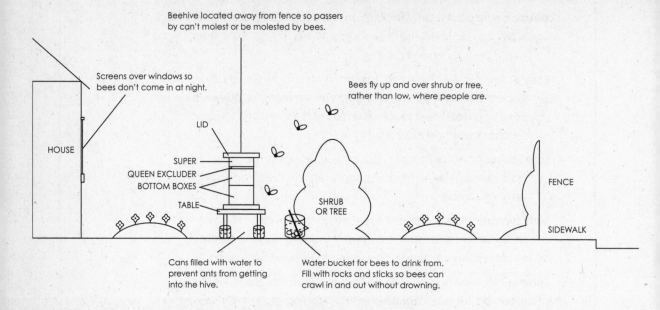

Beehive located away from fence so passers
by can't molest or be molested by bees.

Screens over windows so
bees don't come in at night.

Bees fly up and over shrub or tree,
rather than low, where people are.

HOUSE

LID

SUPER
QUEEN EXCLUDER
BOTTOM BOXES

TABLE

SHRUB
OR TREE

FENCE

SIDEWALK

Cans filled with water to
prevent ants from getting
into the hive.

Water bucket for bees to drink from.
Fill with rocks and sticks so bees can
crawl in and out without drowning.

An ideal bee setup. Place the entrance of the hive toward a shrub or fence to cause the bees to fly up and out of the way of passersby. Placing the hive on a table keeps out pests; cans filled with water create a moat around the table's legs to keep out ants.

SETTING UP

Here is an idealized site plan for your bees.

Windy, cold, shady backyards will never support a healthy hive. The fact is, bees require at least six hours of sun each day year round. They prefer to have their hive facing east, toward the rising sun. In places with a hot summer, they'll need a bit of shade (either a structure or by simply placing a potted tree next to the hive). In places with a cold winter, it makes sense to site the hive near your house, where some heat from your home will leak out to insulate the hive.

If you live in an apartment, some balconies or rooftops will support a box of bees or two—but the location must be sunny and not exposed to too much wind (we once housed a hive on a shady, windy deck—with bad results). It should also be a sturdy location, capable of supporting a honey-filled hive, which can weigh over two hundred pounds. In these cramped locations, consider too that a hive can grow to be 4 or 5 feet tall and needs at least a few feet on either side of it for the beekeeper to comfortably perform hive inspections and do basic maintenance.

BEE BASICS

A word about the caste system of the hive. There are three kinds of bees: males, called drones; infertile females, called workers; and one fertile female, the queen. Drones make up a small population of the hive, and they mostly hang around doing nothing. Their whole purpose for living is to mate with a queen. But that happens only once in a while, during the nuptial flight when a virgin queen is born. The workers have designated jobs in the hive, which change as they get older. Their first job is feeding and caring for the developing brood (their siblings); then they become honey makers and cappers; finally they graduate to becoming foragers who collect nectar and pollen for the colony. The queen's sole duty is to lay eggs which keep the colony alive.

From left to right: drone, queen, worker bees

BUYING BEES AND GEAR

If all is well, you're ready to start collecting your beekeeping supplies.

Unless you know someone who wants to give you bee equipment, or you can find quality used materials, you're going to put out a serious amount of cash. The benefits of buying new are that you avoid getting used equipment that might be infected with bee diseases, and the gear will be in perfect working condition.

You will need:

One bottom box with frames. This is where the bee colony clusters at night. The queen bee lays her eggs here, and you never harvest honey from this box. In common beekeeping parlance, the bottom box is also called the "brood chamber." The number of frames in the brood chamber depend on the size of the box and the beekeeper's preference, but it is usually filled with nine or ten frames. Some beekeepers use two medium supers instead of one deep super to make the brood chamber.

Extra frames. Frames are rectangular wooden objects supported with metal wire and lined with a beeswax (or plastic) foundation. They hang from the top of the super boxes suspended by a slight lip in the frame. Bees store their honey by "pulling wax" on the foundation to form a repository for the honey. After a chamber is made, it is filled with honey, and then capped. Beekeepers, when harvesting honey, remove frames, uncap them with a knife, and then use a centrifugal extractor to spin it out. After the harvest the frames can be returned to the bees, empty of honey but with the wax cells still intact.

One bottom board. The bottom box rests on this solid piece of wood; otherwise, the bees would be exposed to the ground. It is the same size as the bottom box but with a few extra inches in the front that provides a little lip for the bees to land upon. These also come in a screened version, which allows for better ventilation.

Telescoping cover. Basically, the lid to the hive. It fits completely over the top and extends down a few inches, so the wind can't blow it off. The cover keeps the hive dry and draft-free. The outside is usually lined with sheet metal to prevent rot from rain and weather.

Inner cover. A piece of wood that offers ventilation and insulation and reduces moisture. It goes right over the box, and the telescopic cover then fits over it.

Supers with frames. Supers are simply the honey frame-filled boxes that are stacked upon the bottom box as your colony grows. The bees mostly store honey and pollen in the frames. Supers are what you take away to extract honey, because it is where your bees store their excess.

One hive tool. It's like a small crowbar and is used to move frames around the box; they are often stuck together with wax and propolis.

One smoker. It's the classic beekeeping tool. It has bellows connected to a chamber, where you burn burlap or twigs or smoker fuel in order to generate smoke. The tool is used when making hive inspections or to harvest honey without getting stung. The smoke calms the bees. Smokers come with or without a cage that would prevent you from

The hive tool is useful for prying apart sections of the hive that the bees have sealed with wax or propolis.

A Langstroth hive setup from the bottom up: bottom board, brood chamber, queen excluder, super with frames, inner cover, telescoping cover.

A smoker is used to calm the bees when working the hive.

A veil and helmet are essential safety equipment.

367

burning your hands on the hot chamber. If you're klutzy or absent-minded, get a caged smoker.

Veil and bee suit. Although you look like a hazmat worker wearing this ensemble, it's nice to not worry about getting stung.

A basic bee book. Dadant's classic is *First Lessons in Beekeeping*. But *Beekeeping for Dummies* is also good.

All of this will cost you about five hundred dollars. But remember that, if treated well, your equipment could last for twenty years or more, and so it is a good investment.

BUYING BEES

Once you've acquired all the gear, you're ready to buy a package of bees. Many city beekeepers do this online, and the bees arrive through the U.S. Post Office. They will arrive in a "bee package"—a small box, usually made out of wire. There will be around ten thousand bees in the mesh box, with one queen, who will be suspended within her own private chamber. Your postmaster will think you are crazy.

Unless you have a beekeeping store nearby that will order a package for you, buying bees through the mail via the Internet is your best bet. You should put in your order in fall or early winter—any later and the bee breeders might be sold out. A three-pound package of bees in 2010 cost about one hundred dollars.

A few of our favorite bee breeders are listed in the appendix, and there are listings of many breeders in magazines like *American Bee Journal* or *Bee Culture* magazine. You should choose the one closest to you so the bees have the shortest possible journey and will thrive in your climate.

If you do have a breeder near you, ask if you can buy a nuc from them. A nuc—or nucleous—is basically a minicolony of bees, a queen, and a few frames of brood and honey. Nucs allow you to get a jump-start on the packages, because the bees are already developing brood and honey stores.

Another option is to catch a swarm from a neighbor's hive; see page 375.

BREEDS

In America, there are generally two varieties of bees to choose from: Italians (*Apis mellifera ligustica*) or Carniolians/Slovenian (*Apis mellifera carnica*). These two are not different species but are considered two of twenty-five different races of bees worldwide. Beekeepers in the United States debate about which are better producers, and there are hybrids, such as the Buckfast and Midnite, that are marketed as mite-resistant, less prone to swarming, or with any number of other desirable traits. Since the bees' performance mostly depends on the climate, we recommend researching how each breed does in your specific area. The best way to figure out which bee is best for you is to ask a local beekeeper which type of bee they keep and why, and then make your own decision.

INSTALLING A PACKAGE

Bee packages are mailed in April and May, depending on where you live. Once you pick up yours, you'll want to install them into their new digs on a warm afternoon. Setting them up late in the day ensures that they stay the night in their new home, which increases the likelihood that they'll stay. If bees are placed into a housing situation they don't like, they can and will leave. Here's what to do:

1. Handle the package properly. If you pick up your package in the morning, or you can't immediately install them into their hive, it's a good idea to mist one side of their wire mesh wall with a 1:1 sugar water solution. This will give them something to do (eat sugar), which will calm them down. Be sure to keep uninstalled packages in a shady location, so they don't get overheated.

2. Decide on placement. To get ready for the installation, first decide where you are going to permanently locate them. Remember: It's very hard to move a box once it's settled—it's confusing to the bees, and heavy and dangerous for you. Place the empty beehive—bottom board, brood chamber, inner cover, and cover—in this place. We like to set the bee boxes on top of an old desk or table so they're up off the ground. Other people use cinder blocks or old pallets. If your area has an ant problem, place moats of water around

INSTALLING A PACKAGE OF BEES

Step 1:
*Remove the lid and
take out four frames.*

Step 2:
*Pry the lid off
the bee package.*

Step 3.
*Remove sugar feeder
and slide out queen box.*

Step 4:
*After suspending
the queen box in the
new hive, shake the
bees out of the package
into the super.*

Step 5:
*Place inner cover on
hive. Leave the bee
package box at the
entrance of the hive.*

Step 6.
*Place telescoping
lid on hive.*

Step 7:
*If desired,
feed the bees sugar
water from a feeder.*

the legs of the table or desk—set the legs into old metal coffee cans or industrial-size canned tomato containers, then fill them with water. Ants can't swim, and so they won't be able to access the hive as long as you keep the containers filled.

3. Prep the box. Remove the telescoping cover and inner cover from the empty beehive. Pull out three of the frames from the brood chamber and set them aside. It will be easier to unload the bees into an empty space.

4. Install the queen. Place the bee package on the ground near the empty hive. Pry the lid off your package of bees with a knife or your hive tool. Some manufacturers use an inverted container of sugar water as the lid. While it does seem totally insane to pry off a lid to a box that contains ten thousand bees, don't worry: The bees will be fairly gentle because the pheremones of the queen calm them, and because they aren't defending their home (they don't have one—yet). Once the lid is gone, remove the small inner chamber where the queen is housed and put it into the beehive by suspending it on the edge of one of the frames. Be extemely gentle here, as the queen is the mother of the hive; without her, it's a dead hive.

5. Pour bees into the hive. Once the queen is balanced inside their new home (she will be released from her chamber later), the other bees will be eager to join her. To do that, remove the rest of the lid to the bee package and upend it into the empty space where the frames were. Don't worry: This may seem like a surefire way to get stung, but the bees will be glad to have such a nice new home. They will also smell the queen inside and will want to be near her. When the package is inverted, give it a good shake, or even a thump like a ketchup bottle, to get all the bees out. After most of them are out, put the inner cover and the telescoping cover back over the beehive. Any bees that are still clinging to the package will eventually figure it out; place the box next to the opening at the front of the hive.

Following Up

After a few days, check that the bees are getting used to their new home. Put on your veil and gloves (you will be more cautious because the bees have something to defend now), fire up your smoker, and gently open up the top. Blow a few puffs of smoke in; lift out the queen's chamber. Usually, the bees will have eaten away a stopper made out of candy at the bottom of the chamber in order to release the queen. If she's still in the chamber, you'll need to remove the stopper that is sequestering her in her chamber by using tweezers or pushing the stopper aside. Be very gentle! New queens can cost up to fifty dollars. She should stride right down into the brood box and start laying eggs as soon as possible. After you've made sure the queen is released, replace the frames you took out when you first installed the hive back into the box. Be gentle, and move slowly when you do this. The bees should be fairly docile and will move out of the way.

At this point, some beekeepers will feed the bees sugar water. There are side feeders and mason jar feeders; some people drill a hole in the top of the hive and invert over it a sugar water–filled mason jar with holes drilled into the lid. Feeding encourages them to stay at home and work on making honeycomb, so that the queen can start depositing her eggs in there more quickly. Other people think that feeding them makes them weak and less prone to forage great distances. Since the ideal is to get the queen laying and producing brood for the first nectar flow in early summer, depending on where you live, we think it can't hurt to give them a little help. Sometimes bees won't eat the sugar water and will prefer to forage for themselves. Every colony is different. When trees and shrubs and flowers bloom, the bees will want to be in the middle of the action. The more bees a colony has, the better.

MAINTAINING YOUR HIVE

Now that you've installed your hive, you can sit back and let them do the work. After the first two to four weeks, you should open it up again to make sure the queen is laying and honey is being stored. To do this, suit up and light your smoker. Since they are now defending their property, this is when you might get stung. Using your hive tool, pry up a few frames and inspect them for the queen's circular laying pattern. You're looking for white dots curled up at the bottom of each comb chamber. These are the eggs. You may have to tilt the frame to see them. Raised dark yellow–capped areas indicate pupating larvae. You might also see bright dots in the comb—yellow, white, orange. This is their pollen, or bee bread, which they store to feed the young. You might even see some capped, or sealed, honey in the upper corners of the brood frames.

During your inspection, if it's starting to look crowded in the bottom box—more than half filled with larvae, it's a good idea to add a super. These come in three sizes—deep (about 10 inches tall), medium (about 7 inches), or shallow (about 6 inches)—and go directly over the brood box. It's really your choice which size to add. Some people, us included, put another deep super over the brood box and never collect honey from it. The way we see it, the brood box and the deep super will allow the bees to have enough room to raise new bees and store enough honey to make it through the winter.

As the season goes on, continue checking and adding supers as they fill the ones you provide. Besides the two bottom ones, any others we add are for our personal honey use. To dissuade a queen from venturing into your honey stores, where she will contaminate it with her brood, you can place a queen excluder—a metal fence, basically—in between your honey super and the brood box. We've found that we don't need one, because two deep supers are enough space for a normal queen to lay, and so it's a fairly safe bet that she won't venture upstairs. While the queen excluder is a cool idea, it can sometimes get gummed up with wax and honey, and then it prevents normal passage for the worker bees. But, again, this will depend on your bees and your climate.

As the summer progresses, you should monitor your hive. Open it up whenever you feel it might be necessary. Some people inspect their hives every month, some once a year. It's up to you and your temperament. But remember that you should only open up the hive on a warm, sunny day, when the bees are busy foraging and working and so won't be annoyed by your intrusion. Never open up a hive on a rainy day, at night, or during a cold spell—the bees will die!

SWARMING

When you do your hive inspections, another thing to monitor for are queen cells. There are two kinds: swarm and "supercedure" queen cells. Each looks like a peanut. Supercedure queen cells are a normal way for the bees to make sure they'll have a queen if theirs becomes ineffective or suddenly dies. The supersedure queen cells are found on the surface of the frames, usually cropping up on the brood chamber frames. It is normal to have these, and they should not be removed if you encounter them during an inspection.

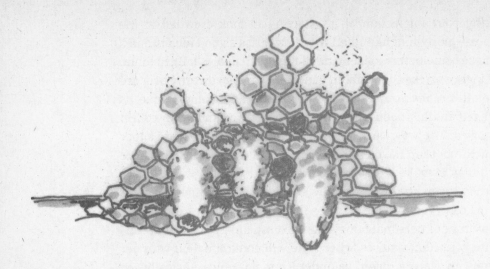

Queen cells in the hive signal intent to swarm

In contrast, the swarm cells are suspended off the bottom of the frames.

These are a sign to the beekeeper that the hive has become too full of honey or brood, and the workers have decided the colony should split off. This is why, as each super gets filled with wax and honey, you should add another super. If you don't, the bees will run out of space which will cue them to build a queen cell, put a worker egg inside, and start feeding it royal jelly to make a new queen. These steps lead to swarming: when the new queen is born, half the hive will leave with the old queen to form a new colony. You can prevent swarming by removing the queen swarm cells—just smear them out with the tip of your hive tool—and then add a new empty super.

If you don't remove the swarm queen cells, eventually you and your neighbors are going to be in for a spectacular, and often frightening, sight. First you'll hear a low murmuring sound, and many bees will come out of the hive and circle. Some decide not to leave and will stay with the original colony. Then, suddenly, the circling bees will reach the tipping point, cluster together, and fly away with their queen. It does look just like a cartoon. What's going on is that the old queen leaves the hive after the virgin queen is born. She takes 50 percent to 60 percent of the workers with her. Secondary swarms can occur too, when the colony is very strong. Another virgin queen is born and leaves with even more of the workers. In this way, swarming reduces the population of a too crowded hive.

The swarms usually land a few yards away from the other hive—in a tree, on a telephone wire, you name it. In the city, swarming bees are quite a sight. Some beekeepers think swarms are natural and let their bees go every spring and summer. But in an urban area they are not a good thing, because they freak out the neighbors.

If your bees do swarm, or if you are catching a swarm, bear in mind that the bees will be docile. They have no territory to defend. It's a good idea to have an extra bottom box, board, and covers at your house in case of a swarm. It's best to catch it directly into the extra empty boxes that you keep around for just such emergencies. In a pinch, however, you can use a cardboard box and relocate the bees later to a proper home. But relocation should be done within a few days or the bees will protest mightily.

To catch it, simply place the bottom box near the swarm cluster and snip whatever they've landed on and place it into the box. In the case of those that settle on immovable objects or large branches, you may have to literally scoop them into the box: Place the box near the swarm cluster and coax the bees in using a stick or bee brush. Swarming bees usually know a good thing when they see it and will settle right in to the offered home. If the swarm is totally unreachable, lures that mimic the smell of queen pheromone can be bought. Douse a tissue with the lure and place it in the back of the empty hive—and place the empty box in the general area of the swarm. Scout bees should be attracted to the box, and hopefully they will lead the rest into this new home.

When you bring your swarm home, locate the new hive box at least 5 feet away from the other colony or the bees may become confused and drift into the wrong colony.

BOXES

In addition to different kinds of bees, there are many different bee houses besides the Langstroth we describe. There are giant willow-woven orbs that uber-natural biodynamic beekeepers use. There are Kenyan Top Bar beehives that are cheap and easy to construct. There are Slovenian boxes that open on the side instead of the top. Our advice is to start with "Lang" boxes and then dabble with other methods if you're so inclined.

EXTRACTING

Extracting is the harvest of the honey crop. It is a joyous, satisfying, sticky time. Most temperate-climate beekeepers extract twice a year—after the main nectar flow in late spring/early summer and then again in the late summer/early fall. Extraction methods vary. Some people will merely sneak out a frame of honey from one of the bees' supers and replace it with an unfilled one. They cut off the wax capping of the honeycomb-filled frame and let the honey gravity-feed into a container. This obviously is a sticky, ant-attracting business, but it works for some.

A more efficient method is to use an extractor.

These are cyclinder-shaped machines that have little baskets into which you place the uncapped frames, and then use centrifugal force—spinning—to dislodge the honey. Extractors come in a range of sizes: from small hand-cranked models that hold three frames to giant electric ones that hold sixty-four. For a hobby beekeeper it makes sense to use a smaller extractor, but even those can be pricey—four hundred dollars—and so consider buying one with fellow beekeepers and sharing it as it's needed. Since you will only use the machine twice

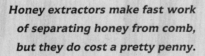

Honey extractors make fast work of separating honey from comb, but they do cost a pretty penny.

a year or so, it makes sense to own the machine collectively. Another options is to rent one from a beekeeping supply store. Not all extractors are created equal: Make sure it is easy to use and simple to clean before buying. Our favorite one so far is the Mann Lake 3-frame hand-cranked model.

Before you can extract you need to get the honey-filled frames out of the supers. The bees won't be happy with your stealing their honey. Because of this many people simply suit up, fire up their smoker, and go in, gently brushing bees off the frames until they can sneak it away for extracting. Other people use a chemical product that is placed inside the hive and causes the bees to flee so the supers can be carried away and extracted. Another, nontoxic, method is a bee escape. This is a wooden structure that goes in between the bottom box and the honey-filled supers.

It is basically a maze that the bees walk through in order to descend into the brood box to cluster around the queen and keep her warm at night. In the morning the bees can't figure out how to get back into the honey supers. If the beekeeper places the bee escape twelve to twenty-four hours before they plan to extract, the honey supers will be completely bee-free and can be carried away. Wait longer than twenty-four hours and at least some of the bees will have figured out how to get back into the honey supers.

Whatever method you choose, once you have some frames to extract, you'll need to set up an uncapping station, which can just be a table covered with plastic wrap (it is a sticky, messy process).

You'll need:

- an uncapping knife
- a pan or dish filled with hot water, to clean the knife off in between frames
- dish towels
- a bowl in which to place the waxy cappings (one which you don't mind getting coated with unremovable wax)
- the extractor nearby
- jars with lids for the honey
- wire-mesh filters, if you want filtered honey (without chunks of wax)

Go through the frames you intend to extract. Bees place a cap of wax over each honey cell only after it has reached just the right sugar level. You might see some empty or half-filled honey combs—avoid spinning these frames in the extractor, because the uncapped honey tends to taste a little off. Just put those frames back into the hive.

After you have what looks to be good, honey-filled frames, lay one on the table, and gently shave off that fine wax cap with the honey knife.

We use a serrated decapping knife dipped in hot water, then wiped dry. (Some people use an electric decapping knife that plugs in and stays hot.) You tilt the frame up and gently saw along the edge. A thin coat of wax will come up in a long white sliver. Wipe the cappings into the bowl, dip your knife in hot water, wipe it off, then uncap some more. When the frame is free of capped wax, you're ready to flip it over and do the same thing to the next side. Don't worry that all the honey will come spilling out when you flip the frame over—some will drip out, but it's a negligible amount.

Once both sides of the frame are free of cappings, place the frame into the extractor. After it's filled with frames (some hold three, others four), start spinning! You will see the honey splattering the sides of the extractor and dripping down the stainless-steel sides and pooling up in the bottom of the extractor.

After spinning, remove the frame to check that the honey has become dislodged—the honeycomb will look empty. Then flip it around so the other honey-filled side is exposed to the outside of the extractor. Spin again until the frame is completely honey-free.

If the honey looks solid or crystallized, it will be tough to extract.

The first step to extracting honey is uncapping the comb

Because of cold weather or a certain nectar crop, the honey can become hardened. Return it to the hive, where it will be eaten by hungry bees.

Most extractors have a tap (called a honey gate) at the bottom that you open to release the honey. When you've spun all your frames, open the tap and let the honey drizzle out into mason jars or other containers.

Some people filter their honey at this point to avoid having wax or other bits in their honey by placing a mesh strainer under the stream of honey. You don't have to filter, though we've found that the bits tend to float up to the top of the jar postharvest and can be skimmed off the top of the honey easily.

To clean the extractor, close the tap and pour water along the sides. This will wash any honey or wax on the sides. Then slosh the water around and release it from the tap. Repeat. We've noticed that this honey water tastes very good, so feel free to save it for drinking, hot or cold. It

A fully loaded extractor, ready to spin. Note that the frames must be flipped so that the honey is removed from both sides of the frame.

also makes a very good version of simple sugar to use in making cocktails or lemonade.

Finally, place the extracted frames back in the supers and put them back on the beehive. You might notice that the girls are slightly annoyed with you—the smell of uncapped honey drives them crazy!— so be cautious: Suit up and use smoke. Once the supers are replaced the bees will get to work salvaging what you plundered, licking up all the honey and placing it into a new comb, and chewing up ragged pieces of wax to build more. They are so tolerant and hardworking!

GIFTS FROM THE HIVE

Honey

Honey is the reason most beekeepers get started. It has healing, antibacterial properties. Some people say that eating local honey can help relieve your allergies. But most of all, your honey will be the tastiest, freshest honey you've ever had. It also makes a great gift for friends and neighbors who are bee curious. Depending on the health of the hive and the weather and nectar flows, expect one to ten gallons of honey per hive per year.

Pollen

Bees collect pollen on their back legs in what are called pollen baskets. Basically, the hairs on their legs are positioned to hold huge loads of pollen; they push the pollen down as they move from flower to flower. You can see bees loaded down with pollen, which can be bright yellow, orange, or even white. They place the pollen inside the comb with a dollop of bee saliva, which turns the pollen into something we call "bee bread." This essentially is fermented pollen. The bees feed this high-protein food to their young and store it over the winter where they can nibble on it as a protein source.

Humans like pollen too. It is said to help alleviate allergies and is considered a health food. In order to gather it from your hive, you can buy a special collector: a reduced entranceway that the bees have to walk through. As they squeeze by some of the pollen falls off into a little chamber that you can reach in and empty occasionally. Store the pollen in a jar and keep it in a dark dry place, enjoy it in smoothies or in baked goods. Depending again on weather and hive health, bee-keepers can collect several ounces of pollen per hive.

Wax

The smell of beeswax is intoxicating. It can be used to make candles and is often an ingredient in soaps and hand salves. To recover it after uncapping the honey, you should set up a double boiler. Boil water in one pot, then place the other (one you don't care will become waxy forever) above the boiling water. Add your cappings and honey. After all of the wax melts, take it off the heat. Let the pot return to room tem-perature. As it does the wax will float to the top, the honey will stay on the bottom. Simply peel off the layer of wax and put it in a jar. It's ready for whatever craft project you have in mind! And your house will be filled with the ambrosial smell of hot honey and wax. Depend-ing on weather, hive health, and the amount of cappings and wax you save, expect a couple quarts of wax per year.

OVERWINTERING BEES

After the fall harvest, it is time to put the bees to sleep for the year. If you live in a cold climate, you'll need to remove any supers, so that the hive is back to its starting point as a brood chamber and super. Since winter is generally a season when the bees are not harvesting nectar, they spend most of their time clustered together for warmth. Extra supers make the hive bigger and more difficult to keep warm; that is why we remove them in the winter. You may also block most of the hive's opening with a hive reducer, which keeps out water, drafts, and pests.

In areas where there is snow, consider building a shelter of straw bales around the hive. In areas with lots of rainfall, consider building a shelter over it to prevent mold and disease.

In spring the whole cycle starts over again. The bees emerge, forage nectar, gather pollen, the queen lays her brood, the beekeeper goes out to do a spring inspection to make sure all is well and to feel glad.

There are pests to be careful of with bees as with any animal husbandry. This is a partial list; it is not in the scope of this book to provide full information on all of the potential bee diseases and pests.

Ants

If allowed to run rampant, ants can and will kill a healthy colony of bees. They are attracted to the warmth and food in the hive and will try to set up a colony inside the beehive. One protective method is to place the hive on a table or desk with each leg immersed in water-filled buckets. Keep all foliage and other objects away from the hive; otherwise the ants will create a path into the hive. Another method is to smear the bottom of the hive with a sticky product called Tanglefoot, but this must be reapplied weekly to be effective. You can also place sticks of boric acid bait near the bees. The ants eat the stuff, take it back to their queen, and die.

Mice

In colder areas especially, mice like to take up residence in a dormant winter beehive. And who wouldn't? It's warm, draft-free, and smells like heaven. But mice are destructive chewers and will gnaw up wax, eat honey, and generally make a huge mess. Best to keep them out by raising the hive off the ground, and, in the winter, place a hive-reducing board at the hive's entrance.

Small Hive Beetles

More common in the South, the small hive beetle can devastate hives. Their young hatch in hives and then tunnel through frames, destroying honey and wax. They can be controlled by bait traps placed in the hive.

Yellow Jackets/Wasps

During the summer yellow jackets and wasps can be tough on a honeybee colony. They will enter the hive to steal honey and young bees to eat. If your beehive is weak, the entire colony can be driven out and taken over by yellow jackets. The best practice is to keep your bees healthy and use an entrance reducer so your bees have a smaller area to defend.

Parasites: Varroa and Tracheal Mites, and *Nosema*

Lately, a rash of parasites have been infecting the honeybee populations. Varroa are tiny red critters that attach to the bees and suck their essence, kind of like fleas. Tracheal mites enter the bees' mouths and suck nutrients. *Nosema* suck on the bees' guts. There are a variety of methods to get rid of tracheal mites—one is to feed the bees a patty of shortening and sugar, which might cause the tracheal mites to pass through the bees' intestines. For Varroa mites, a late-summer treatment is to pour an entire box of powdered sugar over an opened-up hive, with a pan placed in its bottom to catch the falling sugar and mites, which can't stick to a bee covered in powdered sugar. It is incredibly satisfying, as you can imagine. Another tip for getting rid of Varroa is called "drone baiting": A full frame of drone comb—commercially made larger-celled comb—is placed in the colony. Just before the drones hatch, the whole frame is removed from the hive. The idea is that Varroa mites prefer to infect drone larvae cells (which are larger), and so by baiting and then removing them, you'll be removing the mites too. But don't forget about removing the drone frame, or you'll be overrun with good for nothing drones! *Nosema* can be controlled by allowing better ventilation in the hive.

Foulbrood, Colony Collapse Disorder, Chalk Brood, and More . . .

We don't want to start getting you worried. There are hundreds of bee afflictions, too many to list. Consult a bee book for advice on how to manage the specific pests in your region. You'll have to decide for yourself if you want to treat the bees with miticides and antibiotics. But for the most part, healthy bees are the result of proper placement, the attention of the beekeeper, and the genetics of your bees. So buy bees with mite- and disease-resistant genes, place them in a protected place with some sun (away from strong winds, with lots of nearby forage), and do your job as a beekeeper and check on the girls every once in while. You'll be greatly rewarded.

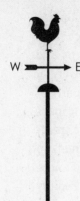

CHAPTER 17

POULTRY: CHICKENS, DUCKS, AND TURKEYS

CHICKENS

So much ink has been spilled in the name of chicken husbandry. There are entire books on the sole subject of chicken diseases, glossy photo books of chicken breeds, and how-to guides for keeping the egg-laying creatures in the backyard. There are Web sites and blogs devoted to hen worship. And we don't think there's anything wrong with that. In fact, we love chickens, too.

Almost everyone buys chicken eggs, and almost anyone with a small backyard can keep a few hens. They are relatively quiet (especially when compared to, say, a yippy dog), they create a wonderful nitrogen source for your garden (their composted poo), and their feed can be supplemented with supermarket scraps and weeds. Most hens lay an egg every day or two, so their fresh eggs will be a source of delight and culinary ecstasy. You can't beat the taste of a fresh egg laid by a happy chicken that has been fed plenty of green forage and allowed to scratch up bugs. Think: orange, pert yolks surrounded by thick, healthy egg whites. Your poached eggs will be a revelation, your meringues a sensation. (We're gearing this section toward egg production, see page 413 for our thoughts on meat birds.)

The eggs are good. But chickens also teach city people about our connection to the natural world. You'll need to pay attention if your chickens are going to thrive. Do they have enough water? Where are they laying their eggs? Is their roost comfortable? Are they fighting? Are you supplying oyster shells to make strong eggshells? Before long you'll find yourself going into the backyard just to observe your hens. Their clucking noises are mesmerizing; their antics can be hilarious. Suddenly you, a city person, will be taking part in an agricultural act every day when you go out to gather eggs and check on your hens. Their happiness—and yours—depends on your ability to gauge their needs and act on your observations. We hope you're up for this mission and that you'll at least consider having a few of them on your urban farm.

Requirements:

- your willingness to accept the responsibility of hen ownership. Not unlike a dog or cat, your backyard birds will need attention and love and should not be considered a casual dalliance. Chickens live for over eight years.
- a small, partially shaded henhouse and yard (collectively, the coop) with at least 6 feet of exercise space per bird
- a place protected from predators where they can roost at night
- nesting boxes where they lay their eggs (two for each five chickens)
- daily feed (usually corn, soybeans, grain, and minerals in a pellet or crumbled form)
- scratch (coarsely ground grains that they eat as a snack)
- critter-tight, metal feed-storage bins for their food
- ground-up oyster shells
- grit (pebbles, sand, and other goodies for the chicken's gizzard)
- fresh water
- access to grass, weeds, and vegetable scraps
- straw, dry leaves, or solid wood sawdust for nesting boxes
- companionship of at least two others of their kind (chickens are not solitary creatures)
- a place to compost the soiled coop bedding
- a plan for what you will do once your hens are beyond laying age (three to five years) or if you get any roosters by mistake (roosters are not necessary for egg laying, a common misconception)
- no dogs with killer instincts on the premises (most cats will not attack full-size chickens)
- a reliable "chicken sitter"—a friend, relative, or roommate who will look in on your girls while you're on vacation

In some cities it is illegal to keep chickens; in others, it's legal to have hens but not roosters. Before you get chickens you should consult your city's ordinances to see whether it's legal or not. The most important aspect of these is usually the distance your coop must be from your own and your neighbors' houses. Because animal control departments mostly respond to neighbor complaints, you might consider approaching anyone living nearby who will see or hear your hens to gauge how they feel about having chickens next door. Promise that they'll be quiet (hens are not roosters!), swear that you'll keep everything clean, and pledge to share the eggs.

SETTING UP

Before you get chickens you should plan where their coop will go. It is comprised of a completely enclosed house and a more open run. The house should have a watertight roof and a door that can be closed to keep out predators. The hens will sleep inside the house at night, perched on a roost. It is also where they will lay their eggs during the day, in a nesting box. It needn't be an elaborate, enormous house—it just needs to have enough space for a roost and the nesting boxes, and it must be fully secured so that predators can't attack your hens. It's best practice to keep the feeder and waterer in the secured house as well, to prevent rodent infestations.

The chicken run is attached to the house and is an open but fenced enclosure where the hens are allowed to get exercise during the day. Runs are often enclosed with hardware cloth or bird netting to prevent your birds from coming in contact with rodents and wild birds. Ideally, a run should have a good sunny spot for the hens to take their afternoon dust baths, where they clean their feathers using dust. The function of the run is to get the birds exercise without enabling them to destroy your garden.

Most coops are made out of lumber, but there are infinite designs for coops, including prefabs and kits. There are endless cute pictures of proud coop designers next to their work on the Internet—great for ideas; see Resources in appendix 11 under chicken housing. Some of these kits can cost hundreds of dollars, though. Since we live in the city, we recommend making your coop out of found scraps of lumber, doors, and other urban detritus that accumulate on street corners and in vacant lots. We've seen chicken houses made out of pallets lined with old cookie pans and made of bamboo and chicken wire, even a small

IDEAL CHICKEN COOP
(Five to Fifteen Hens)

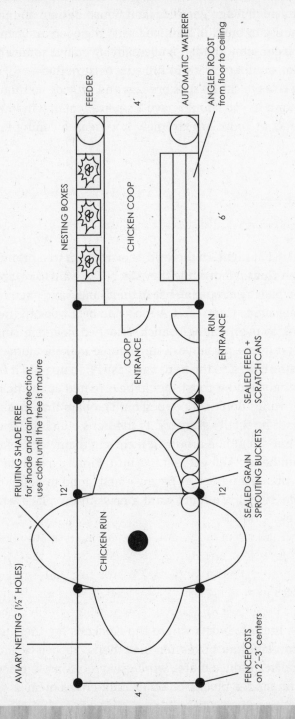

FEEDER

AUTOMATIC WATERER

ANGLED ROOST
from floor to ceiling

4'

NESTING BOXES

CHICKEN COOP

6'

COOP ENTRANCE

RUN ENTRANCE

SEALED FEED + SCRATCH CANS

FRUITING SHADE TREE
for shade and rain protection;
use cloth until the tree is mature

SEALED GRAIN SPROUTING BUCKETS

12'

12'

CHICKEN RUN

CHICKEN RUN (4' x 12')

This is the area where hens get exercise. It is made
with aviary netting, hardware cloth and wood
with sawdust or straw bedding.

CHICKEN COOP (4' x 6')

This is the nesting area where hens sleep at night
and lay their eggs. It is made with wood, hardware
cloth floor to keep out rodents and sawdust or
straw bedding.

AVIARY NETTING (½" HOLES)

FENCEPOSTS
on 2'–3' centers

4'

An ideal chicken coop for five to fifteen hens includes a
run where chickens can get plenty of exercise

henhouse made of recycled milk crates. Start looking around for your city's most abundant street resource and imagine what you can do.

Here is an illustration of an idealized chicken coop layout that gives the general principles of good chicken house design, no matter what material you use to build it. It's nice to site the coop underneath a tree, where it will get some shade—a necessity if you live somewhere with extremely hot summers. If it's in full sun, be sure there are windows on opposite sides to catch cross breezes and provide ventilation. A good rule of thumb is, you should have 1 square foot of window for every 10 square feet of coop. In extremely hot areas, a ceiling fan is necessary.

THE ROOST

Chickens like to roost at night. In the wild, they roost in tree branches, clutching with their clawed feet while they sleep. You want to re-create a version of this by placing a roosting bar in their house and encouraging them at a young age to sleep on it. A roost can be a wooden (never metal) pole, dowel, or even a long branch, attached near the ceiling and secured so that it doesn't move. Some people make a ladderlike roost, with ascending rungs, so the hens can easily hop up onto it from the ground. Note that a ladder roost should be oriented at an angle so hens on the lower rungs don't get pooped on. The bars that make up the roost should not be totally slippery, as the hens clutch their toes on the bars, and they should be around 2 inches in diameter (smaller if you have bantam hens). Allow 2 feet per bird. Finally, it should not be directly above the nesting boxes, because hens tend to poop even when they are asleep, and you don't want a mess in the place where you collect the eggs.

NESTING BOXES

Nesting boxes are literally boxes where the chickens lay their eggs. People often build elaborate boxes for their hens, but we've found that milk and wooden produce crates work great too. Simply place it, slightly elevated on cinder blocks or other milk crates, near a dark wall. Line the boxes with 3 inches of straw, leaves, or shredded paper. To give the hens the right idea, place a ceramic egg, a golf ball, or a marked supermarket egg in the center of the nest. When your hens are ready to lay, they should seek out this nesting box. Since chickens lay around the same time in the morning, you should provide two for five hens—they can share and will work out a good schedule. The biggest problem most people face with nesting boxes is convincing the hens to lay there and not in some dim corner of the hen-house. Be sure to place it out of the sun, in a dark place without drafts, and with some privacy. Be patient too; the hens will eventually learn where to lay.

GATHERING EGGS

The eggs should be in a place that makes them easy to collect. Many people create a door near the nesting boxes that can be opened from the outside of the house. Leave a fake egg in the nesting box to fool the chicken into thinking you aren't stealing their eggs. To ensure eggs aren't broken, you should probably collect them at least every other day. They can be stored out of the fridge, because they still have their protective coating. if they look dirty, just wash them—scrub in warm soapy water—right before eating. If you're distributing eggs to others you will want to integrate washing eggs into your weekly routine—washed eggs should be refrigerated. Some people keep a record of how many eggs each hen lays, or how many were gathered every day. This is a safeguard to monitor a change in behavior that might signal illness.

KEEPING THE HOUSE SAFE

The house should have a door or secure sleeping area that can be closed at night to protect them from predators, such as opossums, rats, raccoons, owls, skunks, and dogs. These predators flourish even in the city and would love the chance to dine on your girls. Raccoons are very intelligent and have been known to open doors, so think about adding a locking mechanism. We can't emphasize enough that you must keep your hens safe. There is nothing worse—or more common—than losing hens to predators because the house is not properly secured. Think Fort Knox. And always remember to shut their door before night comes. Hens will naturally head to their house as the sun starts going down. If you don't shut the door, nocturnal raccoons, rats, and possums will slink in. Automatic doors can be set up with timers for opening and closing, a convenient—if expensive—option.

FULLY ENCLOSED RUN

Some people will build a completely enclosed chicken house and run so the door to the house doesn't have to be closed every night. This is ideal for people who go out at night or want to go out of town for the weekend without hiring a chicken sitter.

To build a fully enclosed and secure run takes a bit of effort and a lot of hardware cloth, a durable fencing material with small openings that don't allow mice or rats to gain access. Its small openings—unlike chicken wire—also prevent raccoons from reaching into the run and grabbing one of your hens and pulling her head off (yes, we saw the aftermath of this once).

To protect from these pests, when you are building your coop either line the entire floor and interior of the house and run with hardware cloth or dig a 1-foot-deep and -wide trench surrounding the house and run and sink hardware cloth around the perimeter. If you choose to bury the cloth, it should be bent in an L shape extending away from the house (critters aren't smart enough to dig down *and* around). Then you connect the buried cloth to the structure of the run. For the roof, use aviary netting, which is lighter and costs about the same as chicken wire but is more effective at keeping pests like mice and sparrows out.

FOOD AND WATER

Hanging feeders are nice because the hens will waste less food and rodents can't access it as easily as if it were on the ground (they still can get to the food, smart critters that they are.) Feeders should be suspended just to chicken-beak level—any lower and the hens will soil the feeders. How much feed per bird will depend on the size of the bird, her age and metabolism, and access to other food sources such as bugs and scraps. A general rule of thumb is 5 to 10 chickens will go through a 50-pound bag of layer feed in one month.

Founts, or chicken waterers, usually come in one, three, or five gallon sizes. We prefer the metal ones, because they tend to last longer than the plastic variety. Chickens often will kick dirt into their waterers; to keep the water clean, raise the waterers up on cinder blocks to beak level. Chickens drink a lot of water—a major requirement for making eggs—during the summer; a good rule of thumb is to have one gallon of water per hen daily.

EXERCISE

Hens need to get exercise. Their favorite pastime is kicking up dirt to find bugs, technically called "scratching." People who don't have a chicken run should let their hens out for exercise. But you need to call the shots: Don't let them get into your vegetable garden. Though some people swear hens in the garden are great bug eaters, we've noticed they will first eat all your greens, destroy your precious seedlings, and dig up entire plants. Which is why we suggest that you build a fence that keeps them out of the garden, or construct a "chicken run" of 50 to 300 square feet (depending on the number of birds) that contains the chickens and their enthusiastic scratching.

If your plan is to keep the hens in a run exclusively, its floor should be sprinkled with sawdust, straw, or dried leaves. The hens will enjoy scratching at such materials. They also enjoy taking dust baths, literally dusting themselves with loose soil or sawdust, in order to clean their feathers and remove parasites. Chickens do need sun too, so be sure the run has at least a patch of sun during the day. Toss fresh greens or grass into the run to keep the girls happy.

Some people will let the hens out into the garden to dig up and fertilize a bed you will soon be planting. To keep them from laying into beds you want to preserve, a temporary fence or bed-size cage with an open bottom works great—some people call this a "chicken tractor." Just be sure to put them back in their house at night.

We've found the following breeds to be the best for city living. They are quiet and high producing, and they don't mind a smaller amount of space.

Australorp: Big, beautiful black hens with green feathers. They are docile and quiet. Brown-egg layers.

Black Star: Black hens with some red feathers at the neck. Extremely productive layers of brown eggs. The chicks are sex-linked, so the black chicks are female, which avoids the problem of accidentally getting roosters.

Rhode Island Red: Reddish hens, very docile, and amazingly productive layers.

Araucuna: The Easter egg chicken. They are a bit batty, but they tend to be smaller and so can handle life in a run. City people love the colored eggs they lay.

Cochin: These come in a variety of colors and sizes, with the common trait of having feathered feet. We think the cochins look like they're wearing leg warmers! One of the most gentle, sweet breeds around.

NOT RECOMMENDED

Cuckoo Maran: A beautiful black-and-white-colored hen that lays dark chocolate brown eggs. But they tend to be quite loud and are a bit high-strung.

Cornish: Usually raised as a meat bird, some people raise these girls for egg production. We've never encountered a more flighty yet somehow stubborn breed in our lives.

White Leghorn: This is the factory-farmed-egg chicken. The hens have white feathers and lay white eggs. Though they are amazingly productive, often laying 365 eggs per year, they just aren't very personable. And who wants to raise a backyard chicken that's been engineered for factory farming?

BUYING CHICKENS

Once their house and scratching area are built, it's time to buy your chickens. Remember that hens are social animals that evolved living in wild flocks: They need other hens around to be happy. There are two ways to acquire hens.

Buying Full-Grown Chickens

One is to simply buy some laying hens from a local farmer or pet store. If you live in a major metropolitan area, farms are usually only an hour's drive away. Internet billboards often have chickens for sale (a quick check on New York City's craigslist brought up several egg-laying chicken opportunities). It's a good idea to buy chickens between six and eight months old if you are anxious to get eggs immediately. Buying local ensures that the breeds you get will do well in the local area (if the farmer or breeder is to be trusted).

The downsides to buying full-grown chickens are considerable. Full-size hens tend to be more expensive. They might be a loud breed unsuitable for city life (see sidebar for the best and worst breeds for the city environment). They might have a disease or parasites that you don't know about. If you're buying from a disreputable breeder, they might be old and toward the end of their egg-laying service (hens usually lay really well for the first two years, then production drops off). It's also stressful for hens to be moved from one farm to another. The key to avoiding these pitfalls is to find a reputable breeder with references.

When you go to the farm or breeder, bring a cardboard box lined with some straw or a medium-size pet carrier (about two chickens can fit in a cat carrier). Don't bother putting food or water in the box; they will be nervous and won't eat or drink on the car ride home. Once they are home, set them in their henhouse with access to food and water. At this point it's a good idea to shut them inside for a day and night to let them get adjusted to their new environment. By the next day, they will have adjusted to the new place and will be scratching and pooping. You should monitor them closely for the first week for signs of stress—watching "chicken television" with a cup of coffee is one of the urban farmers best delights. Make sure they are laying in their nesting boxes and roosting in the designated places. If they aren't, you might need to adjust the boxes or roost to suit their needs. Remember to shut their door at night and open it in the morning—word will travel through the predator underground that there are some tasty new arrivals on-site, so you must be vigilant.

DAY-OLD CHICKS

The other way to get chickens is to buy them as chicks at a feed store. If you don't have a nearby feed store, you can order them directly from a hatchery. It's very postmodern feeling, buying day-old chicks on the Internet and having them arrive by mail at your home in the city. People like to buy day-old chicks, because they can order special breeds, and chicks are rather inexpensive (a couple of dollars each) compared to full-grown chickens. Having said that, it's only a good idea to buy day-old chicks if you are curious and devoted and don't mind having live animals inside your house, garage, or basement. And keep in mind that it takes five to seven months before the chick develops into a full-grown chicken and starts laying eggs.

Making your order is incredibly fun, and you'll have to make some decisions.

Straight Run or Sexed?

Buying a straight run means that none of the chicks have been sexed. If you order straight run, you are guaranteeing that you will have roosters! This is a cheaper option, and many people choose to buy straight runs with the idea of eventually eating the young roosters from the group. Assume half the order will be males. If you order sexed hens, you can choose to buy females only. Since chicken sexing isn't always accurate, you may end up with one or two roosters in your order anyway. If you buy sex-linked chicks, you are fully guaranteed the sex of the chicken based on the color of the chick. This is because sex-linked chickens are crossbreeds whose sex is linked to feather color.

How Many?

Many hatcheries require a twenty-five-chick minimum order. This is because the chicks must be kept warm—85 degrees—which can be achieved only when they are able to cluster together in the box. Since some backyards can't handle twenty-five chicks, the best option is to find a few other people who are interested in raising chicks and divide the order among the group. Always add a few extra in case of death or predators or roosters (in past orders we've averaged about one or two out of twenty-five that are DOA). For example, if you want to have five chickens in your backyard, order seven chicks.

Vaccinated or Not?

Most hatcheries will offer to vaccinate your chicks for Marek's disease and coccidiosis for a small fee. Marek's causes lesions that paralyze chickens; coccidiosis is the common name for a group of protozoan parasites that can kill chicks. We think it's worth it, and these vaccinations are recognized as kosher by organic labeling agencies. Another benefit is that, if the chicks get the coccidiosis vaccine, there's no need to feed medicated chick starter. We've gotten chicks that we accidentally did not have vaccinated, and they died or later developed Marek's: It's worth it to vaccinate!

Which Breed?

There are white-, brown-, and colored-egg-laying chickens; there are full-size chickens and smaller breeds called bantams; there are exotics and show breeds. While looking through the chicken catalog or Web site, it's easy to get swept up with buying Weird Chickens. If you're a beginner, it's probably best to stick to the proven backyard breeds, such as Australorps, Red and Black Sex-links, Rhode Island Reds, and Barred Rocks. See the sidebar on choosing breeds (page 392) for other suggestions.

SETTING UP FOR DAY-OLD CHICKS

Before your birds arrive (the hatchery will tell you when they will be shipped), you should set up a brooder. This is a warm, draft-free area in the house for them to live in until they develop true feathers. Some people use cardboard boxes, others use galvanized washtubs, and others buy wire cages. We like to reuse waste from the city, so we recommend going to a supermarket and asking if you can have a big box—a watermelon box, for example, is the perfect size for raising fifteen to twenty-five chicks. Use whatever you have on hand or can find, but bear in mind that they need 1 foot per bird for the first four weeks.

Here's the ideal setup of a chick brooder.

CHICK BROOD BOX

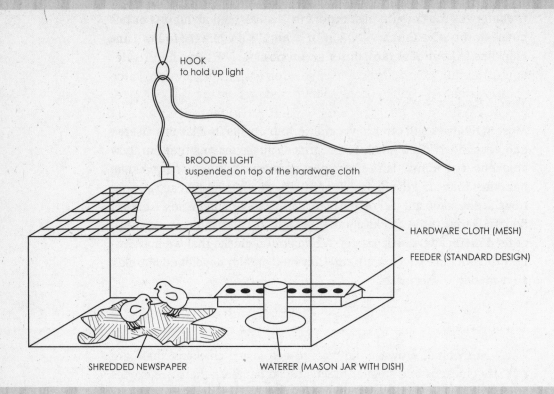

HOOK
to hold up light

BROODER LIGHT
suspended on top of the hardware cloth

HARDWARE CLOTH (MESH)

FEEDER (STANDARD DESIGN)

SHREDDED NEWSPAPER

WATERER (MASON JAR WITH DISH)

A proper brooder keeps chicks safe and warm

Bedding

The floor of the brooder should be scattered with shredded paper, strips of newsprint, or straw. For the first three days it's a good idea to keep shredded paper or straw under their feet, so they don't develop spraddle foot, a weakness in their hocks that prevents them from standing properly. Avoid sawdust at this stage, as the chicks will tend to eat it and get digestive problems. The chicks will poop mightily: If you have twenty-five chicks, we recommend a full cleaning of the brooder every day.

Light

A brooder should have a light source to keep the chicks warm. They need to be kept at 90 degrees to 95 degrees for the first week. We recommend a 250-watt brooder light with a ceramic base (not plastic, which might melt after the light is on continuously for several months) suspended above the box and directed onto the floor of the brooder. Using a red-colored brooder bulb has a more mellow light that has been shown to reduce how much they peck on each other (aka, cannibalism). Whichever color bulb you choose, make sure the brooder light can't fall into the brooder, as you don't want the house to catch on fire. Adjust the height of the lamp based on how the chicks cluster together. If they look like they are huddled under the lamp, it's too cold in the brooder and you might need another lamp. If they are fanned out completely and seem to be panting, the light is too hot or too close and should be moved farther away from the chicks. You must leave the light on all night. If it's warm outside, during the day you can open up some windows to let the chicks get some fresh air, but don't let it get drafty. Chicks chill easily and should be kept at 90 degrees to 95 degrees for the first week, 85 to 90 the second week, 80 to 85 the next, and so on. By the time they are two months old, they will have formed feathers and can handle room temperature.

WATER FOR CHICKS

You should equip the brooder with a small waterer—they come as stand-alone plastic or metal units. You can also make your own using a mason jar, a mason jar lid with about ten small holes in it, and a small

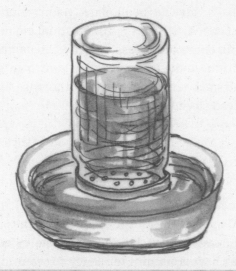

Waterer for chicks using an upside down mason jar

bowl. Simply fill up the jar, screw on the holey lid (you can make the holes with a hammer and a nail), and turn the whole thing upside down on the bowl. Water should trickle out to a point, then stop. When a chick dips its beak into the waterer and drinks, the jar will let down more water. You can also buy metal waterers that screw onto mason jars that use the same principle.

FOOD FOR CHICKS

You should buy a product called chick starter: It's a higher-protein form of regular egg-laying chicken food. It comes in bags and has a powdery consistency that makes for easy eating. Place the food into a trough-style feeder with a chick guard that is available online or in pet stores. Alternately, you can serve their food on a plate or dish, but we recommend mixing it with a little water to make a paste, so the feed doesn't get scattered everywhere in the brooder. In a shallow bowl or tray, sprinkle some food, then add a trickle of water until the feed is damp but not runny. Place the feed tray in the brooder, away from the circle of lamp light. Feed chick starter for the first twelve to sixteen weeks.

SUPPLEMENTS

Chicks should also be given a small plate (yogurt container lids work well) of a product called grit. This is an intestinal aid of small pebbles and sand that will, once eaten, pass into the chicks' gizzard, where it helps to grind up the food. You can make your own grit by digging some pebbles and sand out of your garden; otherwise, buy some online or at your local feed store.

ROOST

After the first week or two, add a thin pole or dowel about the diameter of a pencil to the brooder. Suspend it a few inches above the floor of the brooder and see if any of the chicks are interested. Even at their young age, chicks have a natural tendency to want to roost. There's nothing cuter than little chicks sitting on a pole, and it's a good skill they'll use when they are older.

WHEN THE CHICKS ARRIVE

If you ordered your chicks online, they will arrive in a box at the post office. They are surprisingly resilient but should be given water immediately. You may have to dip their beaks in the waterer before they understand how to drink on their own. Sometimes there may be a dead one in the box due to trampling (parents: Check first to avoid a traumatic memory). These can be disposed of in the compost or garbage can or buried.

After they are given water, you should set out their feed. Chick starter should be freshened at least twice a day, and the chicks should be allowed to eat as much as they want.

The chicks will make a mess. They start pooping immediately. We recommend changing the litter on the floor of the brooder at least every other day. The shredded newspaper and poo makes great bedding for your worm bin. The chicks can also be rather loud for such tiny things. Don't put the brooder in the kitchen or near bedrooms. The laundry room, warm basements, garages, or mud rooms are ideal places. Brooders should be safe places, far away from dogs and cats that might like to eat the little ones. After a few weeks, if you haven't already, you may need to cover the top of the brooder with hardware cloth or some other mesh screen to keep the chicks from flying out but that still allows light and air in.

As they grow, you should supplement their feed with greens, pieces of grass, and small bugs. This stops them from picking on each other and makes for healthy hens. The chicks are tremendously fun to watch. They often fall asleep in midwalk. They sleep in a puffy pile, jumbled up on top of each other. If given a snack, like a pill bug or a piece of greenery, they will chase each other in order to get it. Invite friends over to watch their antics and practice holding them. It's a good idea to get them used to you so they will be tame once they become big chickens.

CHICK PROBLEMS

Most of the time, the chicks will be fine. But problems can spring up.

Paralyzed

If the chicks aren't walking well, or seem to be resting on their haunches or not moving well, it's almost always a nutrient deficiency. A shortage of calcium, riboflavin, vitamins B, K, or many others. The best thing to do is give them a vitamin supplement like Quik Chick in their water. Remember too that chicks will thrive if given greens, grass, or bugs, which provide a full spectrum of needed nutrients. Check with your feed store to make sure the feed producer didn't accidentally make a bad batch of chick starter (it can happen if they forget to add an important nutrient). If you didn't use shredded newspaper, the chick may have spraddle foot, which is incurable. **Never keep day-old chicks on a smooth surface!**

Picking Each Other

If the chicks are acting aggressively toward each other, you're doing something wrong. Perhaps the brooder isn't big enough for them. Perhaps you need to feed them more often. Perhaps the litter needs to be cleaned more often. Perhaps the air is stagnant and they need some fresh air. They might also be too warm. Adjust their environment and see if that changes anything. Unfortunately, once they get a taste for picking at each other and blood is drawn, they will keep at it. You can purchase a special paint-on skin solution that covers over the picked area with a dark pigment, fooling the pickers. Otherwise, isolate the pickees for a few days until they heal and have no scabs showing.

Pasted Vent

You should watch the chicks' fluffy bottoms during the first week or so. They can develop pasted vents, which will look like a cluster of dried poo on their bottoms. If you notice a pasted vent, pick it up and gently tug at or cut off the poo with scissors. If it doesn't come off, you can run some warm water over their backside and try to work it off, or use a Q-tip dipped in mineral oil. You might rip off some downy feathers in the cleaning process, and the chick will cry, but it must be done or the chick will get sick. Wash your hands after doing this!

Dead

Nature works in mysterious ways. A chick is both fragile and resilient. Sometimes, through no fault of yours, a chick will quietly die in the brooder. We like to think that the chick wasn't ready for this world yet.

For a complete book about diseases of the chick, see Gail Damerow's *Chicken Health Handbook*, or consult the Poultry U at University of Minnesota at www.ansci.umn.edu/poultry.

TRANSITIONING THE CHICKS OUTSIDE

After eight to twelve weeks, their soft down will be replaced by pre-feathers—precursors to their real feathers. They will look rather raggedy and tattered, like middle-school students. At this point they can be put out into the coop, as long as the weather is mild. Some people put a light inside their chicken house to keep them warm, especially at night. Provide them with a low-hanging roosting bar in the hen-house. You can use a stick, a small dowel, or whatever you have around they can comfortably perch upon. Place the adolescent hens on their roost every night for a week or so. Eventually, they will be trained to roost on their own.

Chickens are birds and they can fly, and they will perch in trees if they get the habit. If you don't want them to do this, keep their wings clipped until they are about a year old. To clip the hen's wings, hold the bird and spread out the wing.

The feathers near the tip—the flight feathers—are longer than those near the hen's body. Your goal is to clip the wing feathers so that they are the same length as the ones closer to the body—this renders them unable to fly. Just trim an inch at a time—watch that you don't draw blood—then clip another inch, until the wing feathers are trimmed.

Clipping flight feathers from a chicken's wing will prevent it from flying

You only need to clip one wing, as it will make them unbalanced, which will keep them from flying.

It can be very difficult to introduce new chickens to an existing flock. Because hens like to keep their hierarchy, new birds will upset them and fighting will ensue. The best thing to mitigate fighting is to create a newbie area in the chicken house, or an entire area in the house devoted to the young ones. Then slowly integrate the flocks, providing a box or a cage that only they can fit into as a hiding space.

At this point you will already have developed a habit of watching your birds as if they were the most fascinating soap opera. This is good, because it's important to get to know your birds so you will notice any unusual behavior. In this entertaining time of adolescence (yes, teen chickens look awkward too), you will want to keep a close watch out for rooster behaviors. If you plan to keep a rooster (one is the maximum in a flock or you'll have fighting), isolate any others in their own separate enclosures or cages. We vote for fattening them up for the soup pot. To do so, feed them a rich diet of greens, scraps, pellets, and lots of cracked corn until mature (six to eight months is preferable). You should be watching for egg-laying behavior as well. Hens will seek out a nice nesting box, especially if it is equipped with straw and a ceramic egg. When a hen lays an egg, she will make a big deal out of it—cackling like crazy. When you start hearing that noise, you've got your first egg!

FEEDING YOUR EGG-LAYING CHICKEN

After keeping them safe from predators, keeping your hens fed and watered is next on your priority list as a chicken owner. They should have access to water at all times, and be fed once or twice day. At twelve to fifteen weeks, you can start feeding your young hens a product called laying feed or layers mash, which is usually a combination of grains and soy that promotes egg laying. These come in either pelleted or crumbled form. We always feed our chickens some form of pelleted or crumble feed, but purist cheapskates will insist that their chickens live on stale bread and old rice. Maybe. But the feed store–bought food gives you peace of mind that the chickens are getting all the correct vitamins, protein, and minerals they need to be healthy, egg-laying chickens, which is their job. If you give them scraps and kitchen waste, that is fine, just adjust it so that you feed them less feed. Believe it or not, but overfeeding is one of the biggest problems for backyard birds. Since they are often seen as pets that are adorable and always hungry, it's natural that an inexperienced chicken owner would overfeed. Overweight chickens are not healthy though, and will have trouble laying eggs and might die prematurely. Don't overfeed! Ten full-grown, egg-laying chickens will eat about 2½ pounds of chicken feed per day, less if you are supplementing with scraps.

SOY: IS IT BAD?

Many urban farmers get nervous when they see that chicken feed contains soy. It's a key crop that is genetically modified. It's also known to be one that is grown in a vast and horrific monoculture. However, the easiest way to get protein in chicken feed that can be absorbed and assimilated into hen digestive systems is through soy. Highly accessible protein translates into high egg production. Soy makes eggs, basically. If this makes you nervous, ask your feed supplier about the source and quality of their soy. If it's organic, it's not genetically modified. If sourcing local grains is important to you, or if you're allergic to soy, there are a handful of mills that are trying different protein sources, such as peas and fish waste. Ask your feed supplier—it's a great conversation to have!

BEYOND CHICKEN FEED

For people who want to save money on feed costs, or who are wary of feeding animals food (soy, corn) that could feed humans, we encourage them to tap the urban waste stream to feed their chickens. Not only will you save money and earn the moral high ground, when hens are fed scraps and weeds they seem more engaged, and their yolks will turn an amazing orange color from all the extra carotenoids. Think about it: All over your city people are throwing away all kinds of food that chickens like. You can divert it with a little effort.

ASKING FOR PRODUCE

The best places to inquire are grocery stores; just ask the produce guy next time you're in the supermarket. Or ask the person at the farmers' market stand what they do with the discards and blemished fruits. Explain (without seeming too crazy) that you have hens, and you'd like to supplement their diets. If they smile and agree to let you in on the bounty, rejoice and make a plan for how and when you'll pick up these goodies. Some places will want you to ask every time and won't agree to a regular pickup. Others might be ready for long-term sharing. At this point you'll need to be clear about how often and how much you'll need. Don't take five boxes because you feel like you have to—just take what you and your hens need. There's nothing more gross than rotting vegetables and fruit in your backyard!

If you're too shy to ask, there's always Dumpster diving under the cover of night.

DUMPSTER DIVING FOR CHICKENS

If the produce manager won't even talk to you, you might need to go Rambo for your hens. Dumpster diving—also known as going through the trash of restaurants, farmers' markets, and grocery stores—is a great way to supplement the diet of your chicken. It is literally providing local food to supplement your store-bought chicken feed and will save money. In Dumpsters, we've found whole wheat bread, old cheeses, whole cantaloupes and papayas, and piles and piles of greens. All of these things are beloved by chickens!

Scope out the place where the food waste from your favorite grocery store goes. It's usually in a Dumpster in an alley near the store. It's probably best to go at night, so bring a flashlight. You should also bring buckets or bags to carry your spoils. Make sure everyone is gone for the night before throwing open the Dumpster doors. Usually the produce on top is the best stuff—it's just been added, and the fruits and veggies tends to get soggy and crushed near the bottom. Still, don't be afraid to sift around a bit and get all the best morsels.

FEEDING SCRAPS

Whether you are actually Dumpster diving for your hens, or have worked out an agreement with a business owner, or are choosing to supplement from your kitchen scraps, not all food is created equal in the mind of a chicken.

Hen favorites:

- cantaloupe, watermelon, papaya
- lettuce
- kale and greens
- apples, peaches, apricots, tomatoes, pears
- whole wheat bread
- pasta
- rice
- leftover beans
- meat (chickens are omnivores! And need a fairly large amount of protein to set eggs)
- oatmeal
- sour milk/yogurt/cheese
- vegetables like spinach, tomatoes, winter squash

Hen least favorites:

- meat with bones (choking hazard or potential gross-out factor)
- lard or oils (chickens should not eat fatty foods)
- cabbage or onions
- raw potatoes
- coffee or tea
- sour and spicy foods, such as citrus (they don't like them)
- rotten foods (a little bit of mold is okay, but like us, they can get sick from spoiled food)

Of course, not all chickens have the same palate. Some chickens love carrots, others hate them. Experiment, because every chicken is different

WEED AND BUG COLLECTING FOR CHICKENS

Another way to feed your chicken well is to collect weeds for them. Though it still shocks many, cities do grow plenty of chicken food. Before you give them a lot of a weed they haven't tried before, offer them a little. If they eat it enthusiastically and seem fine the next day, offer them more. Chickens know what they like. Here are some of our favorite weeds for chickens, but remember, not all chickens are alike.

- **pellitory.** This weed literally grows out of cracks. It has tenacious seeds, needs almost no water, and has found a good life in cities, growing abundantly.
- **grasses.** Any kind of grass, chickens like. Clippings from a lawn that you know hasn't been sprayed. Grass growing out of cracks. That horrible Bermuda grass. They love grasses with seedheads, too.
- **clover.** Not just for rabbits—chickens like to get some three-leaf action too.
- **dandelions.** In the early spring and late fall, the chickens love their dandelion greens.

Snails and Bugs

You can really cut down on your feed costs if you collect and feed your hens snails and bugs. Snails are found in most vegetable gardens, lurking on artichoke stems, in the sorrel bed, or latched onto shady sides of raised beds. Hunt for snails at night or before the dawn with a flashlight and a bucket—you can gather hundreds at a time. Be sure to cover the bucket till morning (snails move fast) and feed them to your hens for breakfast. You might need to step on the snails to crush their shells. Another benefit: calcium in the snail shells! You can also "raise" bugs by maintaining an active brush pile. Simply gather twigs and branches and pile them in a corner. Every week, move the pile around and gather up all the insects the pile has attracted: millipedes, worms, beetles, and pill bugs. Watch the hens go crazy when you feed them the bug assortment of the week!

SPROUTING GRAINS

A final option for purists who don't want to use pelleted feed and want to save money is to buy whole grains and sprout them. This is an advanced topic and mostly unscientific field, but from our experiments, sprouted grains are full of protein and vitamins and hens favor it over layer pellets. It also tends to be cheaper than milled pellets. We urge you to buy whatever is your local grain—whether that is wheat, oats, corn, or barley. To sprout grains: Fill a clean bucket a little less than half full with your grain of choice and add water until the bucket is almost full. Place a lid with air holes on top of the bucket and let the grains sit for a day or two at 55 to 65 degrees. After a day or two, drain off the water and pour on fresh, stir the grains around, then decant again. Leave the wet grains in the bucket until they sprout. This usually takes a few days. Some people put the soaked grain into grow trays. Periodically check on them and stir them up a bit. You'll see that they absorb the water and will begin to sprout. After germination, you can feed it to your chickens. You'll have to have a few buckets going at different stages to make sprouted grains an everyday feed source. Be sure to supplement with plenty of greens and oyster shells. This is a lot of work but will save money.

OYSTER SHELLS

Chickens go through a lot of calcium. Every time they create an egg (which is almost every day), they lose calcium and need to regain it. The easiest method is to offer them ground-up oyster shells. Don't worry—you don't have to sledgehammer whole oyster shells saved from a trip to the oyster bar. They are an ag product, sold for cheap (ten dollars for a fifty-pound bag) at most feed stores. If you don't have a local feed store, just buy the oyster shells online. If you're wondering how often to feed the chickens oyster shells, let the eggs tell you. If you notice they are getting thin, time to pass out the oyster shells. Or just leave oyster shells, free choice (always available to eat), in the coop.

Instead of oyster shells, you can refeed your chickens eggshells. But you want to disguise them from looking like eggs, because this might encourage a chicken to become an egg eater, a terribly bad habit that, once learned, is rarely kicked. As you use eggs in the kitchen, place the shells in a container until you have enough to fill a cookie tray. Place them on the tray and bake at 300 degrees for a few minutes. Let them cool, then crumble the shells. Sprinkle on their food, when needed, as you would with oyster shells.

While their hilarious antics and nutrient-packed manure are reasons enough to host your lovely chickens, don't ever lose sight of the main product of your ladies: eggs. Homegrown eggs. Each one is a treasure, a jewel, evidence of some serious effort by one of your beloved hens. Because they are special, be sure to collect them each day, or at least every other day; otherwise they may pile up and break (one of the most common ways hens get into the terrible habit of eating their own eggs). We like to collect them in an antique basket and sit them in a bowl in the refrigerator (homegrown eggs will keep fresh for over a month there). Eggs have a natural coating (called the "bloom") that keeps bacteria out, so if the eggs aren't encrusted with chicken poo, don't wash them, and they will last for weeks outside of the fridge. If they are a bit dirty, however, give them a rinse and gentle scrub, and then refrigerate or eat as soon as possible.

Homegrown eggs are different from those at the store, or even the farmers' market. One thing you'll notice is that the egg yolk and white hold together very well and make especially good poached eggs. Fried, the eggs are amazing; the orange yolk looks pert and inviting; the white is sturdy and crispy. Cakes and waffles and all kinds of baked goods prepared using your hens' eggs will take on a deep yellow color and taste divine. Perhaps it is a bit of work to get the hens and shelter and feed them, but they will in turn feed you, and this is a very special relationship.

Sometimes a hen gets sneaky and lays a pile of eggs underneath a bush somewhere. Since you are unsure about the age of the cache of eggs, you might be tempted to just compost them. But there is a good way to figure out which ones are fresh and which ones are old. Fill a large bowl with water. Gently drop one whole, uncracked egg at a time into the water. If it is too old to eat, it will float. This is because, as an egg ages, the yolk and white contract, which causes airspace to form in-between the yolk and egg white and the shell. The older an egg is, the bigger the airspace, the more likely it will float. If it's just older but probably okay to eat, it will partially bob toward the surface. Discard the floaters and eat the rest!

CHICKEN TROUBLESHOOTING

These are the classic problems that chicken owners will encounter.

Food Wasting

They just kick and drop food all over, even from a self-feeder. Because of this, wild birds and other critters (rats) show up and eat their food. If this is a problem, limit their access to food to twice a day. Feed them as much as they can eat in fifteen minutes, then take it away.

Won't Lay in the Nesting Box

As lovely as their nest box is, they lay their eggs on the ground. This is normal for very young chickens, but if they continue, you need to intervene. One thing to do is add a fake egg to the nesting box to entice them to lay there. The nesting box might be too high up to access. The box might be too small for the chicken. It might be in a breezy or too bright area. Adjust the box until it becomes attractive to the chicken.

One Is Picking on the Others

Usually there is one leader of the pack. She will often jump on the others and peck them a little bit. This is normal. What isn't normal is chickens picking feathers off other chickens so much that their entire back area becomes exposed. Picking behavior may signal that the chickens are overcrowded or aren't getting enough food. Some people suggest adding a salve to the picked-on chicken. We haven't found this to be a successful remedy. Once a picker, always a picker is our motto. The picking chicken must be culled or killed (see pg 416 for how to kill a chicken).

If the problem is introducing full-grown chickens to each other (called combining flocks), place the new chickens in a cage in the henhouse for a day or so to let the existing gals get used to their presence. After a few days you can add new girls to the flock at night by setting them on the roost. Since every day is pretty much a new day for a chicken, they might not notice the new members at all. In all likelihood, there will be some squabbling—don't worry, just let nature take its course.

So Many Flies!

Flies are a huge issue, especially in urban areas, where space is at a premium and chicken coops are often very close to our homes. The best way to keep down fly numbers is to keep the coop very dry. Flies need water in order to successfully hatch, so applying a thick layer of sawdust to the chicken area keeps down insects. It's also prudent to clean out the chicken house of droppings every week. Fly strips offer some relief—our favorite stuff is made for barns and comes on white strips called the Sticky Roll. Hang these near the ceiling in the chicken house. Another option is to buy bags of baited fly traps—add water and watch as thousands of flies die a watery death. Finally, you can buy predatory insects that feed on fly larvae. These come as unhatched larvae, which are sprinkled in fly-breeding areas—once hatched, these winged creatures then lay their eggs in the larvae of the fly, effectively killing it. (See Resources in appendix 11 for more information about all of these.)

So Much Poop!

It's true: Chickens poop a lot. It's good practice to take a dustbin around the chicken areas and scoop up their droppings into a bucket. This keeps down flies and keeps your shoes clean. Add sawdust to the droppings, and add the bucket to your compost pile. Sawdust is sold in pet stores or can be acquired at cabinet or woodworking shops. They are often pleased to give shavings away for free.

Colds and Flu

Hens can get colds or flus. She'll look tired and droopy, and might wheeze a bit. In this case, it's a good idea to isolate a sick bird from the others and provide her with water and food in a warm place. Wash any feces that may accumulate on her hind feathers. Usually they just need a little time alone and they'll feel much better. If this doesn't do the trick, adminster a little Vet Rx. If in a day or two she seems sick, call in the antibiotics.

Mites

Hens control mites by taking dust baths, but if an infestation gets out of control (you'll see little red specks crawling under the feathers) use a commercial powder or spray treatment. We've found that food-grade diatomaceous earth (DE), sold at garden centers, is an effective mite reducer. Simply sprinkle it in the nesting boxes and in places where the hens take dust baths. Folk remedy people swear by wood ashes placed in the dust bath area.

Some hens will get mites living in the scales of their legs. If your hen is infested, you'll notice their legs look crusty and their scales are raised. These can be removed by bathing the chicken's legs in warm water with flea/tick shampoo and then applying a thick coat of some oily product—like petroleum jelly, vegetable oil, or mineral oil—to the legs. Repeat every week until the mites are gone.

Prolapsed Vents

It looks like your chicken's butt hole is falling out. Well, it kind of is. Prolapsed vents are a fairly common chicken malady. Basically, the vent where the egg comes out turns inside out. Kind of like a hernia. The best thing to do is don some gloves, lube the chicken's vent with Vaseline, and push the vent back into the chicken's body.

Egg Binding

This is another vent problem that can happen when an egg gets lodged up inside the vent and can't come out. The hen will look uncomfortable and move strangely. Some solutions are to put her in a warm place, massage vegetable oil around her vent, and massage her abdomen toward the vent—this may help her pass the egg. Worst case scenerio: You might have to just break the egg and then pick out the shell.

Egg Eating

Sometimes chickens eat their own eggs, sometimes they eat another chicken's eggs. It's usually a sign that they aren't being fed enough or they aren't getting enough calcium. Adjust their feed accordingly, add some fresh greens. Collect the eggs once or twice a day to discourage this behavior. Sometimes it's hard to get a chicken to stop once they do it. Consider culling—don't pass her off onto some other unsuspecting farmer.

Won't Get Off the Nest

This is called going "broody." She sits on the nest all day and night. If you go near her, she makes a horrible noise and might try to peck you.

Sitting on a clutch of eggs is a natural instinct that some chickens follow. There's no cure except to wait. Usually it takes twenty-one days for them to work through their broodiness, which is also the amount of time it takes to incubate a fertile egg.

Not Laying

When your hens first start laying they may lay outside of the nesting box. Look for their nests and put the eggs in the nesting box. You can even use grocery store eggs to get them started, or stone eggs in the boxes.

Looking Dead on the Ground but Flopping

Don't worry—this is a totally natural behavior, called taking a dust bath. Hens clean their feathers with dust. They will lay in the dirt and use their wings to brush dirt on their feathers—they may look dead or injured when they do this!

Shedding Lots of Feathers/Looking Bald

This is called molting; they molt naturally, or lose and regrow their feathers, once a year in the fall. They may look a bit strange during this time, but it is normal.

Neighbor Complaints

One of the biggest problems with urban farming is convincing your neighbors that it is a good idea. Chickens can be loud and attract flies and rodents, and they can ruin gardens. So if your neighbor complains, don't roll your eyes and wave them off. They have a right to let you know when something is wrong. Even if you don't like your neighbors, listen. If the problem is noise, find out your neighbor's sleeping patterns and let the hens out of the coop after your neighbor wakes up. If your chickens are getting into their yard, make a great effort to stop this from happening. Most complaints are cries to be given respect. Respect your neighbor by listening and taking every action you can.

If their complaint is that urban chicken keeping is weird, take the opportunity to tell them what you like about the chickens, how you feed them from the urban waste stream, and about how they alleviate some worries you might have about food security. Of course, hand them some eggs next time you see them.

Predators Kill Them

Disease and old age are not the main killers of backyard poultry: predators are. There's a whole hoard of city-living predators that would love to eat your birds. Opossums, raccoons, wild dogs, skunks, hawks, and sometimes rats would all like to taste your hens. They will go to extraordinary lengths to get at your birds: We once had a raccoon who slowly, over a period of nights, removed the staples that held down the chicken wire on one of our coops until it could reach into the run and kill the hens. Oppossums will eat chickens alive, munching on their soft oviducts. Hawks will dive-bomb and grab hens. Dogs won't be able to resist their kill instinct and will shake hens until they break their necks. Do what you can to prevent them: The most effective measure is to secure the entire coop with sturdy wire and shut the hens in at night.

MEAT CHICKENS

Many people embark on chicken raising with meat on the brain. We think raising chickens as meat birds exclusively isn't a good idea in an urban environment. The main concern is cost: Meat breeds like the Cornish cross and broilers are huge eaters of feed. In the country these chickens can range as pastured poultry, eating bugs and grasses instead of feed. In most city areas, this isn't feasible. The other reason is that the fast-growing meat chickens actually don't taste very good. They are indeed the same birds that are raised in factory farms—they have enormous breasts, thunder thighs, and not any kind of sense at all. They do grow quickly, though, often ready for butchering at two or three months. They taste bland. And our goal as urban farmers is not to have miniversions of industrial farms. So we recommend meat-motivated folks try their hand at turkeys or ducks.

However, that's not to say that you won't have occasion for eating one of your chickens. If you raise egg-laying chickens, in fact, it is almost a foregone conclusion that you will have to kill a hen or rooster at some point. Not to be a downer, but we think it's a good idea to plan what you will do with your chickens after they stop laying *before* you buy your chickens. Laying hens are productive for two to four years and can live up to eight to ten years or more. What will you do with your beloved egg-laying chicken? Remember that they still need food, water, and attention: Are you willing to continue spending money on a chicken who doesn't earn her keep? We know lots of people who are fine with this—they feel like the chicken earned the right to live her full life span after years of service. Other people, though, make a plan to get rid of the hen once her egg-laying ceases. Since we think it is inhumane to off-load an elderly chicken on the local animal shelter or into the wild, where they cannot survive, the most acceptable way to get rid of a chicken is to kill it. You should be prepared to do this yourself humanely and effectively, by taking a class or having someone who knows show you how to do it, or see page 416.

TIME WITH THE CHICKENS

Many people ask us, How much time will the chickens take out of my day? Here's how we do it.

DAILY
Morning: fifteen minutes
Let hens out.
Feed hens layer pellets.
Fill up water containers.
Collect eggs.

Evening: Ten minutes
Feed scraps or collected weeds.
Close coop door for the night.

WEEKLY
Clean out chicken coop: thirty minutes.
Dumpster dive: one hour.
Sprout grains: thirty minutes.

MONTHLY
Deep clean of coop and run: one hour.
Buy chicken feed and bedding at the feed store: one hour.

A YEAR IN THE LIFE OF A CHICKEN

Late winter: Buy day-old chicks for arrival in spring (page 394).

Early spring: Raise day-olds in brooder (page 396).

Late spring: Place adolescents in coop.

Early summer: Begin feeding layer pellets at twelve weeks old.

Late summer: New chickens start laying eggs; if you have roosters, they begin crowing—harvest as necessary. Be sure to feed oyster shells and a good mixed diet.

Late fall: Egg production may lower because of diminished daylight.

Early winter: Cull any unproductive chickens/other roosters for the table or soup pot (page 416).

Winter: Hens may start molting, or losing feathers.

Early spring: Hens begin laying again in earnest, and are one year old. They will lay steadily for another year at least. Plan to cull in the winter at two to three years old.

BREEDING CHICKENS

If you don't want to continuously buy chicks through the feed store or online hatcheries, there is a way to raise your own. You'll need two things: access to a rooster and a bantam hen or two. Although some full-size breeds of chickens are able to "set" eggs, bantams of any breed make the best mothers. Bantams are small chickens that have retained their ability to raise chicks. Since most cities have ordinances against roosters, you might need to just borrow a rooster for an afternoon of stud service. Choose a rooster of a breed you like and introduce him to your hens. After a session of mating, your hens should be fertile for about ten days. Save all their (now fertilized) eggs for a few days and place them in a separate nest. Encourage the bantam to set on this clutch of eggs, by showing her the eggs, but don't include any bantam eggs in this nest, unless you want more bantam chickens. She should take to this task easily, setting on the eggs, turning them. The other hens probably won't notice. After twenty-one days, the eggs should hatch. Let the bantam care for the chicks, and be sure to lock them up at night to protect them from predators. You shouldn't have to separate the chicks and bantam mom from the other chickens unless you notice the big hens picking on the wee ones. There is nothing more cute than a little bantam hen with her gaggle of chicks. Be sure to socialize the chicks by picking them up regularly and feeding them special treats. The downside of this technique is that it is difficult to vaccinate the chicks against Marek's. Most farm supply stores only sell packages of one thousand vaccinations—making doing your own too expensive. If you've had this disease in your coop before, it is not recommended that you breed chicks without the vaccination, so consider splitting an order of the vaccination with some fellow chicken breeders.

KILLING A CHICKEN

There may be a day when you need to kill a chicken. This is controversial in the chicken-loving world in which we live, but it's something that we think meat eaters should consider doing at least once.

There are a variety of reasons for killing a hen. She's old and has stopped laying. She's a picker who brutalizes her fellow chickens. She turned out to be a rooster and is crowing every hour on the hour. Or she was raised as a broiler and was destined for the dinner table since day one. Whatever your reason, here's how to kill and dress a chicken humanely and quickly. We can't emphasize enough that finding a mentor or taking a class on chicken killing is the best way to learn; these instructions below should act as a supplement for a hands-on experience.

There are various methods for killing a chicken. Some people place them in a "killing cone," a device that holds their bodies upside down and exposes their head for a cut to the jugular. Others pith the chicken with a jab to the back of the mouth. Old-timers wring the chicken's neck. The African diaspora brought us the highly effective method of wringing the chicken's neck by swinging it around holding its head (this is actually quick and humane). Most of these methods emphasize the efficiency necessary to kill many chickens at once. We prefer using a pair of very sharp pruning loppers to remove the chicken's head. It's quick and effective and virtually foolproof.

The act of killing is a somber one. It's a good idea to create a ceremony before killing an animal. It's a time to focus your respect and do a good job moving your animal from the living to the dead. We like to light some incense or burn some tobacco, perhaps say a few kind words to the chicken. Thank the hen for giving its life in service to your sustenance. This will make you calm, and thus will calm the chicken as well.

Setting Up

You'll need the following, preferably in an outdoor area:

- a safety cone with the tip cut to allow a chicken's head through the hole. The cone should be attached to a solid object like a large board, fence, or tree, tip facing down, at 3 feet to 4 feet above the ground.
- a pair of lopper pruners designed to cut a branch 2½ inches thick
- a bucket for feathers, offal, and blood
- aprons
- a bowl of water and soap for washing your hands and knives
- pot of extremely hot water in which a whole chicken can be dunked
- a sharp knife
- a clean pan of cold water to place chicken in
- dish towels
- a four inch piece of string
- a cutting board covered with butcher paper or plastic wrap
- a plastic bag for the chicken and gizzards
- an elevated working surface to keep meat away from the ground
- the bird in question, food withheld for at least twelve hours; twenty-four is preferable

After making your peace with the death of the bird, have a friend hold the chicken upside down by both legs with one hand—this relaxes the chicken so that it will quiet down.

Lower the bird into the safety cone, so that the head appears at the tip of the cone. Have a bucket directly under the cone to catch the blood. You may have to guide the head down and out of the cone. The chicken should be relaxed and will feel secure in it, and its main function is to prevent excess wing flapping and access to the bird's head (if you don't have a cone, you can simply hold the chicken tight or wrap its wings in a towel). Open the loppers and place them around the chicken's neck. Firmly close the loppers—you don't have to do this quickly, just firmly. The chicken's head will be severed and drop into the bucket. The body will jolt quite a bit, and blood will exit from the neck.

Using either a killing cone or by holding the chicken, duck, or turkey firmly upside down, make a cut to the throat using a pair of pruners or loppers. After the cut, let the blood drain into a bucket for a few minutes.

Fully experiencing the life force draining away really helps make it a respectful process, and it does take a while. Some people think it's better to slit the chicken's throat with a knife instead of using loppers to cut off the whole head. The knife method takes a little bit more experience, which is why we suggest loppers for your first time.

Pluck

Once most of the blood has drained out (usually two or three minutes after the kill), the body should be dipped into very hot water (140 degrees) for a short period of time—one minute is usually sufficient. This dipping loosens up the feathers, so you can pluck them off. After dipping, lay the hen on a flat surface or hold it in your lap, and pull off the feathers in the direction in which they lay. If you pull against the feathers, you may rip the delicate skin of the bird. The feathers will come off in clumps in your hand. If you plan on using the feathers, put them somewhere to dry; if you aren't going to use them, put them in the bucket with the blood and the head. After a few minutes of plucking, your chicken will start to resemble a grocery store chicken. Pick off all the little downy feathers as well. The wing feathers will be the most difficult to remove. Just pull hard, or use a pair of pliers to get them off. Note that a rooster is much more difficult to pluck than a hen, especially the wing and tail feathers.

Place the chicken in the pan with cold water for five minutes to clean and cool it off.

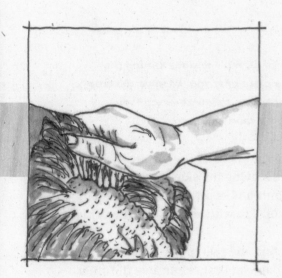

After dunking in hot water, pluck poultry by pulling feathers in the direction they grow. Wing feathers are the most difficult to remove.

Eviscerate

This is the most difficult part. Many people worry about contaminating the meat with fecal matter from the intestines, which is a legitimate concern for a commercial operation. Since you have no reason to hurry, move very carefully. If you do accidentally get some poo on the chicken, you can cut off the part that made contact and wash off the chicken very well with soapy water. Next time will be easier!

The first thing to do is get the chicken on its back, breast bone up.

Cut off the feet by making a cut at the joint right where the scales end and the leg bends. Peel the outer layer of scales off and soak in soapy water if you plan on eating them or using them for stock.

Now go to the vent area—which looks like a butt hole—and you are going to "free the butt hole."

Cut all the way around the vent, leaving the skin intact but making a circle of cut skin around the butt and carefully not cutting any organs. Tie a knot around this circle of skin with a piece of string. This will prevent feces from spilling out.

Next, go to the neck area and find the two tubes—one, the trachea, will resemble one of those bendable milk shake straws; the other is the soft and pliable esophagus. Cut the skin around the tubes about 2 inches down, so they are exposed. The esophagus connects to the crop, an organ located in the neck region that chickens use to squirrel away food for digestion later. The trachea connects to the lungs. With your fingers, follow the esophagus, gently pulling and prodding until you find the crop, which (if you didn't withhold food) will be a bulging little sac of chicken food and greens. Keep in mind that the crop is very pliable, and work your fingers around it until it is free. Cut the end off, careful not to spill any of its contents onto the carcass.

That done, grab the flap of skin just under the breast bone and make a shallow cut with the tip of your knife. With your fingers, tear open the slit by gently tearing and prodding, until you reach down to where the tied-off butt hole is. You'll be able to see the organs at this point. If not, the organs might be blocked with fat or the peritoneum, which you should clear out of the way with your fingers or a careful knife. Don't start cutting away with a knife—you might pierce the intestines and contaminate the meat. Once you have a clear sight of the

 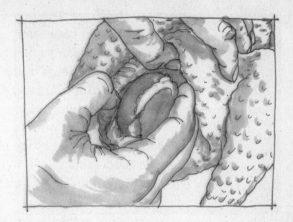

After making delicate cuts to expose the organs, reach your hand into the bird and pull out the viscera.

organs, place your hand or fingers inside the body cavity. Delicately release any of the organs held by the membranes around the body cavity. Then grab a full handful of the organs, trying especially to grab the hard, large gizzard, which is located just below the breastbone, and pull them out of the body of the bird. If successful, the intestines, gizzard, and liver will come out all together. Be sure that the intestines all come out, including the tied-off area. You might have to make a few cuts with the knife to completely dislodge the tied-off vent. Discard the intestines in the slop bucket, but reserve the liver and gizzard.

Cut off the liver. It is a dark-colored organ with two lobes. It should have a greenish-looking sac embedded in one of the lobes. Carefully cut this out or use your fingers to carefully pull it off. It is the bile sac and is filled with greenish bile; it tastes terribly bitter. Interestingly, the bile used to be used as an ink by resourceful pioneers.

The gizzard is the large, hard organ covered with silver skin and sometimes bright yellow fat. Some people eat gizzards. Even if you don't plan on eating it, it's interesting to see what's inside the gizzard. Make the cut along its long edge, like you are cutting a grapefruit. Inside you'll find bits of rock and particles of mashed-up feed. If you or someone else are going to eat it, dislodge all the contents and peel away the greenish-yellow lining.

Place the washed gizzard and liver in a bag if you plan on eating them.

Go into the chicken again with your hand for another bunch of organs—this time the lungs and heart. You'll have to reach far up to get to the heart. The lungs are wedged in little depressions on each side of the rib cage—they are delicate and frothy-looking. Pull down the trachea and any other residual organ matter, which might include some unformed eggs in the oviduct.

Finally, flip the chicken over and locate the oil gland. This is a funny looking bump on the tail about the size of a dime. Cut the whole thing off—you'll notice it contains a yellow-orange colored wax substance. This is used by the hens to preen their feathers, but it does not taste good. Dispose of it in the bucket.

Wash your hands with soap, then wash the chicken well in plenty of cold water.

At this point the chicken needs to rest for at least twenty-four hours. Resting causes enzymes to start breaking down the meat, which makes for a less tough bird. We like to submerge the chicken carcass in a 5 percent solution of salt and water in a bucket in the fridge. Salt draws the blood out of the carcass. After twenty-four hours, remove the chicken from the saltwater, dry it off, wrap it in a towel, and store it in the fridge for before cooking.

Don't expect that you'll be popping this chicken in the oven to make a tender roast *poulet*. Because home-raised chickens tend to get more exercise than store-bought hens, they will tend toward the tough side. We encourage you to braise them in vegetable or chicken stock, or slow-cook them on the stove in wine and herbs until tender. You can also poach them—shred the meat and use the stock in various dishes.

The offal and feathers remaining in the bucket make a wonderful garden amendment. We usually dig a 3 foot deep hole near a tree and deposit the bucket of muck there. It provides the tree with a well-needed snack.

And that's chicken raising, from almost birth to death. Yeehaw!

DUCKS

Many urban farmers get the idea that they are going to raise chickens for meat. So they order meat variety chicks, known as broilers, raise them from day-old chicks, and wait, and wait, and wait until it's time to finally harvest the bird. Even with the freakishly fast-growing ones this can take up to three months (and those tend to taste bland), and up to eight months for standard breeds of chickens. Since time equals feed—you will have to feed the chicken every day until it's slaughter time—you'll spend a hefty sum on food just to get one roasted home-grown chicken on the table.

And so, we turn to ducks, our favorite meat bird on the urban farm. They breed on their own, grow quickly, and reach butchering age in under two months. They are foragers, often surviving on just greens, table scraps, bugs, and some grains. Their meat is delicious and lined with unctuous, tasty fat. Visions of duck confit will become reality. Their soft underfeathers make nice pillows. If you get attached, you can keep the ducks for egg production, breeding, and for slug and fly control.

Requirements:

- a duck area at least 6 square feet per duck
- a duck pen for sleeping at night
- a water-filled pond or pondlike re-creation
- duck food
- a feeding trough
- a water bucket they can't swim in
- grit, such as sand or rocks
- water and critter-proof feeding bins
- access to grass, weeds, and vegetable scraps
- the companionship of at least two other ducks
- a place to compost the soiled bedding
- a garden or some other place to dump the soiled pond water
- straw or sawdust for keeping their duck pen dry
- a nesting box set on the ground if ducks are to be egg layers

In some cities it is illegal to keep waterfowl such as ducks. So before you get any, you should consult your city ordinances to see whether it's legal or not. You might also consider approaching any neighbors who will see or hear your ducks and see how they feel about having them next door. Well-known for their quacking (with the exception of Muscovies, which make a strange hissing noise instead), ducks do make more noise than chickens, so you should get neighbor approval first.

SETTING UP

Ducks don't require the elaborate house with roosts and nesting boxes that chickens do. They'll do just fine living in a fenced-in run, with a small sleeping and laying area. Contrary to popular opinion, ducks don't need a pond, but we've found that they are happiest when they have access to large quantities of fresh water, where they eat, poop, and mate. Don't think you have to build an entire wetland, though. You can get creative with the pond concept: Dig a hole and sink in a washtub or a big plastic bin, then fill with water. Or fill a kiddie pond full of water and provide a ramp for the ducks to have access. The one thing to keep in mind is that the pond should be easy to clean. They get disgustingly swampy, so have a convenient way to drain the water and replace with fresh.

Once you determine the best place for the pond and have installed it, build your duck pen and run. The pen is more simple than the chicken house; it's essentially shelter from the elements. It doesn't have to be huge, and it doesn't need to be tall. But it does need to be predator-proof and a safe place to sleep at night. Some people build the duck pen in the area under their porch. It should have dry straw bedding on the ground, and will need to be cleaned out periodically. In a pinch you can use a large dog crate for their sleeping area.

The run should be just like a chicken run; see page 387 for those plans. Alternatively, some people allow their ducks free range throughout the yard and just lock them up at night in their pen.

Many people want to know if you can raise ducks and chickens together. If you have a big enough space, we do think it is a fine idea. In fact, we've found that the presence of ducks reduces hen infighting. Maybe they are embarrassed to squabble in front of the peaceful gaze of the more intelligent waterfowl. Who knows? Because ducks forage on the ground, they often eat fly larvae, too. But if space is cramped, consider having either hens or ducks but not both. Ducks are quite messy with water and in close quarters will create too much moisture for proper hen health. In rare cases, ducks may attack or rape chickens if they aren't given enough space.

As with chickens, you'll need to keep the ducks out of your garden unless they are supervised. Ducks will eat slugs and bugs, but they also love greens and will strip a collard plant in less than two minutes. They are easier to keep out of the garden than hens: You need only to set up a fence boundary between the ducks and the garden. They usually aren't as motivated as chickens and are less likely to fly over the fence to get to the garden.

You'll find that ducks are very good-natured and curious and explore everything—your hands and pant legs, plants, the ground—with their bills. They love to float around their pond, submerging themselves to get clean and groom themselves.

BUYING DUCKS

Full-Grown Ducks

Once their pen and sleeping area is set up, it's time to buy your ducks. There are two ways to acquire them. One is to buy older ducks from a fellow urban farmer, or a rural farmer. This is a good choice if your goal is duck eggs, or if you're eager to start a breeding program. The downside to this is that you might be sold a geriatric, ill-tempered, or loud one, so buy from someone you trust.

If you buy a grown-up duck, bring a durable pet carrier with you

BREEDS OF DUCK

You should choose breeds based on what you're hoping to get out of your ducks: meat or eggs. Before you get too gung-ho about raising ducks for eggs or meat, you should try eating said duck products. Duck eggs are much larger than chicken eggs and have a richer flavor and a bigger yolk, which is great for baking. Some people think duck eggs taste swampy, so make sure you like them. The same with duck meat, which is more like steak than chicken.

Pekin: This is the famous white duck much grown on Long Island. They are smart and grow fast and, most important: incredibly delicious. Great meat bird, but they can be loud.

Muscovy: This is a large meat breed from Central America. They look really weird, with strange red eye flaps. They hiss instead of quacking, making them ideal for city environs. They will set on eggs and hatch out young. Highly recommended!

Welsh Harlequin: This is the ideal egg-producing duck for the city dweller. They are incredibly calm and will easily lay an egg a day. Highly recommended!

Indian Runner: This is a small egg-producing duck. They are a bit skittish.

Khaki Campbell: Another high egg-producer, sometimes laying 355 eggs a year.

Cayuga: A gorgeous, green-colored, endangered meat- and egg-producing duck. Said to be more quiet than Pekins.

to transport the animal home—ducks excrete wet poops that will soak through a cardboard box. Line the carrier with newspaper and/or straw. You should put a dish of water in the carrier if it's going to be a long trip.

Once they are home, set them in the run near the pond and give them some food. Ducks are similar to dogs in that they will be your friend if you feed them snacks. The ducks will adjust to the new place in no time. You may have to encourage them to retire into their pens for the night by herding them in at dusk. After a week they should know to go into their pen on their own.

DAY-OLD DUCKLINGS

If your goal is meat, you're better off buying day-old ducklings, which are rather inexpensive (a couple dollars each) and will grow fast, compared to buying them full grown. You can also order special rare breeds. Day-old ducklings can be purchased either on the Internet, from the local feed store, or from a local breeder. As mentioned, they grow quickly and will be out in the yard in less than four weeks.

If you order from the Internet, you'll have to make some decisions.

Straight Run or Sexed?

Buying a straight run means that none of the ducklings have been sexed. If you're getting ducks for eggs, you'll want to get females, which usually cost slightly more. If you're in it for the meat, you may as well get the straight run, which is the cheaper option. Note that the presence of one male duck often makes the females more content, so it's worth it to have one onboard if you have the space. Male ducks, unlike male chickens, don't crow or make more noise than the females.

How Many?

Many hatcheries require a ten-duckling minimum in order to keep the ducklings warm from body heat during their flight/drive to your urban homestead. Since some backyards can't handle ten ducks, the best option is to find another person who is interested in ducks and split the order.

Which Breed?

While looking through the waterfowl catalog or Web site, it's easy to get swept away, and you'll want to try some rare breed or try your hand at geese (don't do it!). If you're a beginner, though, and you want meat ducks, stick with the Pekins or Muscovies. If you're after eggs, try the Welsh Harlequins.

SETTING UP FOR DAY-OLD DUCKLINGS

Before your birds arrive (if you ordered from a hatchery, they should tell you when they will be shipped), you should set up a brooder. A waterfowl brooder is the same as the chick brooder; see page 396. The only difference is that the ducks will need more water than chicks, and they will grow faster—requiring a brooder lamp only for the first two weeks. Note that ducklings can't swim! Don't put them in a pond of water until they have developed true feathers. Place a towel under the water source and plan to change it daily—it will become saturated with water and duckling feces very quickly.

WHEN THE DUCKLINGS ARRIVE

Day-old ducklings are exquisitely soft and fluffy. They are also amazingly hearty. They will be thirsty, so give them water immediately when you put them in their brooder. If you ordered from an online hatchery, because of the risks of mailing live poultry, there may be a dead one in the box, due to trampling (parents: check first to avoid a traumatic memory). These can be disposed of in the compost or garbage can, or buried under a tree that will be grateful for the extra nutrients. If this bums you out, buy your ducklings direct from a feed store or breeder.

WATER AND FEED

Ducklings should be fed unmedicated chick starter or duck ration—medicated chick starter will kill ducklings. The unmedicated feed can be served in a shallow bowl. Whenever ducklings eat, they like to drink copious amounts to wash the food down and clear their nostrils. So during feedings, be sure to provide them a large container of water to drink from, but it must be low enough so they can access it—a plastic dish tub works well.

Ducklings will also readily eat scraps from your table, such as wilted lettuce, apple peels, bits of melon or fruit, and greens. If you're feeling motivated, look in your garden for snails, slugs, and pill bugs.

SUPPLEMENTS

Ducklings should also be given a small plate (yogurt container lids work well) of grit. This consists of small rocks and smashed oyster shells that will grind the food up in their gizzards, a large muscle in their digestive system responsible for the initial breakdown of food (ducks don't have teeth!).

The ducklings will make a huge mess. They start pooping large wet turds immediately. We recommend changing the litter on the floor of the brooder at least once a day. The shredded newspaper and poo make great bedding for your worm bin, or compost.

PUTTING THE DUCKLINGS OUTSIDE

The ducklings will form true feathers after four weeks. At this point they can be put out into the duck run. After cleaning up after them for almost a month, it will be a joyous day to take them outside. Fill up the duck pond, make sure there's additional drinking water, and be ready to feed twice a day. Once they are outside, begin feeding them more vegetables and scraps from your table. Ducks are especially fond of mushy watermelon. You can feed them duck or chicken feed, but they will also thrive on a diet of grains, like whole wheat, oats, cracked corn, or millet. Again, put the food near the drinking water and expect the watering dishes to get cloudy with food particles—you'll need to change the water often.

DUCK INSTINCTS AND LIFE CYCLE

Ducks are productive for three years and can live between eight and ten years or more. A flock of three egg-laying breeds will lay about two eggs per day. Ducks are notorious egg hiders, and will often lay them in unexpected places. Like hens, they also can get broody, a state of duckdom that involves sitting on their nest for weeks and weeks at a time without getting up. Their instinct is to try to hatch their eggs. When they aren't broody, they will spend a lot of time sleeping, swimming, preening their feathers, regarding you with curiosity, and eating enthusiastically. If you ordered a straight run of meat ducks, check your resolve and remind yourself regularly why you chose ducks: They grow fast and taste delicious.

Ducks are social animals that evolved living in flocks, so they need other ducks around to be happy. They are birds, and if they figure it out, they can fly. If you don't want them to do this, keep their wings clipped. See page 401 for wing-clipping instructions.

FEEDING YOUR DUCK

If you're growing a meat duck, feel free to just feed them scraps, grains, and vegetable trimmings. They'll troll around the pen eating bugs and grass. If you're raising them for their eggs, we do recommend feeding them some form of pelleted or crumble feed that ensure they are getting all the correct vitamins and amino acids for egg production. See page 403 about feeding dos and don'ts when collecting from the urban waste stream, weed hunting, and collecting bugs for your duck.

DUCK TROUBLESHOOTING

Compared to chickens, ducks are a breeze to take care of. Here are a few of the minor problems you might encounter with them.

Sneaking out

Ducks are pretty smart, definitely more wise than hens. So if they discover a way out of a pen—a hole, flying—they will escape over and over again. Once learned, it's tough to break them of that habit. So make sure the fence doesn't have holes and that they can't fly (by clipping their wings).

Egg Hiding

Again, they are smart. Every day is a new one for chickens, but ducks remember. They tend to notice that you are stealing/collecting their eggs. And if they are motivated to reproduce, they will start laying in secret spots. Luckily, our brains are bigger than theirs: Simply watch from a safe distance and see where she goes to lay.

Being Broody

If ducks figure out that they can't hide the eggs from you, they might simply sit on them and refuse to move. While we admire their passive resistance tactics, you can still just grab the eggs from underneath their soft bottoms.

BREEDING DUCKS

If you have both a female and a male duck, chances are they will breed—often and successfully. We've had great luck breeding meat ducks, and keep a breeding pair around at all times, harvesting their young as they mature. The only requirement is to take a hands-off attitude, letting the female collect a nestful of eggs and sit on them for twenty-eight to forty-five days (depending on the breed), and out will hatch ducklings. At this point, if you are worried about predators, you can bring the ducklings inside and let them grow up under a brooder until they form true feathers. If you trust Mom to guard them from predators, let them grow up in the pen under her care and then harvest when they've achieved a suitable size.

MEAT DUCKS

When they reach six to eight weeks, it is time to harvest your meat duck crop. This can be an emotional time if you've become attached to your ducks and allowed yourself to think of the birds as pets or friends rather than as food. If you can afford the luxury of keeping a pet, or have an appetite for duck eggs, by all means let your ducks live. But whatever you do, do not release domesticated ducks into the wild. They will put an unfair strain on native populations of birds.

If you have considered the ducks food from day one and have been fantasizing about duck confit for weeks while watching the little guys grow, it is time to harvest.

KILLING A DUCK

The method for killing a duck is the same as for killing a chicken; see page 416. Because ducks tend to have stronger wings than chickens or roosters, we always use killing cones to process ducks.

Cleaning a duck follows the same process as cleaning a chicken, but because ducks have fine feathers, plucking can take an hour or more. We recommend that you invite some friends to pluck. You can opt to skin the duck, but then you miss out on all that wonderful fat. Another difference is once most of the blood has drained out, the body should be dipped into very hot water (150 degrees) with a small amount of laundry detergent added. The detergent makes it easier for the water to insinuate itself into the duck's water-repelling feathers. Holding the duck's legs, submerge it for three minutes, bobbing it up and down in the water. Eventually the dipping loosens up the feathers so it's easier to pluck the big feathers off. Do this, and then do a quick second dunk once you reach the down layer.

After the second dunk, lay the duck on a flat surface and pull off the downy feathers. Some people get out tweezers and quickly pluck at the pin feathers; others use a propane torch to scorch off the small feathers instead of plucking them off. Note that these are very soft feathers, designed for warmth. You won't be able to get every single one of the small feathers off the carcass—give up after you've gotten most of them (later they rest will probably come off after they're soaked in salted water for twenty-four hours). If you think plucking a duck is hard, try plucking a goose. It's one of the main reasons we don't recommend keeping geese. If you are getting serious about a duck meat operation, seriously consider buying a mechanical plucker, which uses many rubber fingers mounted on a cylinder and can pry off all the feathers in just a few minutes.

To eviscerate, follow the directions for the chicken on page 419.

COOKING DUCK

Like chickens, the duck should be allowed to rest for at least twenty-four hours in the refrigerator. Home-raised ducks, unlike home-raised chickens, are great candidates for being roasted whole in a slow oven, perhaps drizzled with olive oil and honey and stuffed with an orange. The reason they are so good is their fat. If you process a large number of ducks, you can render the fat, and then make duck confit—salted duck legs slow-cooked in their fat. The duck breasts can be seared like steak or salt-cured and air-dried to make a "duck proscuitto."

TURKEYS

Turkeys are the most American of birds—native to North America, eaten by Indians, Aztecs, and pioneers. Most people, come November, eat a Thanksgiving turkey without really knowing what one looks or acts like, much less all the work that goes into raising one of these birds for the table. An urban farmer might be tempted to give it a go, and we say, Why not? It's fairly simple—if you can raise a chicken, you can raise a turkey. Unlike chickens, though, the final process of raising a turkey is undoubtedly the slaughter. Some turkey-eating people might shy away from the fact that a turkey had to die in order to feed them, but an urban farmer will burst with pride when they brag about the bird they raised from chick to golden-skinned turkey. They'll have a lot of stories and memories of their birds, and somehow that makes the food taste that much better.

Requirements:

- a small, partially shaded henhouse and a yard with at least 10 feet per bird
- a very high place to roost at night
- daily feed (some meat bird feed pellets—soy, corn, minerals—supplemented with restaurant and grocery store scraps)
- fresh water
- straw, leaves, or sawdust for the run area to keep down flies
- critter-tight metal feed storage bins (rats eat through plastic)
- access to grass, weeds, and vegetable scraps
- the companionship of at least one other turkey or a few chicken friends will suffice (turkeys are not solitary creatures)

- a place to compost the turkey poop, which they produce voluminously
- a plan for dispatching the turkey humanely

In any city where it is legal to keep chickens, it's probably okay to keep turkeys as well. Check your local ordinances. The most important aspect of these ordinances is usually the distance your turkey yard must be from your own and your neighbors' houses. You might also consider approaching any neighbors who will see or hear your turkeys. If you have hens already, most neighbors won't notice the difference except in one crucial way: tom turkeys (the males), when mature, tend to gobble. It's a loud, chortling sound made in response to any city noises: an ambulance siren, a passing bus, even a loud radio. It's cute and novel at first, grating and horrible after the twelve hundredth time. If any of your neighbors are sensitive to noise, don't get any male turkeys. Females, on the other hand, make relatively soft chirping and clicking sounds.

SETTING UP

Before you get your turkeys you should decide how many you want to raise. Three turkeys will need as much space as three medium-size dogs—they get big! In an ideal situation turkeys would be enclosed in a giant run (10 square feet per bird) with a tall, protected roost where they would sleep at night. If you have trees in your backyard, consider building the turkey roost and run around a tree or two. The turkeys will most likely want to perch on the trees at night, and wander the run during the day. They should be allowed out of their run only when you are home and monitoring them. This is because turkeys love to range free and, in a city, that might mean they will want to walk into a busy street and risk getting hit by a car. They are very curious and not frightened by anything. We know someone who had her roaming turkeys impounded by animal control; if you want to avoid that, keep them penned when you are not home. The roosting area where they sleep should be fully enclosed, with hardware cloth or aviary wire to protect them from predators, which include opossums and raccoons.

With these thoughts in mind, decide which approach you want to take and where the turkey setup will be. If you have chickens already, it's a safe bet that they will make themselves at home with the hens, usually becoming the dominant members of the flock because of their size. You should know that commercial operations always separate turkeys and chickens, because they fear a disease called blackhead which chickens carry but are unaffected by but can pass on to turkeys which are affected. If you are very worried about this, separate the chickens and turkeys, never allow them to come in contact with one another, and disinfect any waterers or feeders that were to be shared by both species.

BUYING TURKEYS

To start, you should buy day-old turkeys, called poults, via the Internet or at the nearby feed store. There are a number of good hatcheries with online catalogs that ship day-old poults through the U.S. post office. You may be able to find poults at a feed store as well, but they won't have as wide a breed selection as a hatchery.

There are some exciting heirloom breeds of turkeys to choose from. We like heirlooms because they taste better than the standard white turkey, which was engineered to grow quickly and develop enormous breasts. On a small scale, where taste and quality is more important than quick production, heirlooms make sense. The Bourbon Red, Broad-Breasted Bronze, Royal Palm, Narragansett, and Rio Grande are just a few of the heirloom breeds. Turkeys are native to North America, from where they were taken to Europe, domesticated, and then returned as a handful of distinct breeds. These breeds are now recognized by the American Livestock Breeds Conservancy.

SETTING UP FOR DAY-OLD POULTS

Once you make your order, you'll need to get your turkey poult brooder ready. The setup for them is exactly the same as for chicks; see page 396.

PUTTING POULTS OUTSIDE

At eight weeks, the poults will be ready to go out into their run area/roost. By this time they will have grown quite large and will have some real feathers peeking out. It's not an attractive time for the poults (think: middle schoolers), and you'll be ready to shoo them out into the real world.

If you are introducing the turkeys to, say, your flock of laying hens, you'll probably want to draw out their introductions. The best method is to sequester the poults in an area of the coop where the hens can see and smell them but can't actually peck them. Keep them separated like this, with their own food and water and light, for two days. After that, release them. You might be surprised how assertive (some might say reckless) the poults will act toward the much larger hens.

If you aren't combining groups, just place the poults in the outdoor turkey run and roost and monitor their behavior. You can leave a light in the area if they seem very needy. Watch to see whether they can reach the roost you've put up (start low and increase the height as the poults grow up); make sure they are eating and drinking. At this stage the poults will consider you their mother, and it will be normal for them to run over to you when you venture outside. Spend time with them and enjoy these strange creatures. Please don't forget to shut the door to their house at night, or they will certainly become some other beast's dinner.

TURKEY NUTRITION

Most turkey-growing operations recommend feeding poults 28 percent protein for their first eight weeks of life. This is the ratio found in meat-bird pelleted food. There's nothing wrong with pellets, except that they can be expensive, and are often not organic, if that's a priority for you. After eight weeks, and for the rest of the bird's life, it can be fed a normal chicken feed, which is about 16 percent protein. If you are looking to tap the urban waste stream to feed and grow your turkeys, remember this ratio. Turkeys will gladly eat Dumpstered food, such as yogurt, bread, fruits, and leafy vegetables. They will also enjoy protein-based foods, such as meat scraps, beans, and restaurant leftovers.

Follow the same guidelines for turkey nutrition as for chickens; see page 402.

TURKEY TROUBLESHOOTING

Like hens, you might have issues with too much poop and flies and them picking on each other (see page 409), but hands-down, predators are the most likely way something horrible will happen to your turkeys. Predators are most likely to attack when the birds are small, and they'll usually come at night. Raccoons, stray dogs, opossums, and skunks are the most likely critters that will try to kill your turkeys. The only surefire way to protect them is to make their run and roost as inaccessible to predators as possible, bearing in mind that a raccoon can climb quite high and will reach its paws through chicken wire and fish around for little turkeys. Besides strong fencing, a protective dog, motion-sensing light, or an electric fence are some other options.

TURKEY BREEDING

Breeding turkeys is a step up from simply buying poults and raising them in a brooder. It requires a hen and a tom that are mature and a safe place for the little poults to grow up under the care of their mother.

Turkey breeding usually begins by selecting a male and a few females from your initial poult order that you'd like to keep. When deciding which ones to keep for breeding stock, be sure to select birds with traits that you want to pass on, like size, personality, and hardiness. Sometimes the Darwinian aspect of backyard poultry-keeping will lead to the self-selection of the hardiest survivors, who will then go on to breed. Both male and female turkeys must be nine months or older in order to breed successfully. Since the standard white turkey isn't able to mate, you'll need to choose heritage breeds. You really don't need to help much; just allow the tom and hens to be together. You'll notice that the hen often initiates breeding, when she crouches and spreads her wings out. The tom will make some huffing noises and circle around her. After a while he'll jump onto her wings, balance, and run his ovipositor on her oviduct. Yep, tom turkeys don't have an actual penis. But sperm is passed into the hen's oviduct, where they eventually meet up with an ovum and fertilize it. The hen can store the sperm for many days, and numerous eggs become fertilized. After laying a clutch of eggs, the hen will settle in for the big hatch. If you check on her, she will look immobilized, just sitting. But every couple of hours she will rearrange and turn the eggs, so that they develop properly. Her warm feathers keep the eggs at the exact right temperature.

After twenty-eight days of this, the eggs will hatch. Little poults slowly work their way out of the shell over a period of hours, or a whole day. Once they are all hatched, proud Mama will take the poults out for a walk to show them what to eat. There is nothing like a mother's love—turkey hens actively protect their young by shielding them from predators or other dangers.

HARVESTING YOUR TURKEY

Killing a turkey is like killing a really big chicken. Follow the instructions on page 416, but you'll need to use a large-size killing cone and a larger pot of water than for a chicken. Also, recognize this fact: You can't do it yourself, so invite a friend or two over to help out with holding the bird and plucking the feathers. And remember later to invite them to the Thanksgiving feast. So many things that we do on the urban farm are enjoyed much more when one or a few other people gather together for the task. It's a way to—dare we say?—create community.

COOKING A HOME-RAISED TURKEY

If you raised a heritage breed turkey, you should know that the bird cannot be cooked like the Butterballs at the grocery store. Usually factory-raised standard whites are cooked slowly, at a low temperature. This works fine if your bird has an enormous breast and is pumped full of water. A heritage bird, though, needs to be treated like a game bird. That is to say: cooked quickly at a high temperature. Like an hour at 450 degrees, turned every fifteen minutes. They tend to be more tasty if they are not stuffed with the usual bread stuffings but instead larded with butter and garlic that is inserted under the skin. If the bird is large, pour a few cups of chicken or vegetable stock and white wine in with the turkey after the first hour, and then tent it with foil. A thermometer inserted into the breast should read 165 degrees. Take the whole turkey out, remove the legs and thighs, and finish cooking them in the oven for about half an hour in the remaining liquids. Place the legs back near the body for presentation.

After raising the bird and slaughtering it yourself, your Thanksgiving turkey might possibly be the best thing you ever tasted. Be sure to invite all your friends who helped raise the bird so they can celebrate the turkey as well. Use the carcass for making a rich soup stock—simmer with onions and herbs (but do not boil or the stock will turn cloudy). This extraction of the best of the turkey is the final way to celebrate the whole animal.

CHAPTER 18

RABBITS

Rabbits might just be the new chicken. More and more urban farmers are discovering the benefits of raising rabbits for meat, fertilizer, and hides. Bunnies have a very low impact in a city environment: They're quiet, prefer to be kept in shady locations (thus not taking up space in the vegetable garden), and reproduce quickly. Rabbits provide high-quality, hormone-free meat that is low in fat and delicious when properly prepared. Two so-called waste products—their fur and poo—should not be ignored. You can tan their hides, which can be used to make clothes, blankets, or toys for pets (dogs love rabbit hide). Rabbit manure is an ideal amendment for the garden and can be added directly to your vegetable beds and fruit orchards.

To start out on your rabbit venture, we recommend the bare minimum of animals: two does and one buck. If bred three times a year, and the does give their average litter of eight kits (baby bunnies), you will have fifty rabbits for your table in an average year—that means you'll be eating rabbit almost once a week! So starting small doesn't mean you'll be missing out.

Requirements:

* motivation for growing rabbits: whether to sell the meat to someone else or to eat it yourself. If it is for your table, make sure you like the taste of rabbit meat.

THE RABBIT LIFE CYCLE

Female rabbits are called does. They can become pregnant as early as four months old, but in order to be a good parent to their kits, or offspring, they shouldn't be bred until they are at least six months. Does can breed for about three years and live for over six years; most breeders cull the older, unproductive does.

Male rabbits are called bucks. They don't become potent until they are six months old but will be more successful at eight months, and the most able to reproduce and sire large litters at one year. Bucks can live to be ten years old, and generally remain potent for five or so years.

Rabbit babies are called kits. They can be accurately sexed at three weeks old. Many people wean them from their mom at eight weeks, and they can be kept together until they are three months old; then they will begin to exhibit sex behavior. Young rabbits, sometimes called fryers, are harvested at eight to twelve weeks and are very tender. "Roasters" are twelve weeks to six months old and typically do better braised or slow-cooked.

- a partially shady area to house your rabbit pens, fattening cages, and yard area (collectively known as the rabbitry)
- protection from predators such as dogs and raccoons
- metal cages or pens no smaller than 3 feet by 3 feet for each rabbit
- at least one doe (female rabbit) and one buck (male rabbit) of breeding age (at least six months old)
- kindling boxes for rabbit mothers to give birth in
- daily feed of high-quality rabbit pellets or concentrates (usually made of alfalfa and barley)
- critter-tight, metal feed-storage bins (mice love rabbit food)
- fresh water in a drinking crock, watering bottles, or self-watering system
- straw, leaves, and/or sawdust to line the pens and sprinkle on the ground to reduce flies, odors, and make the rabbits comfortable
- access to grass, weeds, and vegetable scraps
- a place to compost soiled bedding and rabbit poo
- willingness to experience killing a rabbit at least once, so you know you will be able to be a meat rabbit farmer

AVERAGE COST OF EVERYTHING YOU'LL NEED TO START RAISING RABBITS

Cages: $50 each

Breeding age rabbits: $20–$50 each

Bag of feed: $15 for a fifity-pound bag; $25 for organic

Water bottles/crocks: $6 each

Metal feeders: $12

Kindling boxes: $11

SETTING UP

So you have a shady spot in your garden. Or a deck that is well ventilated and fairly shady. Or an area on the side of the house that is never used and isn't terribly close to your neighbor's house. Any of these places make an ideal spot to raise rabbits. When you are considering the right placement for them, remember that bunnies are quiet, and so they won't disturb the neighbors with loud noises. However, rabbit urine is quite smelly, and so you must be diligent about cleaning out cages regularly or your neighbors might complain.

HOUSING REQUIREMENTS

When considering placement you want to give the rabbits as much space to run around as you are able, but the bare minimum is 9 feet by 9 feet for a herd of three rabbits (two females, one male), each adult rabbit must have its own cage. An ideal situation is that they live in 3 foot by 3 foot wire pens but are occasionally allowed into a larger area to get exercise. This area must be secure from predators or escape. Though some people think bunnies should be roaming around free, consider the natural habitat of a wild rabbit: They spend the majority of their time holed up in a burrow, where they like to take naps and feel safe from predators. At night they become more active, and this is the time of day they prefer to roam a little, eat, and play with other rabbits. With this way of life in mind, when you set up your rabbit area, be sure to provide private areas (a milk crate is perfect) within their cage that they can retreat to. You should have a plan to get them out of the cage into the exercise area every once in a while to run around a bit. Some people use a "rabbit tractor," or a big cage that they place over grass in their yard: The rabbits can nibble, hop around, and mow and fertilize your lawn. Others build a nighttime run for the rabbits, and let a different rabbit out depending on which night it is. Whatever you plan, remember that you don't want to put mature males and females together unless you want them to mate.

For cages, we recommend using wire pens with bottoms made of a fourteen-gauge wire mesh with ½-inch-by-½-inch holes or cages specifically designed for rabbits. The holes in the bottom of the cage should be small enough to prevent the rabbits' feet from getting tangled up but large enough to allow the poo to fall through the floor for easier cleanup. Open cages also allow for good air circulation, which is important for rabbit health. Some people make their own cages, but our eyes glaze over at the designs for such plans, and we wonder if we will really save any money by building it ourselves on such a small scale. The best cages to buy, because they are long lasting, come from the Bass Equipment Company, which specializes in rabbit supplies (see appendix 11, page 544). Their sturdy metal cages come in a variety of sizes, and we've seen them in good shape even after ten years of use.

Another way to go for rabbit cages are through pet shops—get the largest cage they sell for pet rabbits. Alternatively, an all-metal dog cage can be used. If buying a dog cage, make sure the floor has small diameter holes so their feet won't fall through, and that the rabbit can't squeeze through the holes on the sides. Whatever you do, don't use wooden cages! They will absorb urine and are not easy to clean. The rabbits will chew on wood too, and might injure their mouths. We also urge people to avoid using plastic dog or cat crates for rabbits— the plastic bottoms tend to accumulate urine and feces; then it festers and creates an unhealthy environment for your bunny.

The cages should be raised off the ground—we've balanced them on milk crates, built wooden platforms, hung them with nails to a back wall, or just hung them off large hooks attached to the ceiling with heavy rope or chains, so they were suspended off the ground. The nice thing about having them raised above ground is twofold. One is, you don't want to introduce disease, and having rabbits on the same patch of dirt will likely cause coccidiosis, a deadly rabbit disease that is caused by a protozoan (see page 459). The other is, it's easier on your back to feed them and give them water at waist level.

IDEAL RABBIT AREA

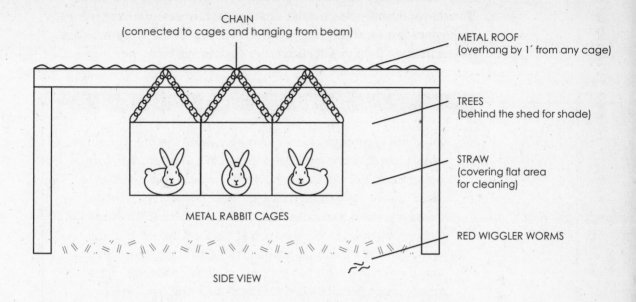

CHAIN
(connected to cages and hanging from beam)

METAL ROOF
(overhang by 1´ from any cage)

TREES
(behind the shed for shade)

STRAW
(covering flat area
for cleaning)

METAL RABBIT CAGES

RED WIGGLER WORMS

SIDE VIEW

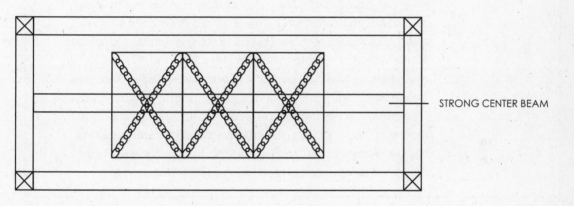

STRONG CENTER BEAM

TOP VIEW

CAGES

- Lounging crate (milk crate/box)
- Water bottle
- Feed bowl

SHED

- Should be sturdy and have a wall to protect against prevailing winds and a roof for rain
- There should be trees or shrubs on one side to provide shade and block the wind

An ideal rabbit set up keeps the rabbits off the ground, protected from rain and too much sun, and allows for easy cleanup of their droppings. Breeding stock should have their own separate cages.

The rule of thumb is, each adult rabbit must have its own cage or area. If they are put together in a pen, they will reproduce willy-nilly, and does will often fight or kill each other's offspring.

RABBIT FARMER PROFILE

Shirley has been raising rabbits for twenty years. She keeps about seven sable does and one California buck in a shed with a dirt floor in her suburban backyard in northern California. The cages are raised up 4 feet off the ground and are attached to the sides of the shed using hooks. The cages are open-bottomed, so the rabbit manure falls down onto the ground of the shed. Instead of mucking out the rabbit poo once a week, Shirley has added red wiggler worms in trenches underneath the cages. The worms compost the rabbit droppings quickly and cleanly. Shirley turns the litter every now and again, and adds a little bedding for the worms. When she's ready to harvest the composted rabbit poo, she scrapes a quantity of the finished compost off to the side, leaves it for a day for the worms to migrate to the fresh bedding, and then adds the now worm-free finished compost to her garden. Shirley finds that "red worms mean less work than no worms." Her town gets very hot weather in the summer, so she doesn't breed during the hottest months of the year. During a heat wave she adds frozen water bottles to the rabbit cages, so they can chill out. She harvests the rabbits for her dinner table, and sometimes sells breeding stock.

In addition to being off the ground, the cages should be protected from the sun or rain. Many people build a three-sided structure with four legs connected with a roof made of tin or some other waterproof material that extends a foot or so from any cage. The back wall is built against the prevailing wind to keep the rabbits warm, but bear in mind that rabbits prefer cold over hot, so the area should have plenty of ventilation—an old metal shed without windows would *not* be an ideal place. In colder areas an enclosed shed should be built or retrofitted to protect against snow and ice and chilling winds, but again, it should provide ventilation. The open side of the structure is where you'll access the cages to feed and water the rabbits.

PREDATORS

The rabbit's biggest predator is dogs. Untrained dogs often see rabbits as an irresistible plaything and will want to pick them up and give them a shake. This often breaks the rabbit's neck. Rabbits can also actually die of fright, and nothing scares them more than just even seeing a dog. Another predator is the raccoon. We've actually had raccoons pry open a rabbit cage in order to feast on one of our prized bucks, and we've heard of them reaching through the bottom of cages to snatch babies or body parts for eating.

The only recourse for avoiding these predators is to build very sturdy cages that can't be opened by or accessed by raccoons, and to keep them away from all dogs. Suspending the cages in the air avoids

BLACK GOLD: RABBIT POO

If you can't or don't want to use worms to break down the rabbit manure like Shirley does (see page 444), you should muck out the rabbit area at least three times a week, using a stable shovel to scoop manure, wasted feed, and other scraps that have fallen from the bottom of the cage. Mucking out is much more convenient if you have hanging cages, or an uplifted area, so you can easily access the floor with the shovel. Some people place a bin below the cages that catches the droppings. Scraping out the floor or removing droppings from the bins below the rabbit area reduces odors and flies; it also harnesses the amazing potential of bunny poo to help growth. We scoop our rabbit manure into buckets, then work it directly into the garden soil, taking care not to put the manure near leafy greens like lettuce. If you don't have a garden, we've heard of people who actually sell their rabbit droppings to gardeners and marijuana farmers! If you don't want to add the poo directly to the garden, you can compost it, considering it as a "green" that must be mixed with layers of carbonaceous material like dried leaves, straw, shredded newspaper, or cardboard. Add lime to keep odors down, and keep it moist for quick breakdown. After it has fully composted down, add it to your garden beds and watch your vegetables get huge! See page 147 for more information about composting.

445

dog-on-rabbit confrontations. The best-case scenario is to have the rabbits on a back porch or deck, where predators are unlikely to venture.

BUYING BREEDING STOCK

Once you have your cages and rabbit area set up, you'll need to buy your breeding stock. This is the most important decision you'll make, because every offspring will express the traits of its mother and father. So buy carefully, and consider the following.

Breed

There are many different breeds of rabbit; some of them grow to be quite large, like the Flemish Giant; some are known for their hide production, like the Rex breed; others, like the Angoras, are fiber animals, which has soft hair that it is combed and spun into yarn. You'll need to decide what your goals are before buying the bunnies. If pelts are your main focus, get a sample of a tanned hide from the breeder: Is it the correct thickness? Color? Feel? If you are looking for meat production, find out the meat-to-bone weight ratio. The Flemish Giants are huge (up to twenty pounds) but have large bones that make them less desirable (they eat more yet produce less meat), and as a practical matter, come slaughter time bigger is not always better. We tend to favor the meat breeds like the New Zealand or California. Combo meat and fur rabbits like the Rex, American, and Silver Fox are also attractive, because their meat is delicious and their fur is thick and beautiful. As urban farmers, we can't recommend keeping Angora rabbits, because they are quite prone to illness, require pristine conditions, and don't yield enough fur to make small-scale farming of the breed a practical choice unless you are a totally dedicated fiber geek.

Breeder

The best idea is to buy your rabbits locally, from someone who knows what they are doing. This person will let you know why the breed they raise is best for your area's climate. Take a tour of their facility to see how the rabbits are being raised. The place should be clean and the rabbits should be calm. We also like 4-H shows and state fairs as places to buy high-quality breeding stock that is appropriate for your region.

Health

When buying your rabbit, check its ears (should be clean and mite-free), its eyes (bright and healthy), and its paws (no sores). The rabbit's fur should be clean, soft, and glossy. There should be no impacted poo or droppings around the tail. Its teeth should be straight and even.

Age

Young rabbits, ages two to five months, aren't sexually mature. Since they are young, you can train them to be friendly and get them accustomed to being held and used to your particular rabbit setup. They will also be less expensive than mature breeding stock. However, you won't be totally sure of their exact adult size, which can be a problem. We once bought a young buck that grew up to be a total runt—not a trait we wanted to pass on to his offspring. Buying breeding rabbits that are six months or older will be more expensive, but at that age they still will be relatively flexible about housing and holding, and they'll be ready to breed immediately.

Cost

Don't be surprised if a doe with registration papers costs fifty dollars and likewise for a buck. Cages and breeding stock will be the biggest investment you'll make, but bucks will be able to provide their services for over five years, and does can produce young for three to five years (they will live much longer, so have a plan for your does' retirement scenario).

Taste

If you are buying rabbits for meat, it makes sense to try the meat of the breed and type of rabbit you are planning on raising. It is surprising how rarely people actually do this, despite the clear advantages. Most rabbit breeders will happily sell you a killed and cleaned rabbit so you can get a sense of the flavor of the meat. If you don't like the flavor, or something else about the meat, it's possible that the breeder isn't right for you. Rabbit meat should be mild and firm and white; it is slightly sweet. It's highly versatile, and can be poached like fish, fried like chicken, or braised or roasted like chicken.

BRINGING YOUR RABBITS HOME

When you go to the rabbit breeder or state fair, it's fine to bring pet carrier crates to transport them back to your house. In a pinch, a cardboard box lined with straw works fine too.

HANDLING RABBITS

If the rabbits are very young, they can be carried like kittens, by the scruff of their neck. But if they weigh over a pound, it's best to just scoop them up from behind in a bundle and cradle them against your body. They feel most safe when they can burrow their heads under your folded arms. But even with the most diligent handling, your rabbits will scratch you. They have very sharp back claws, and when you place them into their cages, their tendency is to kick off your body with their hind paws, which then gouges your arms. When we started breeding rabbits we would be covered with scratches on our forearms, to the point that we looked like cutters. To avoid this unsightly and painful fate, when you place them into their cages, put their back end in first, then let go. As an added precaution, wear a long-sleeved shirt when handling them. Similarly, pick them up from the back end—that way their toenails don't get caught in the cage, which can seriously injure the rabbit.

When you first get your rabbits home, watch them for the first few days for at least half an hour each day. Their body language says a lot about how they are feeling about their new home. They should be lounging around, sleeping, eating, and grooming themselves. If they are constantly hunkered up, ears back, there is something wrong. Check to see what might be causing their angst: Too much sun? Place a bit of cloth over part of their cage. An uncomfortable cage? Place a handful of straw or leaves along the bottom. Not enough privacy? Add another room by placing a plastic tub or milk crate for them to hide in. Too much breeze? Move the cage out of the direct wind. It could be a million things, so tune in and experiment.

WATER

Rabbits need to have drinking water at all times. They hate being hot, and drinking water cools them down. It's also essential to them for digesting food. If the rabbits aren't eating, the first thing to check is if they have run out of water. It can be placed in a crock or a bowl, but it might get filled with turds and urine, depending on the rabbit and what they're used to. We recommend using rabbit water bottles that can be suspended from the outside of the cages. These are inexpensive and easy to find at pet food stores. The downside of water bottles is they can build up algae during hot weather (not a healthy thing) and must be cleaned regularly with a scrubber and soap or bleach. Larger rabbit operations employ a form of regulated drip watering that involves hooking a hose to PVC water lines and emitters that go into each rabbit cage. This is really only appropriate for people keeping six or more rabbits, and it can be a time saver for large urban farming operations. Some rabbits are used to drinking out of bottles, crocks, or emitters and might have trouble learning a new method, so ask your breeder which system they prefer.

FEED

Most commercial operations feed their rabbits pellets. The main ingredients of pellets are generally a combination of alfalfa, barley, and minerals. Pellets are about 16 percent protein. Feeding with pellets is very efficient and clean—they are dry and are added to a feeder daily. Because they can get dusty, buy feeders with a dust screen at the bottom. On average, bucks and non-lactating does will eat about a cup of feed a day. Be careful not to overfeed as a fat doe does have trouble conceiving and birthing. A lactating doe should be offered feed free-choice (as much as she wants to eat). Making milk takes energy, and soon the kits will be sharing the feed with her. Pellets are great for busy people, but after pouring them into a hopper week in and week out, you'll notice the rabbits seem kind of bored—and so are you. So we encourage you to feed them scraps from your garden—carrot tops, flowering lettuce, chicories that are too bitter, broccoli greens, apples that didn't make the cut, weirdly shaped cucumbers. Introduce the greens and produce slowly, and remember that these are just snacks—a fun supplement to pellet feed. There are some people who believe feeding rabbits greens and such will make them sick—especially younger bunnies. We have never experienced that problem, because we introduced the green snacks slowly. We notice that the rabbits seem to look forward to feeding times when they are offered some fresh snacks, and literally pounce on the yummy produce.

If your garden isn't kicking out extra produce, consider visiting the Dumpster. See Dumpster diving on page 156.

Please note that all rabbits are different, some adore carrots and celery, others will only eat fennel. So experiment. Here are the general guidelines:

Do's:	Don'ts:
apples	cabbage (too much will cause diarrhea)
bananas	
beets	chard
carrot tops and carrots	garlic
celery tops and stalks	leeks
daikon radishes	moldy fruits or vegetables or bread
dried alfalfa	onions
fennel stalks and bulbs	
fresh lettuce leaves, not scummy	
greens like collards, kale	
herbs, like parsley	
peaches	
plums	
radicchio	
spinach	
stale, hard baguettes and loaves of bread	
weeds like malva, fennel	

You'll notice that most rabbits will prefer to eat greens over pellets, but you should still offer pellets to them, because they include mineral, fiber, and salts that they must have. When trying out a new food, offer them a little at first. If they don't get sick (diarrhea, lethargic), then you can feed them more the next day.

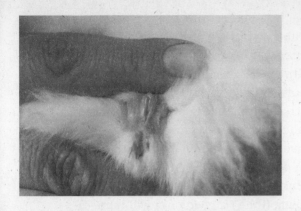 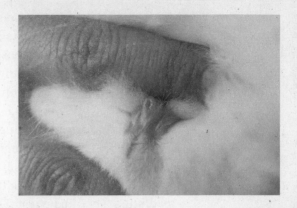

The female genitalia of a rabbit, left, looks like a slit; the male genitalia is more of a circle.

SEXING

You can sex a rabbit after they are three months old. Simply cradle them backside down and press gently just above their genital opening. Males, upon pressing, will protrude the head of their penis—it will look like a little circle. Female rabbits, upon pressing, will look like a slit, not a circle. Rabbit sexing can be tricky; the female and male genetalia look very similar.

BREEDING

Rabbits do breed like rabbits: early, often, and with large litters of five to eleven kits, or baby rabbits. Does become fertile at around five months old but typically aren't good mothers until they reach six to seven months. Their estrus cycle lasts thirteen days, and they go into estrus twice a month. Basically that means they are capable of becoming pregnant almost every day. Following mating, the gestation period is only thirty-one days. By the seventeenth day after breeding, you can gently palpate the doe's stomach to feel if she is pregnant. The babies will feel like marble-size lumps in her belly. Another way to tell if she's pregnant is to put her in the buck cage—if she runs away from him and grunts, she's probably preggers.

As a rabbit farmer, your goal is to orchestrate their nuptials in a way that keeps the rabbits healthy and productive. There are many different breeding patterns—intensive breeding methods have the doe breeding four or five times a year. We think that's a little too hard on the does, and it can contribute to their early death. We prefer a more mellow breeding cycle, such as three times per year—it's easier on the doe and her offspring, which you will be eating.

To breed a rabbit, simply place the doe into the buck's cage, usually in the early morning. If you do the reverse—buck into the doe's cage—the doe might attack the buck as an invader. But at his bachelor pad, she's a curious guest. The buck (if he's a good one) will immediately start smelling the doe and introducing himself. After a few minutes, and if the doe is receptive, the buck will mount her, thrust for a minute, and then fall back in a fuzzy heap, spent. We like to leave the doe in the buck cage for at least an hour before moving her back into her own cage. They will mate repeatedly, and if all goes well, you'll have babies in a little over a month. If she doesn't seem receptive and tries to escape or fights with her suitor, take her out of his cage and try again in a few hours or the next day.

Around twenty-eight days after mating, if it was successful, the doe will begin to build a nest with straw and fur pulled from her chest and belly. This nest building is called kindling, and the result is quite beautiful: a large ball of fur about the size of a football. This nest keeps the kits warm. On the twenty-eighth day (you'll be reminded by her frantic nest building), you will want to place a nesting box into the doe's cage. Some does will want to build their nests early—say, on the twenty-first day—and will run around gathering straw and seeming frantic. If this happens, it's fine to put in the nesting box early. Note that you don't leave the nesting box in the cage all the time, because eventually the rabbit might start using it as her toilet box. The purpose of the nesting box is to create a safe, warm space for the kits that will also prevent them from falling through the cage floor. A nesting box can be a commercially purchased item (they cost around eleven dollars), or you can build your own from scratch. The easiest nesting box to "build" is an extra long milk crate or large plastic Tupperware storage bin lined with straw, with half the top covered with a piece of plywood.

The plywood top provides privacy for the doe and her kits and keeps the wind out. The straw should be tamped down so the baby bunnies don't fall to the bottom and suffocate. Resist the urge to put a soft sweater or some clothing item in the box—it will suffocate the

NESTING BOX FOR RABBITS

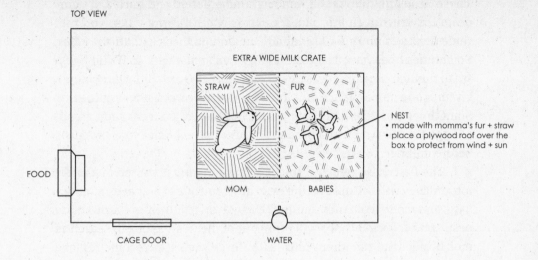

TOP VIEW

EXTRA WIDE MILK CRATE

STRAW

FUR

NEST
• made with momma's fur + straw
• place a plywood roof over the
  box to protect from wind + sun

FOOD

MOM

BABIES

CAGE DOOR

WATER

MILK CRATE DETAIL

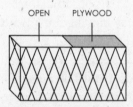

OPEN   PLYWOOD

__Nesting boxes should be placed in a pregnant doe's cage twenty-eight days after breeding; they should be lined with straw and protected so they are not drafty.__

babies! Straw only. Place the nest box in the corner of the cage, where the doe doesn't normally pee.

On day thirty-one, the kits will be born. It's a good idea not to disturb them for at least three days. Then you might want to check on them: count the young, make sure they are warm. If a doe has a small litter, it's a sign of stress; they can abort their young spontaneously if there's some factor they don't like in their area—or an inadequate buck. The young are born with their eyes closed and are covered with a very light coat of fur. They remain warm in the nest with their brothers and sisters. The doe nurses the kits throughout the night and day, but mostly sits outside the nesting box. Sadly, sometimes a kit will die.

They will appear outside the nesting box or in the nest, cold and dead. These dead ones should be removed immediately and buried or composted. Even though it is sad, it's part of nature's way. When you find some dead kits, you should evaluate the doe, as she is usually the killer. Sometimes does have large litters and can't cope with it. If she had a small number of young, or is not a good mother (e.g., killing her young), give her one more chance. But if it happens again, consider culling her. Sometimes the does kill their young because of a stress such as not enough food or water, or a perceived threat, so evaluate your practices before culling.

The first three weeks of the baby bunnies' life is passed in a milky torpor. They will roll around the nest, grow fur, open their eyes after ten days, and nurse from their mother. They begin to adventure around the nest, and precocious ones will jump out of the nesting box in search of food besides their mother's milk. After three weeks you should remove the nesting box from the cage. Depending on the size of the cage, the mother will be fairly annoyed with her young. If you can, provide her with some alone time by moving her into a separate cage for a few hours a day, or by giving her a high platform to sit on, undisturbed. After five to eight weeks, it's okay to wean them by moving the kits into their own caged area.

The kits are ready for harvest at twelve to sixteen weeks. It's a broad time frame, and you should let your needs dictate this decision, asking yourself: Are the kits big enough? Is your rabbit area getting too crowded? Are they starting to become sexual? You want to slaughter any kits that you don't plan to keep around for breeding. Depending on the breed, they will weigh three pounds to eight pounds at slaughter. Note that a four-pound fryer (young rabbit) will weigh around two pounds after skinning and gutting.

SLAUGHTER

Killing a rabbit is not fun, even for a seasoned farmer. They are mammals—quite soft and beautiful, and they can resemble a cat. However, they are also very good eating, and such voracious breeders, that if you didn't kill some you would soon find yourself overrun with rabbits. And so, we kill them with respect and a good appetite, in the most humane way possible.

There are lots of ways to kill a rabbit: bash its head with a blunt object, slit its throat, shoot it in the back of its head, or break its neck. We prefer the last, because it doesn't require special tools and bashing seems a bit too gruesome. Slitting the throat makes for the cleanest carcass but also allows the rabbit to scream, which is a sound no one will like. Some practiced people can easily grab a rabbit by the head and break its neck. We think the most foolproof method is to break its neck using a dowel; following are the illustrations and step-by-step instructions for that process.

Setting Up

You'll need the following, preferably in an outdoor area:

- strong wooden dowel
- an extremely sharp boning knife
- a pair of lopper pruners designed to cut a branch 2½ inches thick
- nail a 2 foot by 4 foot piece of lumber across a tree branch or barn wall, then tap two nails halfway into the wood, about 4 inches apart. Tie a piece of heavy duty string or twine from each of the nails, making a hangman's noose.
- a bucket for offal, and blood
- aprons
- towels
- bucket with ice

1. Saying good-bye.

First select the rabbit you are going to kill, preferably making sure that it hasn't eaten in twelve hours. Bring it to the place where the act is going to happen. Never kill rabbits directly in front of other rabbits: They can smell blood and will get freaked out. It's important to show respect for the life of the rabbit that's about to be slaughtered. We like to burn some ceremonial plant—sage, tobacco—before a slaughter. This calms us and the rabbits—it puts us into a different place than the everyday. We then thank the rabbit—with words, by telling a story, with an embrace or a kiss.

2. Preparation.

Place the rabbit on the ground and lay a dowel (it can be a broom or a rake handle, as long as it lies flat) across the rabbit's neck. Step onto the dowel, placing your feet on either side of the rabbit's head. This will apply pressure to the rabbit's neck.

RABBIT PROCESSING

Step 1:
Place the rabbit on the ground.

Step 2:
Place the dowel across the rabbit's neck.

Step 3:
Stand on top of the ends of the dowel so that the rabbit's neck is pinned. While standing on the dowel, reach behind and grab the back legs of the rabbit and pull up with a snapping motion. This will break the rabbit's neck.

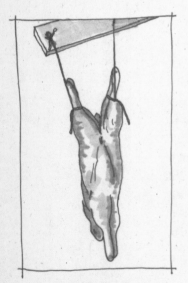

Step 4:
Hang the rabbit by its back legs and remove the head.

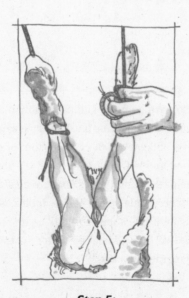

Step 5:
Make delicate cuts into the fur around the ankles until the pink flesh is revealed. You can then use your fingers to work the fur off.

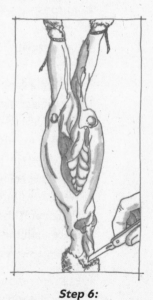

Step 6:
As you pull, the hide will begin to easily peel off the carcass. Guts can be removed before or after taking off the hide.

3. Break the neck.

While standing on the dowel, firmly grasp the rabbit's back legs—the left leg in your left hand, the right in your right. Pull the back legs straight up at a 90 degree angle, being mindful to not fall off the dowel, and give a firm yank with intent to kill. You will hear the neck break, and the rabbit should thrash around a bit. Holding the back legs, step off the dowel. If it's clear the neck didn't break yet, the rabbit is injured; you probably didn't yank hard enough. We've seen this happen once: The rabbit was stunned but not dead. The easiest thing to do is get out the loppers and cut the rabbit's head off, finishing the job quickly.

4. Head removal.

After the neck is broken, and *as soon as possible*, with a pair of loppers, cut off the rabbit's head and hold the body over a bucket to catch the blood. Alternatively, you can slit its throat with the knife. Either way, the goal is to get the blood out of its body while it is still thrashing a bit. When left in meat, blood can carry disease, and so we want to remove as much as possible. Blood by itself isn't a bad thing, however; the French, British, and Filipino peoples save the blood and make it into a sauce or delicious sausages. They will place a bowl over the rabbit to collect the blood. To prevent coagulation, they stir it vigorously, or add a few drops of vinegar and stir.

5. Skinning.

Hang the rabbit from its feet by slipping the back feet into the hangman's noose. Hang the rabbit up so its belly faces toward you. (This isn't the only way to skin a rabbit: some people use hooks passed through the tendon on the back of the rabbit's legs; others skin it on the ground or a table.) It is much easier to skin while it is hanging, in our opinion. To make the work of cleaning the carcass easy, adjust the position of the rabbit so its feet are at your eye level. Once hung, get out a small but sharp knife and make a cut at the rabbit's back ankle, just below where the string is tied. You will see a bit of pink flesh underneath the fur. The goal is to remove the fur without cutting into the flesh of the rabbit, so in most cases you will make small superficial cuts, and then use your fingers to slowly open up a hole.

After you've cut around both ankles, revealing the pink skin below, insert your fingers and slowly rip the fur off the flesh, revealing the back haunch of the rabbit. Slide your fingers along the back of the rabbit, easing the fur off around the tail area. Come up to where the tail is, and using the knife, cut upward to the tail. Then slide your fingers along to the front of the rabbit, and make a similar cut to release the fur from the front. Be careful here, as the bladder and rectum are in this region. Those cuts made, the fur and skin should start insiding itself out as you pull it down to the rib cage, like when you take off a piece of clothing. This is called "case skinning": As the French say, you are pulling off the rabbit's pajamas. Cut off the front paws before entirely pulling the fur coat off. Pull strong and steady downward with both hands. You will need to yank very hard around the shoulder and neck areas, sometimes even using a knife to work the hide off the muscle attachments.

6. Gutting.

After you've pulled off the coat entirely, you will be left with the carcass, guts intact. At this point, some of the intestines or the bladder may be visible. If they are not, make a small incision at the belly button with the tip of your knife. Careful not to cut into organs, cut a shallow line up toward the rectum. This will expose the internal organs. The bladder is the yellow fluid-filled bag; cut this off, being careful not to spill the urine. Your next goal is to remove the intestines without getting poo on the meat. You must open up a passageway to work the intestines away from the meat. With a heavy pair of scissors or kitchen shears, make a cut at the meeting of the two back legs. Then, with each hand holding one side of the rabbit's haunches, crack the pelvis along the cut you made. With a knife, cut again along the area you cut with the shears and your hands, until you have cleared a direct passageway to snake the rabbit's rectum and connecting intestines out. At this point, mindful of the end of the rectum, you can pull harder and ease the rest of the rabbit's guts into the bucket. The one organ from this pluck you might want to save is the liver, which is delicious—but be sure to carefully remove the gallbladder, which is the dark green bag attached to the liver. If the gall breaks, bitter-tasting green bile will leak out, ruining the liver and any other meat it touches, so be careful. The kidneys can remain attached on the back or cut out. Now make a cut through the rib cage area, which will expose the lungs and heart. Pull out and reserve the heart; place the lungs and trachea in the gut bucket. Cut off the back feet, using a lopper or strong pair of pruners or scissors and catching the carcass as you do so.

7. Burying.

We usually bury the intestines and blood in the garden, under a hungry tree. We save the heads and skins for tanning—they can be stored in plastic bags in the freezer for up to a year. We like the "hair on" section of Tamara Wilder's *Buckskin: The Ancient Art of Brain-tanning* for learning how to tan rabbit hides.

8. Brining.

Place the rabbit in ice-cold water and wash out the inside of blood, and any leftover bits. Take the rabbit into the kitchen and then submerge in a 10 percent salt solution. We use a clean bucket filled with water and dissolved salt. Store in the fridge for twenty-four hours. The salt draws out any remaining blood. After a day, wash off the rabbit, pat it dry, and it's ready to cook! Unlike chicken or duck, the rabbit doesn't have to be rested and can be cooked on the day it is killed. Some people, mostly British, prefer their rabbit hung for a number of days before eating it. To each his own.

9. Cooking.

Many chefs cook an entire rabbit in the oven, larded with bacon, or braised with wine or aromatic vegetables. Another good way is to cut it into pieces, soak it in buttermilk overnight, and then fry it like chicken. We also like to poach a whole rabbit in boiling water with onions and herbs. Once it's cooked and cooled, we pull off the flesh and use it for making stir-fries or add it to salads like chicken. The bones make a very nice stock.

COMMON BUNNY PROBLEMS

We consider rabbits to be fairly easy to keep, but they are more sensitive than other critters on the urban farm. Here are a few things that we've noticed that can go wrong when keeping rabbits.

Coccidiosis

A protozoan causes this disease; its biggest symptom is diarrhea. Often young rabbits pick it up because of overcrowding and unsanitary conditions. The best way to avoid coccidiosis is to keep their area clean and dry, change the bedding with fresh straw at least once a week, and maintain a disinfecting regime in which the cages are cleaned with bleach or borax and then dried in the sun. If you notice your rabbits are growing slowly, they may have Liver Coccidiosis; process one and examine its liver, if riddled with spots, it is infected. The liver should not be eaten, but the flesh is fine. Sterilize the cage with bleach, water, and sunlight.

Heat

Hot weather can be a serious problem, as rabbits have a very poor cooling system. During a heat wave (temperatures in the nineties and up), rabbits can and will die from getting too hot. To prevent this, set up fans, misters, drape wet towels across their cages, place frozen plastic water bottles in their cages, move them to a cooler spot—do whatever you can to keep them cool, or they will die. Does with kits during a hot spell will have the most difficult time, so consider not breeding during the summer months. Along those lines, a buck exposed to hot temperatures for an extended few days might go sterile for up to six months.

Ear Mites

The rabbit's ears become encrusted with a scary-looking brown, crusty gunk. You'll notice your rabbit scratching at its ears. If left untreated, the rabbit can go deaf or get an infection. Treat with ear mite medicine, which you can buy at most pet stores. Since mites migrate from the soil up to the rabbit cages, you can prevent them from entering the rabbit's area by coating the legs of their shade structure with Tanglefoot, that sticky glue used to prevent ants from getting into fruit trees.

Red Urine

Not to worry—a reddish-colored urine is totally normal. It often appears after rabbits eat carrots, during a cold snap, or while on antibiotics. Bloody urine would form speckles of red, while red urine is a consistent red color throughout.

Long Toenails

As your rabbits age, pay attention to their toenails. They might begin to curl and injure the rabbit. Using toe clippers, trim off too-long nails, keeping in mind that the blood vessels do go into a rabbit's toenail, so if you cut too much, they will bleed.

Bad Smells

No matter how much you clean, the rabbits will smell a little bit. This is because their urine is incredibly pungent, especially that of the males, which can be projected several feet. This is why we raise rabbits in metal cages. Every month or so, when odors start to become noticeable, take the rabbits from their cages and give the cages a good scrubbing with borax and warm water. Let them dry in the sun or towel them off, replace the bedding in the cages, and return the bunnies to their new clean homes. The area beneath the rabbits will become smelly very quickly. Unless you are doing a worm system, plan on mucking out the rabbit area at least once a week. Compost the straw, droppings, and urine or spread it directly on the garden. With extremely urine-soaked straw, we recommend sprinkling some sawdust or dirt over it, to prevent a major stink session.

So Many Flies!

Other people have worked out systems for different animals to live beneath the rabbit cages to deal with flies and excess manure. Long ago, the people at the Integral Urban House (an experimental 1970s urban farm in Berkeley) put their rabbits in the chicken house, so that the hens mix up the poo with the sawdust and straw and eat any wasted feed the rabbits spilled. Others put worm bins under the cages; the red wiggler worms digest the poo and other wastes, and create another cash crop: worm castings.

Dead Babies

Some rabbits are not good mothers, some are good mothers but become stressed out postbirth for some reason, such as being moved, overcrowding, or the introduction of a new rabbit to the warren. These moms will kill their young and throw them out of the nesting box. You should consider her motivations—and if you were complicit in the death—before deciding to cull the murderous doe.

Low Birth Rate

A litter of eleven is outstanding, eight is fine, and six is just okay. If a doe has any less than that, you might consider culling. She may be overweight, so adjusting her food might up the number of her offspring. If all of your does are birthing in the low numbers, you might consider switching the buck. Interestingly, pregnant does can absorb their young and not give birth if they are stressed-out. Another reason to keep them happy.

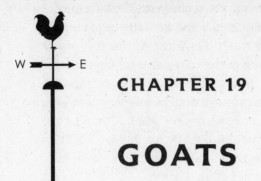

CHAPTER 19

GOATS

Goats are taking the urban ag scene in America by storm. Once seen as exclusively rural livestock, more and more people are keeping them in their backyards for milk, meat, hides, and natural fertilizer. Besides being work animals, they are also enormously entertaining. Depending on the breed, goats have similar space requirements as large dogs, and if one keeps an all-female or castrated male herd, they will not create the strong odors that goats can be notorious for emitting. Most breeds tend to be fairly quiet, with only a few nickering sounds now and then throughout the day.

They are, however, a lot of work. No one should enter into dairy-goat husbandry without full knowledge that goat ownership is an all-engrossing hobby that will suck up your time and money. Their feed is expensive. When in milk, they must be milked at least once a day and fed twice a day. You have to find (and pay for) a breeding buck. They need shots. They might gnaw on something you like. Their hooves must be trimmed regularly. Their area should be mucked out once a week and dusted with fresh sawdust and straw. All we're saying is: Don't romanticize goat ownership! It's more work than having a dog. On the positive side, they are wonderful, steady creatures that have the ability to change your life for the better. Hanging out with goats after a tough day at work is a sure stress-buster. Morning milking is a special, intimate way to start the day; it makes a primal connection to your morning cappuccino. Our lives have been enriched by backyard goats. Note that one learns a great deal of information while keeping goats over the years. This chapter is just enough information to get you started if you are dairy-goat curious.

LEGALITY

If you're wondering whether you can have goats at all, first consult your city's animal ordinances. There was a dustup in Seattle a few years ago when a goat owner had to fight for her right to keep them in her backyard. She ended up winning and started the Goat Justice League. You may have to do the same. Some cities ban goats outright, and so you will have to start by getting it legalized via your city council.

Here's a checklist for what you need if you want to keep goats:

- at least two goats (they need company!)
- a backyard to exercise in with at least 100 square feet of outdoor area per goat for dwarf breeds, 200 square feet per goat for standard breeds
- a loafing shed they can comfortably fit under to take shelter during rain storms
- a structure where they sleep at night and will probably give birth to young, protected from wind and weather
- access to baled hay and a dry place to store it
- a weather-protected manger or feeding structure that the goats can access
- bags of grains, such as corn, oats, barley, sunflower seeds
- bags of minerals like salt, kelp, and mineral mix
- waterers or buckets, filled daily
- a sturdy pair of hoof trimmers or pruners
- a curry comb and hair clippers if you want to show your goats
- medicine, supplements, and vaccinations
- copper bolus tablets (in areas where copper deficiency is a problem)
- fly prevention devices
- access to climbing structure/ exercise
- straw/sawdust for bedding
- a place to compost soiled bedding and manure
- a dry place to milk during rain or snow
- a stanchion for milking
- pails to milk into
- a mechanical milker if you don't want to milk by hand
- udder wipes/teat dip
- a milk straining device/ pasteurizer
- access to a disease-free male goat with strong milk genetics to breed with
- a truck or car you own or borrow to buy supplies or for transportation

THE GOAT LIFE CYCLE

Female goats are called does. They can become pregnant as early as nine months old. Male goats are called bucks and can be fertile as young as four months old. Male goats tend to be smellier than the females. Does can be breed for about seven to ten years, although they live to be about fifteen years old. The does usually go into estrus in the fall and give birth to their young, kids, in the early spring. The gestation period is five months and multiple young are common.

After weaning the kids, does can be milked for up to a year or longer. The lactation curve is such that more milk is made in the first few months of milking and then production tapers off. A good dairy doe will milk out almost a gallon of milk per day (with two milkings); about half the amount for miniature breeds. Usually milk production is small the first year, reaching its peak of production in the doe's third year.

Goats are herd animals and need other goats around in order to live happy lives. They must be provided hay to eat and given shelter from the elements.

SPACE REQUIREMENTS

Assuming it is legal to keep goats in your city—or you are feeling like you can get away with it even though it isn't legal—assess whether or not you have enough space. For two goats (the bare minimum number to keep them content), a law in Seattle suggests that you need a backyard that is 1,000 square feet (that's 100 feet by 10 feet) or larger. On the other hand, some urban goat herders keep their goats in an area as small as 400 square feet and take frequent walks around the neighborhood, and their goats are quite happy, thank you. Instead of throwing out a number, we think it's probably easier for people to wrap their minds around what a dog would need. So ask yourself, when you look at your yard space: Would two medium-size dogs be happy in my backyard? If the answer is yes, then you probably have enough space.

No matter what the size, the entire area should be surrounded by 5 to 6 foot tall fencing. Trees or shrubs you don't want the goats to eat will have to be protected, because they will denude a tasty shrub

IDEALIZED GOAT AREA

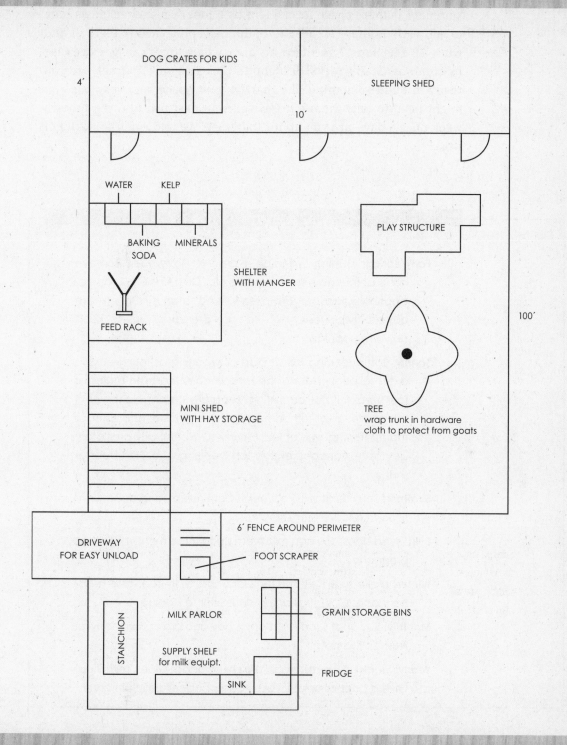

An ideal goat set up will allow for space for the goats to play, sleep, give birth, and get milked.

in moments. Is there a shady, cool spot for them to rest during the summer? Is there space for a 6 foot by 6 foot rain shelter (goats hate to get wet)? Is there an area to build a sleeping cave (4 feet tall and wide, 10 feet long)? Is there an area where they can get exercise by climbing (stairs, rocks, or a child's play structure)? If not, are you ready to build something like that? If your space isn't very big, you might need to take them out for walks at least two to three times a week once a day—are you up for that? See: We told you it was a lot of work!

A YEAR WITH GOATS

Early spring: Kidding in March if bred in November. Remove horns at four days to two weeks old. Give kids CD/T (a vaccination against enterotoxemia and tetanus) shots at ten days to two weeks old. Release predatory flies to keep down fly populations.

Spring: Bottle feeding kid at birth or weaning at four weeks in order to begin milking the mama does; castrate males at four weeks. If not castrating, separate before eight weeks old or they might impregnate a doe (including a sister or mom); trim hooves at least every six weeks; buy predatory flies for fly management. At ten weeks, give a CD/T booster shot.

Summer: Sell kids at twelve weeks. Continue milking does. Continue to release predatory flies.

Fall: Breed does. You can milk them up to three months into their pregnancy.

Winter: Make sure their shelter is dry and draft-free; give pregnant does CD/T shot one month prior to kidding.

Monthly: Buy goat supplies, such as hay and grains. Trim hooves. Freshen fly strips.

Yearly: Stock up on minerals, kelp meal, udder wipes, milk filters. Give a copper bolus if you are in a copper-deficient area.

A DAY WITH DAIRY GOATS

MORNING:

> **Let goats out of sleeping area; place hay in their manger; make sure they have clean water:** fifteen minutes.

> **Brush and milk goats:** fifteen minutes per goat.

> **Make cheese/yogurt with milk:** several hours.

EVENING:

> **Spend time with goats, either on a walk or just sitting together:** one hour.

> **Clean up goat area:** fifteen minutes.

> **Put goats away; check that they have water; place hay in their manger:** fifteen minutes.

WEEKLY:

> **Muck out goat area; put down fresh layers of sawdust or straw:** one hour.

> **Make sure fences are in good working order and that there are no dangers in the goat yard where they could hurt themselves:** thirty minutes.

> **Make sauerkraut:** twenty minutes.

> **Scout for branches and leaves for goats; cut up branches and compost or chip denuded branches:** one hour.

Like every hobby, there's a secret language goat people use. Here are some of the more elusive words defined.

Buck: Male goat.

Burdizzo: A castrating tool that will make any man wince. It acts to clamp blood vessels on the testes, so that once the testis is clamped and cut off, the goat won't bleed to death. Anesthesia is recommended, but most people go to a vet for this process.

COB: Corn, oats, barley. A favorite snack of goats, usually given as a motivator during milking.

Disbudding: Removing the small horns from a young goat. Usually involves burning the prehorns off with a hot disbudding iron when it is less than a week old. Some people use acid to remove the horns, but this can be dangerous. Again, anesthesia is recommended, or most large animal vets will disbud for a nominal fee. Overdoing disbudding can cause brain damage, while underdisbudding might cause bone scars that can grow into the brain.

Doe: Female goat that has had at least one set of kids.

Doeling: Female goat that has not been bred.

Estrus: When the female goes into heat, or is receptive to a male goat for breeding.

Flagging: Tail-wagging that signals estrus or happiness when kids are milking.

Freshening: When your female goat has her kids, and thus starts producing milk. Her udder is "freshened."

Manger: Where the hay goes for the goats to eat.

Mastitis: A contagious bacterial infection that causes painful swelling of the udder, which makes the milk clotty and sometimes bloody. Mastitis is life-threatening for a goat!

Polled: Genetically hornless goat.

Scouring: Literally, having diarrhea. Can be caused by coccidiosis, unicellular internal parasites, or diet.

Stanchion: The platform where goats stand for milking and other grooming; it usually has a way to lock the goat's head and a feeding tray or bucket to keep them occupied.

Teat: One of two nipples of the udder from which milk squirts.

Udder: A swollen gland that is filled with milk; it has two compartments.

Wean: To cease allowing the kids to nurse from their mother or bottles.

Wether: A male goat that has been castrated, often kept as pets or companion animals or as a work animal.

SETTING UP

If you've looked at these requirements and haven't run screaming to the hills, we think you are ready to start building structures, buying supplies, and scouting for your goats. Below is an illustration of an idealized goat area.

In order to transform your back, front, or side yard into a goat paradise, the first thing you must do is build fences. Goats are notorious escape artists and will jump or squeeze to freedom if they are allowed. Don't let them. For regular-size goats, we recommend a fence 6 feet tall; for dwarf or pygmy breeds, 5 feet tall. The fence should be extremely sturdy, because the goats will regularly jump up on their back legs and "stand" with their front feet against the fence. There should be no holes or gaps in the fence either, as goats—especially kids—can slip through even small openings. Cyclone fencing is best, with cemented posts at 5-foot intervals. Another option is to use 6-foot-tall sections of welded livestock panels and attach them with wire to T-posts or wooden posts sunk into the ground with concrete or mortar mix. Using panels is nice, because it makes the fence modular—you can move it around to open up new sections of your yard. We've seen some ghetto-fabulous fences constructed out of pallets standing on end and held in place with rebar that had been driven into the open ends of the pallets. There are a million ways to do it—just make sure it's sturdy enough that if a goat leans her entire weight against it, the fence will hold.

Assume that shrubbery in the goat area will be devoured immediately. On that point, don't let the goats eat plants that are poisonous to

IDEAL GOAT SETUP

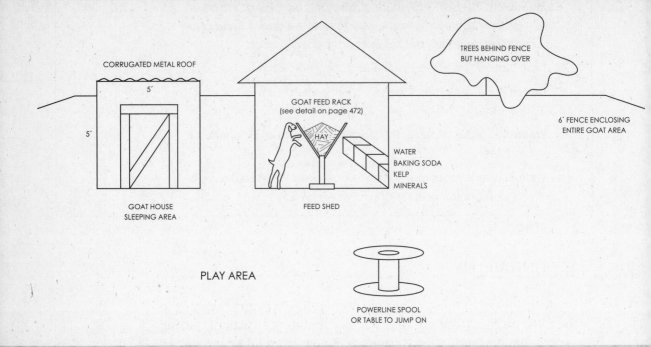

Goats will need a dry place to access their feed and minerals in addition to an enclosed sleeping area.

them, which include rhododendrons and azaleas, bay leaf, eucalyptus, brugmansia, yew, and oleander, just to name a few. Some goats know instinctively not to eat plants toxic to them; some get bored when penned and so will eat anything that is within reach, even the toxic plants. Best practice is to only allow them to eat plants that *you* have positively identified as nontoxic to goats.

If you have trees in the goat area, even large ones, the goats might chew off the tree's bark, thereby killing it. To avoid this fate, wrap the tree trunk with two layers of hardware cloth up to the height the goats can reach by standing on their back legs; or, even better, construct a box around the tree so they can't access it at all.

SLEEPING SHED

Once the fence is in place and your plants protected, it's time to build the structures for the goats. We've found that goats like to sleep in long rectangular sheds that approximate a cave, their ancestral sleeping quarters. Lucky for us, it's pretty easy to build a sleeping shed. Here are directions for one way.

Materials:

4 8-foot-long 2x4s

8 4-foot-long 2x4s

1 roll of 4-foot-tall hardware cloth

3 4 foot x 8 foot pieces of unpainted plywood

1 4 foot x 4 foot piece of unpainted plywood

Roofing material, such as tar paper, corrugated metal, or some other rainproof material

Power drill

Outdoor screws

Optional 4 foot x 4 foot door of your choice

1. Create two rectangular frames out of the 4 8-foot-long and 2 4-foot-long 2x4s by screwing them together. Connect the bottom and top of the frame with the 2 remaining 4-foot-long 2x4s.

2. Roll out the 4-foot-tall hardware cloth, stapling or nailing to each side of the rectangle, leaving one end open, which is where the goats will enter their "cave." Hardware cloth has smaller holes than chicken wire, so it will prevent rodents from bothering the goats, but it still allows for good ventilation.

3. Screw the pieces of plywood along the sides and the back.

4. Place the third 4 foot x 8 foot piece of plywood on the top of the box.

5. Attach the roofing material of choice to the top of the box.

6. If you live in an area with mountain lions or coyotes, or especially vicious racoons, make a door that fits the 4 foot x 4 foot entryway. But note that most goats don't like to be locked into an area, so they can go outside if the urge hits.

7. During the winter, leave the plywood sides intact to keep out drafts and rain; during the summer, remove all or a few sides of the sleeping structure to allow for good ventilation.

Instead of building a sleeping den, some people opt to put their goats in big plastic dog crates or igloos. These are cozy, especially when two or three goats cuddle in the pen. But they need to be cleaned more often, sometimes daily, as ammonia can build up in the dog crate faster than in a wooden shed. Dog crates are great to have on hand while weaning kids; more on that on page 493.

FEEDING MANGER

Another item to build or buy is the hay manger. The majority of a goat's diet should be hay—alfalfa, orchard grass, oat, or rye hay. This is usually given in the morning and again at night by placing flakes of hay into the manger or feeder. Ideally the goats would be allowed access to it without allowing them to pull all of it onto the ground, where they will trample and waste it. There are a variety of different ways to prevent a massive pull down: One is to make a big wooden box with keyhole shapes cut into the side of the box that just allow a goat's head to reach inside. A goat will lock its head in the hole, munching away, and won't spill it using this method. Another is to mount the manger up high so

GOAT FEED RACK

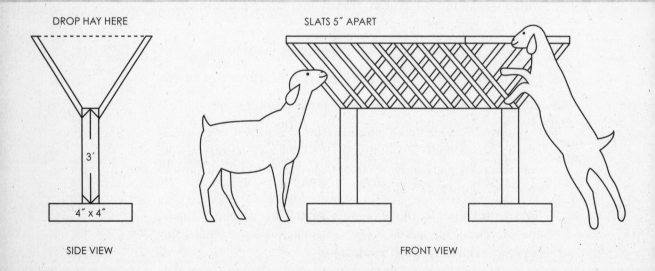

DROP HAY HERE

SLATS 5" APART

3'

4" x 4"

SIDE VIEW

FRONT VIEW

A sturdy feed rack prevents hay wasting

the goats can only access it from the bottom by standing on their back legs. Though this sounds a little mean, it's the position a goat eats in naturally. We do a modified version of both: a V-shaped manger held up high, with slats they have to reach for, and a series of side-access areas. Whichever way you design it, you should put a roof on it; otherwise the hay will get wet on rainy days, and goats shouldn't eat wet or moldy hay.

LOAFING AREA

Goats need a hangout besides their sleeping shed. We've had success inviting them onto the back porch, where they have a good view of the city and are protected from the rain by an awning. Other people might not want goat turds on their back stairs. If so, put a goat-proof gate on your porch and consider building them a play area instead. This area should have a roof but doesn't need walls like the sleeping quarters. It's where you will put their hay manger, water buckets, and small containers of minerals, salt, kelp, and baking soda that you want to offer them free choice or available any time.

Other things to include in the goat loafing area: tables, blocks, big rocks, or milk crates that they can climb and jump on. Design an area with goat fun in mind. A bored goat will overeat and cry. You want your goats to scamper and prance!

STANCHION AND MILKING AREA

The biggest motivation for having backyard goats is usually milk; if that's yours, you'll need to set up a milking parlor. This doesn't mean you have to build an all-stainless-steel milking room modeled after an industrial dairy. You just need to find a place in your house, garage, or shed where you can be as hygienic as possible when handling the milk. The space should be large enough for shelves to store all your milking supplies—such as disinfecting teat dip, udder wipes to clean udders, and milk pails—and a stanchion, which is usually about 4 feet long, less than 2 feet wide, and 1½ feet tall.

The stanchion is a necessary tool to have on the farm, whether you are milking or not, for performing tasks that require the goat to be stationary, as with hoof trimming, administering shots, worming, or

GOAT MILKING STANCHION

SIDE VIEW

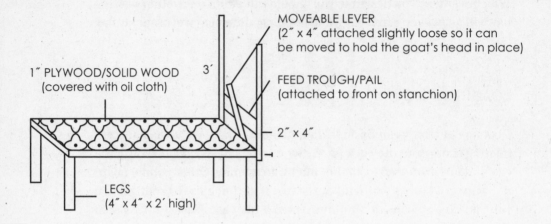

MOVEABLE LEVER
(2″ x 4″ attached slightly loose so it can be moved to hold the goat's head in place)

1″ PLYWOOD/SOLID WOOD
(covered with oil cloth)

3′

FEED TROUGH/PAIL
(attached to front on stanchion)

2″ x 4″

LEGS
(4″ x 4″ x 2′ high)

FRONT VIEW

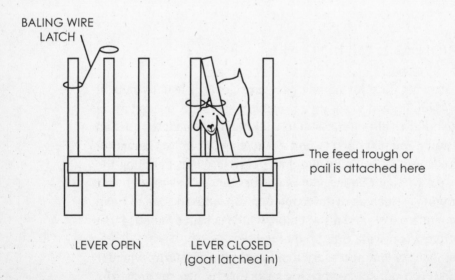

BALING WIRE
LATCH

The feed trough or pail is attached here

LEVER OPEN

LEVER CLOSED
(goat latched in)

Milk stands, or stanchions, hold goats in place for milking, hoof trimming, and administering shots.

shearing. It is basically a long narrow table for your goat to stand on with its head locked in place. Often people place a bucket of grains or other treats for the goat to eat as a distraction. An ideal stanchion will be low slung, with a step-up for the goat, a sturdy head-locking device, and a bucket or trough where the goat's head will be. We like to put a sheet of oilcloth over it, because that material makes for easy clean up if the goat poops or its hooves make the stanchion dirty. For our milk goats, we place a bath mat over the oilcloth, so it is comfortable and not slippery. The height of the milk stand depends on how you will be milking and your own personal ergonomics. You can buy a stanchion from a goat supply house (see appendix 11 on page 543) or custom-build one to fit your exact needs. The illustration on page 474 shows how our homemade milk stand works.

We do our milking in the laundry room just off the back porch. It's convenient, because we didn't have to build a separate milking parlor—we just lead the goats up the backstairs into the laundry room. It's also right next to the kitchen, so hand washing and cleaning the milking equipment is convenient. The fridge—where all milk should go immediately after milking—should also be close by. If you have a large backyard and own your house, it would be ideal to build a milking shed or parlor. This enclosed structure would contain the stanchion, a small fridge, a place to wash hands and milking gear, feed, and supplements. It's critical that the milking parlor be a place where the milk and goats can be handled with cleanliness in mind. The cleaner the hands of the milkers, the teats of the goats, and the milk itself provides for better-tasting milk and protects the goats from mastitis.

BUYING GOATS

Now that you have the infrastructure setup, it's time for the fun part: scanning craigslist and goat-breeder Web sites in search of the perfect goat. Most major metropolitan areas have a breeder within a two-hour drive. They will have Web sites with photos of does and doelings for sale. Increasingly, 4-H is focusing on goat husbandry, so its fairs and auctions are good places to find them for sale. State fairs often have animals for sale too, and are a great place to see the wide range of breeds that are available, and to talk to experienced goat owners.

Animals for sale can range from three-month-old kids to pregnant does to older does to wethers (castrated males) to bucks (intact male goats). You need to be clear about what you are looking for before you start calling breeders or going to auctions. Basically, there are three main categories of goats: for dairy, meat, or fiber/fur.

Dairy goats come in a variety of colors and sizes, depending on their breed, which is the first decision you should research extensively if you are considering them for this purpose. The quality and amount of milk varies depending on the breed. There also tend to be personality differences. Before diving in, the first question you should ask is: What are my goals for keeping dairy goats? If producing a lot of milk is your main focus, and you have the space and/or willingness to walk them, then you should consider all the heavy milk-producers: Oberhasli, Alpines, LaManchas, Nubians, Saanens, and Toggenburgs, which can sometimes provide up to two gallons per day. Each breed has their own stereotypes (e.g., LaManchas are clever, Nubians cry often), but we've found this can vary on an individual basis. Also, don't forget that crosses, or "grade," goats are great options for city keepers who aren't planning on showing their goats.

If you want a little bit of milk but don't have enough space, or have small children who might be intimidated by larger goats, Nigerian Dwarfs have worked out well on our small urban farms. Nigerians are recognized by the American Dairy Goat Association as milk animals, with top milkers producing eight pounds of milk (approximately a gallon) per day. Crossed with larger goats, they are becoming more popular too, and breeders are working on mini-LaManchas, mini-Nubians, and more.

If you're after meat, there are Boers, which are a fast-growing, heavy breed from South Africa, or the New Zealand Kiko, or pint-size Pygmies, for the backyard goat-meat rancher. If you're after fiber production, know that you must provide an absolutely pristine environment, and be prepared to brush your goat often; breeds include the Angora and Cashmere goats, or a cross called the Cashgora.

Once you zero in on your motives for keeping goats, you'll be able to find the right breeder. Many raise mixed breeds; if you want to show your goats at 4-H or any other livestock fairs, they will need to be purebred and have papers.

Temperament is very important when considering keeping goats in the city. Will they settle and be happy at your place? Will they threaten or dominate you, your family, or guests? Our goat hero, Jim Montgomery from Green Faerie Farm in Berkeley has found that bottle-fed kids are usually the easiest to work with and will accept people more readily than those raised by their moms.

When you meet with breeders, make sure to sample their products before buying. For example, taste the meat, drink some milk, or feel the fiber produced by the goats you might buy. We can't emphasize this enough! Milk tastes different from breed to breed, goat to goat. The protein levels, fat content, and other components depend on what it eats and its genetics. So don't buy a goat if you don't like the sample of milk from the breeder! Likewise with chevron (goat meat).

OPTIONS FOR BUYING

There are a variety of options, remembering that you must have, at bare minimum, two goats. Here are a few scenarios, going from the most expensive to the least expensive.

Senior Doe and Her Recently Born Female Offspring

Pros:

Will give milk immediately; you'll know from the breeder about the quantity and quality of the milk; you know the sex of the offspring and can breed her.

Cons:

This is probably the most expensive option, and the most difficult to find.

Works best for:

People who are anxious to start milking immediately and don't mind paying a high price.

Pregnant Senior Doe and Companion Goat

Pros:

Will birth and give milk immediately; you might get some good doelings.

Cons:

A good senior doe will be expensive; might have a difficult birth or give birth to bucklings; companion goats are usually wethers, who will eat but not provide milk; sometimes breeders sell pregnant does with a "buyback" policy where you have to give them the offspring of their choice—watch out for this.

Works best for:

People who are anxious to start milking now, want to witness birth, and have room for extra goats.

Pregnant or Recently Kidded Junior Doe

Pros:

Not as expensive as a good senior doe.

Cons:

Won't have optimum milk production until she is bred again; sometimes breeders sell pregnant does with a "buyback" policy where you have to give them the offspring of their choice—watch out for this.

Works best for:

People who can wait to get milk in the next year but don't want to wait two years.

Two Doeling Kids

Pros:

You will raise them to be friendly; the least expensive option.

Cons:

Doelings will have to reach breeding age, go through a five-month gestation, and usually don't provide a large quantity of milk until their second or third pregnancy.

Works best for:

People who have more time and patience than money.

Regardless of your scenario, know that if you want milk, you'll need to breed your goats every year: You don't get milk unless you "freshen" your doe with some new kids. One aspect of breeding for milk that is often ignored is that your goats may give birth to twins or triplets, thus doubling or tripling the size of your herd. Have a plan for the offspring before they are born. Many farms will harvest any male offspring for meat. Others will castrate the males (making them wethers) and then sell them as meat, pets, companions or pack animals. Doelings can be kept or sold to become milkers, but unless you have superior genetics (which you would have paid for in your stud fees), don't expect to make a lot of money by selling them.

THE SELLER

If you're a first-time goat buyer, you'll need to find a goat breeder who is willing to hold your city-slicker hand a bit. Goat breeders can be found on craigslist, goat breeder Web sites, at 4-H auctions, state fairs, and through goat fancier magazines. Choose a breeder who lives nearby: For logistical reasons we do not recommend buying goats from out of state. Since you will have lots of questions to ask the person you're buying from, choose wisely: They will be your mentor, so you want someone who is knowledgeable yet accessible and understanding.

Questions to ask goat sellers upon initial contact:

1. Are the goats disease-free? (Goats can harbor contagious diseases and STDs! A negative test for Johne's and Caprine Arthritis-Encephalitis—CAE—is important.) If the entire herd isn't disease-free, the goat you're buying might be carrying the disease. It's not rude to ask to see a copy of the test on paper from the lab.

2. How long have they been breeding goats? (Longer is better, though not necessarily a guarantee that they know what they're doing.)

3. What breed are they focusing on and why? What are the main traits they are breeding for? If the answer is blue eyes, run for the hills. You want a breeder who is carefully breeding for milk production, or high-quality meat, paired with friendliness and excellent conformation and body structure.

4. Are the goats registered or do they take their goats to shows? (If yes, it is a good sign that they are raising quality stock and are careful about disease.)

5. Do they offer boarding and stud services? How much do these cost? (For when you want to go on vacation and breed your goats.)

Before first contact with the breeder, you should do some research—carefully read their Web site to find out about their policies, look at photos of their goats for sale, and figure out what is most important to you. Remember to value their time: They probably show their goats to hundreds of people, so don't expect them to spend an entire afternoon devoted just to you.

Just like dating, it's a good idea to check out a few different options before committing. Visit a few farms with goats for sale to get a sense of your options and the methods they're using that you can replicate on your farm, and to have a wide spectrum of goats to choose from. Watch and listen very closely. If you notice that a goat is loud on their farm, it will also be loud on your farm. Ask yourself how you feel about having a few of these animals in the backyard. Since goats have personalities, ask about how they behave, whether they are friendly, if they like kids or dogs (if this is a concern). Be a good guest: Offer to buy some milk or meat or other farm products in exchange for their time.

THE GOAT

Just like people, every goat is different. They have personalities and traits that vary from breed to breed. Their milk and meat tastes different from breed to breed, goat to goat. That's why we recommend tasting these products before making the commitment. For dairy goats, you should also try milking the goat—or one of her relatives—before you buy. Some goats have unusually small teat openings which will make milking a major pain. Some will be ill-mannered on the stanchion. Some will be loving and friendly, others standoffish or aggressive. These are important things to consider! The breed of goat you choose will depend on what appeals to you, kind of like getting a dog or a cat. If you feel drawn to the gopher-eared LaManchas, go for it. If you liked the demeanor of a farm's Nubian, do it.

Once you find The One, make an offer to buy the goat(s), and make arrangements to spend some time on the farm to learn about the goat(s) before you take them home. The key things to know are:

- What is the goat eating? Take detailed notes and continue to feed it exactly what it has been eating. You don't want to change their feed dramatically—this can cause sickness.
- What is the goat's routine? If they milk in the morning, you should milk in the morning; if they feed grains at night, you should do the same.
- What shots or maintenance do they need in the next few months? The breeder should have an idea of when to next trim their hooves and give them wormer, booster shots, and the like. Ask for a copy of their care log, which should detail when these things have been done.
- In the case of registered goats, you'll need to make arrangements to transfer ownership into your name.

Take notes!

FEED AND WATER

Goats are not lawn mowers. Many a person has been disappointed when they brought their goats home, expecting them to trim the grass, and instead the goats denuded the shrubs and trees first. It's in a goat's nature to want to eat "browse," aka branches and leaves. But they will also eat grass, just not as diligently, as, say, a sheep. They also like some grains. But the main feed should always be hay.

Hay is not straw; it is a broad term for fodder used to feed ruminants. Hay includes alfalfa, orchard grass, timothy grass, oat, rye, and barley hay. Depending on the kind, it might have seed heads (oat and barley) or green leaves (alfalfa). For dairy goats, we feed a combo of alfalfa, oat, and orchard grass. Alfalfa is preferred for dairy goats, because it contains a lot of protein (about 16 percent), which gets converted into milk. Goats also like alfalfa because it appeals to their browsing nature—they can nibble the little leaves off the stalks. Oat hay or orchard grass keeps things interesting and isn't as high in nutrients—this is the primary feed for goats who aren't in milk. We tend to offer our goats hay free choice, meaning it should always be available for the goats to eat. We have a hay manger that we fill every morning and top off at night. This way the goats can decide when they want to eat. Other goat keepers feed measured amounts at certain times to prevent excess weight and wasted hay. The best idea is to observe your goats and come up with a feeding plan that makes them (and you) happy. Note that if you have a wether, you do not want to feed them alfalfa exclusively: It will cause urinary stones. If you're raising meat goats, they should be fed hay and a pelleted ration specific for meat animals.

GRAINS

Most dairy goat owners give their does COB (corn, oats, barley). It's a grain mixture usually sold in premixed bags at feed stores. Sweet (sometimes called wet) COB means it has been saturated with molasses to make it even more tasty and to bind powdered nutrients; dry means it doesn't contain molasses. Each goat is different—some love dry, others like wet. We used to buy bags of the premixed stuff until we figured out you could just buy bags of crimped corn, rolled oats, and barley and mix it ourselves (duh). We've found that this is the only way to get organic COB. How much COB you feed is dictated by the cycle of the goat's life. Dairy goat males and wethers should not be fed grains except as a treat. As a general rule of thumb, goats in milk should be given a pound of grain per pound of milk produced. Kids and teen goats should not be fed COB, except as an occasional snack. Pregnant does should be given half a cup of grain a day and limited alfalfa, as the kids can become oversized if she eats too much. We feed the grains in small buckets or pans once a day—each goat gets her own bucket or fighting will ensue. While milking, we place a bucket of grains on the stanchion for the doe to eat: This keeps her occupied, and stimulates the letdown of milk. Bear in mind that feeding goats is like feeding children: Everyone will have a different opinion, and every goat is different.

MINERALS

Depending on the area where your hay is grown, it might not have all the minerals needed for a healthy goat. Enter the mineral blocks, bags, and buckets, which contain iron, copper, zinc, salts, selenium, calcium, phosphorus, and a wide spectrum of vitamins. We buy a bucket of mineral mix through the Internet about once a year and sprinkle a bit on their grains at feeding time. Some people offer it free choice—in buckets or in block form—so the goats can decide whether they want to eat it.

OTHER FOOD

Hay, some grains, and minerals are the bare minimum. Most goat owners, like parents, think their feeding plan is the best. If your goats' eyes are bright, coats shiny, and you're getting good-quality milk, you're doing it right. There are supplements that you can add to their diets if you want. Here's a quick rundown; experiment with your goats!

Black sunflower seeds: Sold in big bags to put in bird feeders, this is a yummy snack that can improve coat quality.

Kelp meal: Just like what it sounds like, this is a powdered kelp. Goats love it! It probably has minerals in it they can't get otherwise. Some people feed this instead of mineral mixes. Goats love seaweed and salty things, and kelp meal is an inexpensive way to go for these: It costs less than a dollar per pound and is widely available at feed stores and on the Internet.

Sauerkraut: Making 'kraut for your goat makes you a total goat geek. But they love it—salty and crunchy with probiotic properties. A hefty handful per day keeps them happy. You can also buy powdered probiotics from goat supply stores.

Baking soda: Many goat people offer baking soda free choice. Since the goats are very in touch with their stomachs, they know when even a slight acid imbalance is starting to take place and will eat baking soda to quell the problem.

Dairy pellets: These aren't necessary, but if you feel like your dairy goat isn't getting all the proper nutrients, these are a palatable snack for her and contain minerals, yeast, probiotics, and vitamins.

Garden snacks: A few nibbles of lettuce or an apple or kale or broccoli leaves make great snacks. But feed them too many of these goodies, and you'll have a goat with some serious diarrhea. Keep fresh-vegetable snacks to a minimum.

Branches/leaves/browse: Since this is a goat's favorite food, you should make this a priority. But since you live in a city, you'll need to forage on their behalf. One of our favorite things to do is carry pruners with us, and when we see a wayward branch or root suckers, we trim it and bring it home to the goats. You must be very careful that the tree or shrub is *not* poisonous to them. Good candidates include: maples, liquid amber, acacia, ash, apple, some oak, loquat, and pine. There's a huge list of no-no plants on the Internet: Read before feeding your goat (www.ansci.cornell.edu/plants/goatlist.html). And remember: You'll need to positively identify your tree or shrub, so get a tree ID book. Over the years, we've made friends with arborists and gardeners who will drop off fodder. This is a lovely way to help each other out: Your goats get their favorite food, your friend doesn't have to take the branches to the dump. Beware: The goats will just eat the leaves, so you will have a large pile of branches to deal with. We recommend trimming them into small sticks and using them for kindling.

WATER

Goats must have access to clean drinking water at all times. For dairy goats especially: Water makes milk! There are goat waterers that connect to a hose and provide clean water on demand. These might get knocked over, so we also place a bucket or two of water around the goat area just in case. Some people claim it should be warmed, but we haven't noticed that goats care about the temperature, as long as it isn't frozen over. Clean is the operative word: If you let their water get funky, you'll taste it in their milk, or they simply won't drink it. Whatever you do, keep the buckets clean and fresh.

MILKING GOATS

We raise dairy goats because we love their fresh milk for drinking, and for making cheese, yogurt, and other fermented dairy products. There is something liberating about not having to buy milk at the grocery store: We can guarantee its purity and know that it is fresh. It is a close relationship, that of a milk goat and her keeper. On one hand, she considers you to be like one of her nursing babies; on the other, you are in charge and give her hay and grain. It's a give-and-take relationship, and it's a very special one, going back thousands of years in human history.

Many people milk twice a day—once in the morning and once in the evening, ideally separated by twelve hours. There are some who choose to milk once a day—you get less milk doing this, but the benefits (a social life) are wonderful.

Basic equipment list for milking:

- stanchion for the goat to stand on while being milked
- udder wipes (homemade cloths or paper ones bought from a store) saturated with soap or other kind of cleaning agent, like diluted teat dip
- teat dipping cup
- strip cup
- stainless-steel milk pail
- stainless-steel milk filter with one-time-use paper in-line filters

1. Loading up.

In the morning and/or evening, at a regularly scheduled time, get your first goat onto the stanchion. If you have more than two milk goats, the dominant one will insist on being milked first. Most goats want to be milked, and will trot right up into the stanchion. To sweeten the deal, most goat owners give her some grain as an incentive. We pour the grain into the trough right when the goat comes into the milk parlor. She then jumps onto the stanchion. We then lock her head into it by moving a slat that holds her head in place, but she can still move her head up and down, which allows her to eat and stand comfortably.

2. Clean the udder.

A goat's udder is made up of two separate compartments, each with its own teat, or nipple. Since goats sleep on the ground, their udders can get soiled and should be cleaned thoroughly before milking. It's obvious but a dirty udder will contaminate the milk, so it's important to get the udder really clean. If the goat had diarrhea or slept in very dirty bedding, you should fully bathe the udder with hot soapy water. Usually, though, it's sufficient to wipe the udder down with a wipe or a wet cloth dipped in soapy water. When washing it, you are also stimulating the letdown of milk. As you rub, push up on and gently massage ot. Don't forget to clean off each of the teats. Wash your hands with soap and water.

3. Milking.

This will take time to learn, and again, each goat is different, and everyone has a different method. We like to sit directly behind the goat to milk. It gives better leverage and equal access to each teat. However, if you have a goat that likes to poop while being milked, this position isn't a good idea. More often, people like to place a stool at the edge of the stanchion and reach across sideways to access the teats. Either way, you'll need a stainless-steel pail to milk into. Stainless steel is preferred because it can be sterilized.

Grasp one of the teats with one hand, pushing slightly up on the udder. Close your index finger around the neck of the teat and gently squeeze. This traps the milk in the teat. Move the rest of your fingers down along the teat until a stream of milk is released into the pail. Do the same thing with the other hand and the other teat. Repeat this again and again until all the milk is released from both sides of the udder. You can be sure to get all the milk possible by making several gentle bumps to the udder during milking—it is similar to something you'll see the kids do when they milk. Milking out a goat can take five minutes, half an hour, or an hour, depending on the doe, your level of experience, your and methodology.

4. Cleaning up.

Postmilking you'll want to dip each teat into iodine or teat dip. Usually this is done with a teat cup, a refillable plastic cup that is the size of a teat. Dipping keeps bacteria from getting into the teat opening, where it can cause an infection called mastitis. Let the goat off the stanchion with a gentle pat and a thanks.

5. Filtering.

Take your milk pail to the kitchen for filtering. There are many kinds of milk strainers; we like the stainless-steel one that has a screen and an opening that fits a wide-mouth mason jar the best. A paper in-line filter is wedged into a screen and then the milk is poured through it, which drains into the mason jar. After filtering, screw on a lid, label the milk with the date and any other marks (some people weigh their milk and add the info to a milk log), and put it in the fridge immediately. Throw away the paper filter and wash the metal strainer and screen with hot soapy water.

MILKING A GOAT

Step 1:
Allow the goat up
onto the milk stand.

Step 2:
Wash the udder.

Step 3:
Milk into a milk pail.

Step 4: After milking, to
prevent infection, dip each
teat with a teat dip.

Step 5:
Place in-line filter
into milk strainer.

Step 6:
Pour milk through milk
strainer to remove any hairs
or dirt that may have fallen
in during milking.

TROUBLESHOOTING WITH MILKING

To some extent, the happiness of your goat corresponds to her milk output. When we first got our milk goats we were nervous wrecks, wondering if we were doing everything right. Remember that this early stage is part of the process of becoming a goat owner. You'll need to learn and adjust to meet your goats' needs. For example, one of our first goats was a shy milker; she hated it to be too bright in the milk room and hunched over if she heard a loud noise. This corresponded to lower milk output. We changed things so that she was happy and comfortable with her environment and she became confident and developed into a very good producer. Here are a few problems you might encounter.

Won't Get On the Stanchion

If a goat won't get on the stanchion, if they are a small breed like a Nigerian Dwarf, you can lift them into place. When you lift, remember to bend your knees and to not lift with your back. Goats don't like to be lifted, so they'll probably walk up onto the stanchion after this insult to their independence. If you have a full-size goat, lifting is too hard on your back, and you'll have to resort to psychological methods—usually treats. When the goat comes into the milking parlor, give her a treat. Place a treat on top of the stanchion. Then show the goat that there are more treats in the bucket on the stanchion. This should be enough to convince them in. If it's not, invite a stanchion-loving goat into the milk parlor to show the hesitant one how it is done. Once you have a shy goat in the stanchion, be especially gentle and offer more treats than normal. Make it a point to never trim a shy goat's hooves on the stanchion—most goats hate getting their hooves trimmed and you don't want this negative association.

Kicking

You sit down and start milking. When the pail is almost full your goat kicks the bucket over, spilling the milk everywhere. This scenario has happened to us more than once. Kicking the bucket can be caused because of a startling noise (they freak out especially when you sneeze, because it sounds similar to a warning snort they make in the presence of a predator), they are impatient with you, or you accidentally pinched her teat. Remember that your goat can sense your mood and state of mind during milking. If you're in a rush she'll pick up on that, and will be impatient. It's not a good idea to yell at or hit your goat. If they start to get kicky, keep your hands on their udder—most goats will give up and let you proceed. If they do manage to kick the bucket, don't cry over spilled milk; just clean it up and try to figure out what happened. Some goats are just kickers, and next time you might have to try strapping their back legs down with a goat hobble. These make it impossible to kick, and eventually the goat learns not to. Another trick recommended by Jim at Green Faerie Farm is to squirt a rebellious goat with water. Goats hate this indignity and will take you seriously after a dousing. Having a squirt bottle handy also allows small children to fend off a pushy goat.

Pooping

You're milking dutifully, and your goat drops a load on the stanchion. This is, for some goat owners, totally fine. We try to avoid this behavior, because it is unhygienic and can be avoided by prepping instead. Before bringing your goat in for milking, train her to go pee or poop by simply waiting at the door and letting her do her business. Goats like to have a warning that they are going to be milked anyway. We also will brush or stroke their sides and the backs of their legs to stimulate excretion. Once they poo or pee, bring them in immediately and give them a treat. Some new mama goats can't hold their bowels for a few weeks after giving birth, so be gentle with these new mothers if they have an accident on the stanchion.

Nervous or Fidgety

Moving around or acting nervous while on the stanchion from the very moment of being placed on it is usually reserved for a first kidder or a goat with an injury. Avoid this by training young does to use it from an early stage—even before they are pregnant, as young as a few months old. Massage their udders to get them used to the process, and reward them with a treat. This will make them have only positive associations with it. Note that goats find full-body contact to be calming and reassuring, so be sure to lean into them while milking or trimming their hooves. If bad behavior on the stand is a chronic condition, you may have to get a hobble, a strap that attaches to their back legs and prevents movement. Sometimes this demeanor can't be broken, and you might consider culling the goat.

Breeding

If you're raising goats for milk and meat, you'll need to come up with a breeding program. Milk goats have been known to provide a reasonable amount for as long as two years after kidding (in some cases longer), but most farmers will breed once a year to ensure consistent milk production. It's considered safe to start a doe at nine months old (they can get pregnant much earlier, but it's bad for them). Some people wait until they're twelve or eighteen months old. Wait any longer than that and you run the risk of having a difficult first kidding.

Does will start to go into heat as the day-length decreases, in the fall, which ensures spring-born kids. Nigerian Dwarf goats are a tropical breed, and so go into heat all year round. Signs of heat include bleating, flagging or fanning their tails, a red and swollen vulva, and a discharge of mucous.

Since urban farms usually can't welcome a smelly male goat into the fold, we have to travel outside of the city to get stud service; we've paid fifty to one hundred dollars each time. Always make sure the buck has a clean bill of health (they can carry STDs) before breeding. It's a good idea to find out the buck's family history as well, so you are sure to get optimum milk genetics. If the doe is in heat, they will usually get busy the moment you put them in the pen together. The doe will "stand," meaning she'll let the buck mount her. If she isn't in heat, sometimes leaving them together for a day or two helps move things along, as the presence of a male will stimulate estrus. However, if you arrive with the doe just as her estrus is ending, you'll have to return in around twenty-one days when she next comes into heat again.

When our girls come home from a stud visit, they always smell like a funky old buck, and we've noticed the milk doesn't taste as sweet as normal, and sometimes has a strong goaty odor—which is called buck taint. Once successfully mated, the gestation period is five months, give or take a few days, or sometimes weeks. You can continue to milk a pregnant goat up to three months into her pregnancy. But since she is using energy to create new life, the doe will be very hungry, and having milk taken may exhaust her if she is not properly pampered and fed.

Kidding

Pregnant goats are the cutest thing. They swell up like watermelons and waddle when they walk. They seem especially satisfied with themselves and love to sit in the sun, chew their cud, and look ridiculously fat. You can gently palpate their stomachs in order to guess how many babies there are starting at three months. Since there is a margin of error in the five-month gestation, you must look for other signs of imminent birth. One is that the doe is acting abnormally friendly. We've had goats lick us, chew on our armpits, and nuzzle our pants in order to let us know the big moment is near. Goats will also develop a sort of hollowed-out–looking area around their tail region; this is a sign that kidding will begin in a few hours or less. This hollow is showing that her bones are softening to allow the baby to pass through the pelvis. Another sign can be the udder filling completely and looking a bit shiny, but note that some goats fill up weeks early, and some don't get their milk until just after giving birth.

Although you don't have to be present for the birth, it's best to try to be there, with a vehicle ready to go to the vet in the unlikely (but possible) event that a goat gets into a life-threatening situation. Jim from Green Faerie reported that he has had to help the birthing doe three times out of the twenty births he's attended. If he hadn't been there the goat probably would have died. The best case scenario is to have an experienced goat person at the first few births. That's why you want to buy your goats from someone willing to be a mentor for you during these times.

We like to keep a kidding kit handy for the big day.
It contains:

- 10 percent iodine solution (an antiseptic)
- gloves
- Vaseline or K-Y Jelly
- a small cup for dipping the kids' navels
- towels
- warm beet pulp or grain mash or molasses tea in a bowl

Gather up your kit and sit with Mama. The first stage is a white discharge coming out of her vagina. After a time she will begin laboring, actively pushing and standing, repositioning herself to allow the kid out. This is the exciting part, and you should watch for the little hooves, followed by the head of the kid. If you see, say, just the head, or only one hoof, you might have to "go in." Meaning scrub up, smear your hands with Vaseline, and prod your way around the birth canal. Only do this if it is totally obvious something is wrong. Usually the kid just needs a minor adjustment in order to present correctly. We advise you to have a few other goat books or a seasoned goat farmer on hand when preparing for delivery. The general rule is that if a goat has been laboring for hours and doesn't seem to be dilating, you should get her to a vet. Once the first kid comes out, make sure the doe licks the embroyonic sac to get the kid to air. After only a few minutes, the doe will have licked the kid clean, and it will stand to nurse from his or her mother's teat. If there are twins or triplets or quadruplets or more, the doe will sometimes give birth to them first before licking. Birth is over once the placenta comes out. This bloody sac is often eaten by the doe. It's a strange sight to see a vegetarian wolfing down a bloody snack. If you want, you can take it away from her and bury it under a tree. After she's done birthing, offer her a long drink of warm water and a sustaining snack, like some beet pulp infused with warm water or molasses tea or warm grain mash. She'll be hungry and thirsty, and this is a great boost for her.

After the kids are all nursing and seem healthy, it's a good idea to check on the kid's sex. A rule of thumb: Lift their tail and count the holes; a male will only have one opening, a female will have two. We are always so happy to get girls, because it means they will eventually become milk producers. Males are a joy too, but we usually eventually eat them. You should also scan for abnormal teats (some have more than two, a defect) and look for any hermaphroditic traits. Hermaphrodites are a rare occurrence and are usually kept as pets or are culled like the bucklings.

KID CARE

Goat kids are ridiculously cute. They can walk, run, prance, and jump from day one. They also like to sleep snuggled-up together. Put a shallow box in the goat loafing area and watch the kids instantly curl up inside it. Kids don't like drafts or moisture—make sure they are warm and dry. Some people in cold climates have to put a heat lamp in the goat area to keep the kids warm. You'll know they are cold if you see them shivering or crying. If a kid gets chilled, take it immediately into your house and wrap it with a heating blanket, and force it to drink some warm milk. Mostly, though, mama goat will make sure the kids are all right.

WEANING CONSIDERATIONS

You'll need to have a plan for how you want to raise the kids. Some farmers take them off their mothers from day one and keep them in crates inside their homes. They milk the moms and then bottle feed the kids this milk/colostrum with the idea that it will create a better bond between human and kid, and will prevent weaning problems later. Other people let the kids nurse for four days, and then take them inside, where they are bottle fed and raised in crates. These two methods are how professional dairies operate. It guarantees control of milk production and the kids. It is also a lot of work: The kids must be fed every two hours for three weeks. Note that the separation of babies can be done without them losing contact with the mom: You can house them in an adjacent pen with an opening for the mom to put her head through to nuzzle her kids or groom them while they nurse on a bottle.

We take a more hands-off approach, letting the kids nurse for three to four weeks before we start to separate them from their mothers. We do want to socialize the goatlings, however, and so make it a point to hold the kids every day for at least an hour. The idea is to get them used to your presence and to accept you as a member of the herd. After a month we start to lock the kids into dog crates at night. This way you can milk the mother in the morning before letting the kids out to nurse. Since goats sleep at night, this seems like the least traumatic way to separate the kids from their moms. We leave the crates in the main area where the mom sleeps, so she can see and smell them; they just can't get to her to nurse. In the morning, after milking Mom, we let the kids out, and they can spend the rest of the day with her, nursing, and bonding. Eventually, the mom will naturally wean the kids, at around three months. If she doesn't, however, you will have to separate them both day and night until they stop nursing. We usually sell the kids at three months, anyway.

BASIC GOAT CARE

There are a lot of small details that you'll need to keep track of on an urban goat farm.

Shots

Kids will need CDT shots after they are a month old, with a booster shot a month later. Does should be given one shot a month before they are due. CDT is a vaccine against clostridium perfringens Type D and tetanus.

Disbudding

Most goats naturally develop horns. At a few days old you can locate two small pimplelike buds that will become horns. Horns are a way for goats to regulate their temperature, guard against attackers, and bring down forage from trees. Though it is more natural to let the goats keep their horns, there are plenty of reasons to disbud—remove—them. The main concern is safety for both humans and the goat. A goat with horns can gouge out a person's eye or cause some other serious damage. Or they can hurt themselves by tangling up their horns in a fence or some other man-made objects and seriously hurt or kill themselves.

The most humane way to disbud is to do it early. At four days old either take the kids into a vet or to an experienced goat tender, or do it yourself on the farm. Bucklings are especially difficult to disbud correctly, so experience is a necessity! The best method is to use a disbudding iron, a very very hot copper plug-in device that heats up to 500 degrees. The iron is placed on one horn at a time and pressure is applied, and then it is moved to the next horn. After a few passes, the horns are effectively burned off. If you don't use anaesthetic, the goatling will scream, and it will tear out your heart—but the procedure is very quick. Apply iodine or antiseptic to the burned-out horn area after disbudding. Some goats are what is called "polled" and will never form horns; these goats will form not the pimplelike buds but more like round flesh–covered bumps.

Hoof Trimming

Every six weeks you'll need to do some hoof trimming. Goats are accustomed to climbing on rocky ledges, which wear down their hooves. However, living in your soft backyard probably isn't enough to do the same. Having overgrown hooves can seriously compromise your goats' health, so they should be trimmed regularly. For pregnant goats, trim six weeks before they give birth. Using proper trimmers,

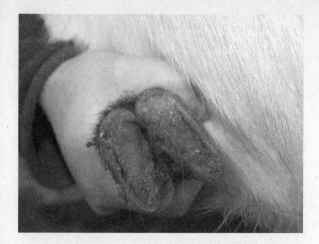

Hoof trimming must be done every six weeks.

Step 1: Begin by trimming the tip of the hoof.

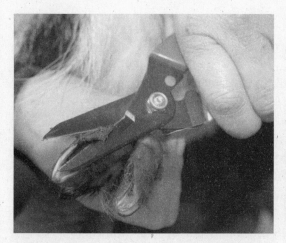

Step 2: Even up the sides of the hoof.

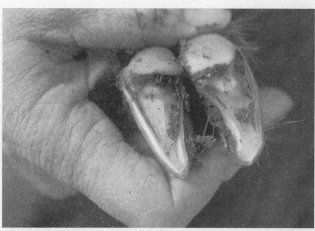

A neatly trimmed hoof keeps the goat healthy and comfortable.

cut off excess bits and make sure the hoof lies flat and straight instead of hooking up or folding under. You've cut too close if you draw blood. The key is to first cut the tip, then trim the sides so they match up. The line of the cut should run parallel with the hairline. Put the foot down and observe how the goat stands on the newly trimmed hoof.

Intestinal Worms

Worms may or may not be a problem on your urban farm. Symptoms include diarrhea and wasting. If you suspect worms, take a fecal sample to your vet. We like to treat with an herbal wormer sold by Fiasco Farm (see page 543) twice a year. It's just a bitter blend of herbs, made palatable by molasses.

Clipping Hair

If you ever show your goat, she must be clipped in dairy style—which is to say, high and tight. The idea is that you don't want goat hair to fall into the milking pail, so udders, flanks, sides, and beards are trimmed every few months, especially in the spring and summer. A pair of good clippers cost over two hundred dollars. We also like to keep a good curry comb around for brushing the goat. They seem to like it, and it helps keep the shedding down to a minimum.

MILK PRODUCTS: CHEESE, YOGURT, AND MORE

The point of having goats and dealing with weaning and other issues is to get milk. If you have successful breeding and kidding, you will eventually have more milk than you ever thought you'd have. Although fresh, clean goat milk will taste as mild as cow's milk, it also can be suitable for people who are allergic to cow's milk. Once you get bored of simply drinking it, you'll want to try making some goat milk delights. Here are some of our favorites:

> **Cappuccino.** Fresh milk, especially goat milk, froths up more easily than any other milk we've worked with as amateur baristas. Our favorite morning snack involves steaming some fresh goat milk and adding a few shots of espresso. It's the most tasty cap we've ever had!

> **Yogurt.** Buy a small container of your favorite live yogurt, making sure that the main ingredients are milk and live cultures (not tapioca or corn starch). Add a tablespoon to a quart of goat milk warmed to 100 degrees, stir well, and keep the jar at 100 degrees for twelve

hours. To keep the jar warm, you can place it in a warm water bath or in an insulated cooler. Alternately, if it is hot outside, place the inoculated milk in a warm area of the kitchen. Once it has firmed up, place in the fridge. Eat with honey or fruit, and use the last of it to make the next batch of yogurt.

Cheese. Warm up a gallon of milk in a double broiler or pan to 86 degrees. Add chèvre starter from New England Cheesemaking (see appendix 11, page 543). Stir for a minute, then let set for twelve hours. In this time the curds and whey will separate. Ladle the curds into cheese molds, sprinkle with salt, and let drain for two days in the fridge. Remove from the molds, sprinkle with more salt and some herbs, and serve.

Kefir. Obtain some kefir grains and add to milk at room temperature. Let sit for twenty-four hours or longer if your house is cool or if you like it more sour. The kefir grains will go to work, creating a thick, yogurty texture drink. Strain out the kefir chunks and drink. Reserve the grains, stored in the fridge with a little bit of the kefir, for your next batch. Kefir fans say yogurt kefir is a probiotic wonder food. It can be pretty strong tasting; to cut it, mix in a blender with orange juice and frozen bananas for a healthy shake.

KILLING A GOAT

Finally, a note about killing goats. We aren't going to cover killing goats in detail here. We feel that the best method, because of the size of the animal, is to find an experienced person to come over to do the deed, or to take them to a small slaughterhouse. If you are raising goats for meat, then you know there will be death, and have probably figured out how you feel about meat. On the other hand, if you are a soft-hearted dairy person with a male goat on your hands, you might feel more conflicted. Sadly, most male offspring can't be used for breeding purposes—they simply aren't perfectly formed or they don't have champion bloodlines and milk genes. And so, unless you plan on housing a smelly, randy goat, they should be slaughtered before they are seven weeks old. They can also be castrated and kept for six months to a year to grow and fatten before slaughter.

Over the years we've found the best method is to find a knowledgeable person who has killed goats before to come to your farm and slaughter the little buck on the farm. One good method is to slit its throat. It is fast and painless. The goat can then be hung to bleed out, eviscerated, and its hide removed. Though it is sad to see your little guy go, the good news is that goat meat is delicious, tanned goat hides are very attractive, and it will mean more milk for you.

HOW TO USE THE APPENDICES

Welcome to the back matter. While the main text of *The Essential Urban Farmer* describes the general principles and specific plans for growing vegetables, fruit, and livestock on an urban farm, this is where you'll find detailed information about plant culture and rotation, in addition to where to buy specific items and how to get involved with an urban farm in your region.

First you'll find important resources on soil remediation to make sure your produce will be safe to eat. Next you'll find tables that describe vegetable families and yield charts that will help you plan how many vegetables your family will need to plant, as well as plant spacing suggestions. Next, there are templates for sowing and planting charts in appendices 5 to 7. These will become invaluable in your plan! appendix 8 illustrates how, depending on your region, you can extend your growing season. Appendix 9 goes over pests that haunt fruit trees and methods to battle them. Appendix 10 is a beast: a complete list of how to save seed from your urban farm. This appendix empowers you to take control and make your own seed!

More broadly, and thanks to the Internet (where every urban farmer can order gear online), we've also included a list of where to buy supplies, in addition to some of our favorite reference books, in appendix 11. Next is a list of urban farming organizations by state. Finally, last but not least, is our answer to a question we get asked all the time: How do I start my own urban farming business?

W   E

APPENDIX 1

SOIL TESTING AND REMEDIATION RESOURCES

SAFE PRACTICES FOR GROWING IN LEAD-CONTAMINATED SOILS

http://cwmi.css.cornell.edu.soilquality.htm

http://cwmi.css.cornell.edu/soil_contaminants.pdf

http://counties.cce.cornell.edu.schenectady/Master%20 Gardener%20Website/projectdocs/factsheets/vegetables/Lead%20 in%20Urban%20grown%20veggies.pdf

SOURCES FOR COLLOIDAL PHOSPHATE/MARINE FISH BONEMEAL

Peaceful Valley Farm Supply

www.groworganic.com

calphos soft rock phosphate sold in 50 lb. bags

PIMS NW

www.pimsnw.com

apatite II (made from fish bones)

sold by the ton, in super sacks (1,650 to 2,200 lbs.), and by the pound

Operation Paydirt

www.fundred.org

New Orleans project testing colloidal phosphate and remediation

Environmental Protection Agency Region and West Oakland Lead Remediation Project

contact Steven Calanog: calanog.steve@epa.gov

APPENDIX 2

PLANT FAMILIES

Amaryllidaceae Family	
Genus	Vegetable
Allium	Chives
	Garlic chives
	Green onions/Bunching onion
	Leek
	Onion
	Scallion
	Shallot

Brassicaceae Family	
Genus	Vegetable
Brassica	Broccoli
	Broccoli rabe
	Brussels sprouts
	Cabbage
	Cauliflower
	Chinese cabbage
	Chinese mustard
	Collards
	Kale
	Kohlrabi
	Mustard greens
	Rutabaga
	Turnip
Eruca	Arugula/Rocket
Raphanus	Radish

Chenopodiaceae Family	
Genus	Vegetable
Beta	Beet
	Chard
Spinacia	Spinach

Compositae Family	
Genus	Vegetable
Chichorium	Chicory
	Endive
	Escarole
Cynara	Cardoon
Helianthus	Sunflower
Lactuca	Lettuce

Cucurbitaceae Family	
Genus	Vegetable
Citrullus	Watermelon
Cucumis	Armenian cucumber
	Cantaloupe
	Casaba
	Cucumbers
	Honeydew
	Muskmelon
Cucurbita	Acorn squash
	Banana squash
	Buttercup squash
	Butternut squash
	Cheese squash
	Crookneck squash
	Golden cushaw squash
	Green striped and white cushaw squash
	Hubbard squash
	Scallop squash
	Spaghetti squash
	Turban squash
	Zucchini

Leguminosae Family	
Genus	Vegetable
Arachis	Peanut
Glycine	Soybean
Paseolus	Common snap beans and dry beans
	Lima bean
	Runner bean
Pisum	Edible podded pea
	Pea
Vicia	Fava bean
Vigna	Cowpea
	Yard long bean

Solanaceae Family	
Genus	Vegetable
Capsicum	Chili peppers
	Sweet peppers
Lycopersicon	Tomato
Physalis	Cape gooseberry
	Ground cherry
	Tomatillo
Solanum	Eggplant

Umbelliferae Family	
Genus	Vegetable
Anethum	Dill
Apium	Celeriac
	Celery
Coriandrum	Cilantro/Coriander
Daucus	Carrot
Foeniculum	Fennel
Pastinaca	Parsnip
Petroselinum	Parsley
	Parsley root

Gramineae Family	
Genus	Vegetable
Zea	Corn

Labiatae Family	
Genus	Vegetable
Ocimum	Basil

Malvaceae Family	
Genus	Vegetable
Abelmoschus	Okra

Valerianaceae Family	
Genus	Vegetable
Valerianella	Corn salad/Mache

PLANT NUTRIENT NEEDS FOR CROP ROTATION

Heavy Feeder	Light Feeder	Nitrogen Fixer	Nitrogen Fixing Cover Crop
Artichokes	Beets	Beans	Bell beans
Arugula	Carrots	Peas	Berseem clover
Asian greens	Chard		Common vetch
Bitter greens	Cilantro		Cowpea
Broccoli	Dill		Crimson clover
Broccoli rabe/rapini	Fennel		Fava beans
Brussels sprouts	Garlic		Field pea
Cabbage	Green onions		Hairy vetch
Cardoon	Leeks		Lupin
Cauliflower	Onions		Purple vetch
Celery	Parsley		Rose clover
Collards	Turnips		Woolypod vetch
Corn			
Cucumber			
Eggplant			
Kale			
Lettuce			
Melons			
Mustard greens			
Okra			
Peppers			
Spinach			
Squash, summer			
Squash, winter			
Tomatillos			
Tomatoes			

W E

APPENDIX 4

CALCULATING HOW MUCH TO GROW

Crop	Number of Plants[1] per Person	Plant Spacing[2] (in inches)	Average Approximate Yield (per square foot in pounds)
Artichokes	3	36–60	½
Arugula	3	6	2
Asian greens	4	8	2
Basil	2	6	½
Beans, bush	8	6	¾
Beans, fava	20	8	¼ (shelled beans)
Beans, pole	6	8	¾
Beets	20	4	1⅛
Bitter greens	4	8	1¾
Broccoli (heads)	4	12–18	½
Broccoli rabe/rapini	8	12	1¼
Brussels sprouts	4	18	½
Cabbage	4	12–18	1¾
Cardoon	1	36	3
Carrots	30	4	1½
Cauliflower	4	12–18	¾

Crop	Number of Plants[1] per Person	Plant Spacing[2] (in inches)	Average Approximate Yield (per square foot in pounds)
Celery	6	8	4½
Chard	8	12	4
Chayote	1	2 plants per large trellis	1
Cilantro	3	6	¾
Collards	8	12	2
Corn	12	12	½
Cucumber	1	12	3
Dill	2	6	¼
Eggplant	2	18	1
Fennel, bulbing	10	8	½
Garlic	12	4	1¼
Kale	8	12	2
Leeks	20	4	4
Lettuce, cutting	4 square feet	2–4	2
Lettuce, head	6	8	1
Melons	4	18	¾
Mustard greens	6	12	2
Okra	2	18	½
Onions	30	4	3
Onions, green	20	3	4
Onions, shallots	20	4	2
Parsley	2	6	½
Peas, edible pod	50	4	1
Peppers	3 sweet 1 hot	12–18	¾, sweet ¼, hot
Potatoes	15	12	1¾
Sweet potatoes	4	12–18	1½
Radishes	20	3	2
Rutabaga	4	6	3
Spinach	10	6	1
Squash, summer	1	24	2
Squash, winter	2	24–36	1
Tomatillos	2	24	½
Tomatoes	3	18–24	2
Turnips	6	4	2

1. If succession planting, plant this number per person per planting.
2. After thinning if direct seeded; assumes trellising for appropriate crops.

APPENDIX 5

SUCCESSION PLANTING

Succession plantings should be done only within the growing season of each crop. End succession plantings in time for the last planting to mature within its growing season.

Don't Plant in Succession	Replant Every Two Weeks to Two Months	Replant Every Two to Three Months	Replant Every Three to Six Months
Artichoke	Basil	Arugula	Chard
Beans, dry	Asian greens	Beets	Collards
Cardoon	Beans, lima	Bitter greens	Kale
Chayote	Beans, snap	Broccoli	Parsley
Garlic	Broccoli rabe/rapini	Cabbage	
	Brussels sprouts	Carrots	
	Corn	Cauliflower	
	Cucumber	Celery	
	Eggplant	Cilantro	
	Fennel	Cut-and-come-again salad greens	
	Leeks	Dill	
	Lettuce	Mustard greens	
	Melons	Onions (green)	
	Okra	Potatoes	
	Onions (bulb)	Rutabaga	
	Onions (shallots)	Turnips	
	Peas		
	Peppers (sweet and hot)		
	Radishes		
	Spinach		
	Squash		
	Sweet potatoes		
	Tomatillos		
	Tomatoes		

SUCCESSION PLANTING AND SEEDING CALENDAR TEMPLATES AND SAMPLES

SUCCESSION PLANTING CALENDAR TEMPLATE

Use this template to create your own Planting Calendar

> **D** = Plant seeds directly in the ground

> **T** = Transplant seedlings

> **#** = Number of plants (see the table on page 507 to help with choosing the number of plants)

If you wish, break each month into weekly sections.

JAN.	D/T	#	FEB.	D/T	#	MARCH	D/T	#	APRIL	D/T	#	MAY	D/T	#	JUNE	D/T	#

DECEMBER	D/T	#	NOVEMBER	D/T	#	OCTOBER	D/T	#	SEPTEMBER	D/T	#	AUGUST	D/T	#	JULY	D/T	#

SUCCESSION SOWING CALENDAR TEMPLATE

Use this template to create your own sowing calendar for starting seedlings.

S = Seeding container type

T = Transplanting container type

F = Open flat

6 = six-pack

4 = 4-inch pot

N = No transplanting necessary

= Number of seeds needed (should be 10 percent to 20 percent more than needed in the garden)

If you wish, break each month into weekly sections.

JUNE	S	T	#
MAY	S	T	#
APRIL	S	T	#
MARCH	S	T	#
FEB.	S	T	#
JAN.	S	T	#

JULY																	
S	T	#															

AUGUST																	
S	T	#															

SEPTEMBER																	
S	T	#															

OCTOBER																	
S	T	#															

NOVEMBER																	
S	T	#															

DECEMBER																	
S	T	#															

SAMPLE SUCCESSION PLANTING CALENDAR

This sample planting calendar can be altered to fit your growing region; note that this is for the Northern California area.

D = Plant seeds directly in the ground

T = Transplant seedlings

= Number of plants (see the table on page 507 to help with choosing the number of plants to plant)

If you wish, break each month into weekly sections.

JAN.	D/T	#	FEB.	D/T	#	MARCH	D/T	#	APRIL	D/T	#	MAY	D/T	#	JUNE	D/T	#
			Bitter greens	T		Chard	T		Basil	T		Tomatoes	T		Tomatoes	T	
			Cilantro	T		Brussels sprouts	T		Cucumber	T		Snap beans	D		Lima beans or cowpeas	D	
			Turnips	D		Parsley	T		Fennel	T		Eggplant	T		Corn	D	
			Peas	D		Cutting salad greens	T		Green onions	T		Melons	T		Cucumber	T	
			Green onions	T		Broccoli or Broccoli rabe	T		Lima beans or cowpeas	D		Sweet potatoes	T		Okra	T	
			Fava beans	D		Cabbage	T		Okra	T		Summer squash	T		Winter squash	T	
			Radishes	D		Carrots	D		Peppers	T		Basil	T		Peppers	T	
			Fennel	T		Head lettuce	T		Potatoes	D		Cauliflower	T		Green onions	T	
			Potatoes	D		Radishes	D		Radishes	D		Asian greens or mustard	T		Celery	T	
			Tomatoes	T		Beets	D		Tomatoes	T		Bitter greens	T		Cabbage	T	
			Asian greens or mustard	T		Peas	D		Turnips	D		Head lettuce	T		Fennel	T	
			Collards	T		Celery	T		Winter squash	T		Spinach	D		Broccoli or Broccoli rabe	T	
			Brussels sprouts	T		Snap beans	D		Other:			Radishes	D		Turnips	D	
			Kale	T		Spinach	D					Beets	D		Radishes	D	
			Cauliflower	T		Summer squash	T					Cilantro	T		Cutting salad greens	T	
			Other:			Arugula	T					Other:			Carrots	D	
						Other:									Arugula	T	

517

JULY	D/T	#	AUGUST	D/T	#	SEPTEMBER	D/T	#	OCTOBER	D/T	#	NOVEMBER	D/T	#	DECEMBER	D/T	#
Beets	D		Turnips	D		Carrots	D		Turnips	D		Cauliflower	T				
Brussels sprouts	T		Peas	D		Celery	T		Garlic	D		Head lettuce	T				
Carrots	D		Green onions	T		Radishes	D		Leeks	T		Radishes	D				
Collards	T		Cauliflower	T		Garlic	D		Fava beans	D		Asian greens	T				
Corn	D		Cilantro	T		Leeks	T		Bulb onions	T		Bitter greens	T				
Eggplant	T		Brussels sprouts	T		Fava beans	D		Green onions	T		Other:					
Head lettuce	T		Asian greens	T		Head lettuce	T		Radishes	D							
Kale	T		Radishes	D		Bulb onions	T		Shallots	T							
Melons	T		Parsley	T		Peas	D		Fennel	T							
Potatoes	D		Fennel	T		Cabbage	T		Potatoes	D							
Radishes	D		Bitter greens	T		Spinach	D		Other:								
Snap beans	D		Other:			Shallots	T										
Spinach	D					Broccoli or Broccoli rabe	T										
Summer squash	T					Cutting salad greens	T										
Sweet potatoes	T					Chard	T										
Other:						Beets	D										
						Arugula	T										
						Other:											

SAMPLE SUCCESSION SOWING CALENDAR

We provide here an indoor sowing calendar that dovetails with the above sample planting calendar. You can alter it to fit your growing region. Some crops take longer to mature and are given more time than others. We assume you will be starting all onions, shallots, and leeks from seed rather than sets. If you choose to direct seed crops such as lettuce, or indoor seed crops such as root vegetables, peas, or corn, you will need to adjust the calendar accordingly.

S = Seeding container type

T = Transplanting container type

F = Open flat

6 = six-pack

4 = 4-inch pot

N = No transplanting necessary

= Number of plants or seeds (should be 10 percent more than needed)

If you wish, break each month into weekly sections.

JAN.

	S	T	#
Collards	F	4	
Kale	F	4	
Cauliflower	F	4	
Bitter greens	F	6	
Asian greens	F	6	
Fennel	6	N	
Celery	F	6	
Parsley	F	4	
Cilantro	6	N	
Other:			

FEB.

	S	T	#
Chard	F	4	
Summer squash	4	N	
Brussels sprouts	F	4	
Cabbage	F	4	
Head lettuce	F	6	
Broccoli or broccoli rabe	F	4	
Tomatoes	F	4	
Tomatillos	F	4	
Peppers	F	N	
Green onions	F	6	
Cutting salad greens	F	6	
Arugula			
Other:			

MARCH

	S	T	#
Cucumber	4	N	
Winter squash	4	N	
Fennel	6	N	
Tomatoes	F	4	
Eggplant	F	4	
Basil	F	6	
Other:			

APRIL

	S	T	#
Melons	4	N	
Sweet potato slips	4	N	
Summer squash	4	N	
Cauliflower	F	4	
Bitter greens	F	6	
Asian greens	F	6	
Head lettuce	F	6	
Tomatoes	F	4	
Tomatillos	F	4	
Peppers	F	4	
Celery	F	6	
Green onions	F	N	
Basil	F	6	
Cilantro	6	N	
Other:			

MAY

	S	T	#
Cucumber	4	N	
Winter squash	4	N	
Cabbage	F	4	
Fennel	6	N	
Broccoli or broccoli rabe	F	4	
Cutting salad greens	F	6	
Arugula	F	6	
Eggplant	F	4	
Other:			

JUNE

	S	T	#
Collards	F	4	
Kale	F	4	
Melons	4	N	
Sweet potato slips	4	N	
Summer squash	4	N	
Brussels sprouts	F	4	
Head lettuce	F	6	
Green onions	F	N	
Parsley	F	4	
Other:			

JULY	S	T	#
Cauliflower	F	4	
Bitter greens	F	6	
Asian greens	F	6	
Fennel	6	N	
Bulb onions	F	N	
Shallots	F	N	
Leeks	F	N	
Celery	F	6	
Cilantro	6	N	
Other:			

AUGUST	S	T	#
Chard	F	4	
Cabbage	F	4	
Head lettuce	F	6	
Broccoli or broccoli rabe	F	4	
Bulb onions	F	N	
Shallots	F	N	
Leeks	F	N	
Green onions	F	N	
Cutting salad greens	F	6	
Arugula	F	6	
Other:			

SEPTEMBER	S	T	#
Fennel	6	N	
Cilantro	6	N	
Other:			

OCTOBER	S	T	#
Cauliflower	F	4	
Bitter greens	F	6	
Asian greens	F	6	
Head lettuce	F	6	
Other:			

NOVEMBER	S	T	#
Green onions	F	N	
Other:			

DECEMBER	S	T	#
Tomatoes	F	4	
Other:			

APPENDIX 7

PROPAGATION
INFORMATION

NUMBER OF SEEDLINGS IN A TRAY

Type of Container	Number of Seedlings in a Tray of Containers
Open flat (varies)	200 to 1,500
Speedling plug flats (varies)	32 to 338
Six packs (varies)	48 to 240+
4-inch pots	16

RECOMMENDED PROPAGATION METHOD BY CROP

Method	Crop	Number of Seeds per Cell or Pot
Seed in open flats and transplant up to six packs	Asian greens	
	Basil	
	Bitter greens (endive, escarole, radicchio, frisee)	
	Celery	
	Lettuce and salad mix	
Seed in open flats and transplant up to 4-inch pots	Artichokes	N/A
	Broccoli (heading and sprouting)	
	Brussels sprouts	
	Cabbage	
	Cardoon	
	Cauliflower	
	Chard	
	Collards	
	Edible flowers	
	Eggplant	
	Ornamentals (from seed)	
	Parsley	
	Peppers	
	Perennial culinary herbs (from seed)	
	Perennial medicinal herbs (from seed)	
	Potatoes (if starting from seeds)	
	Strawberries (from seed or runners)	
	Tomatillos	
	Tomatoes	
Seed in open flats	Green Onions	N/A
	Leeks	
	Onions (if starting from seeds)	
	Shallots	
Seed in six packs	Arugula	2
	Beans	1–2
	Broccoli rabe/rapini	2
	Chives	8–12
	Cilantro	2
	Corn	1–2
	Dill	4–8
	Fennel	2
	Mustard greens	2
	Peas	1–2
	Spinach	2
Seed in 4-inch pots	Cucumber	1–2
	Melons	1–2
	Okra	1–2
	Squash (summer and winter)	1–2
Propagate from cuttings in open flats, transplant up to 4-inch pots, and then possibly up to half-gallon or bigger pots	Perennial culinary herbs	N/A
	Perennial fruiting vines and shrubs	
	Perennial medicinal herbs	
	Perennial ornamentals	
	Sweet potatoes (please see page 190 to learn how to start slips)	
	Tree collards (also known as Jersey kale)	

W — E

APPENDIX 8

REGIONAL YEAR-ROUND GROWING GUIDE

Please see page 234 for instructions on how to use season-extending techniques.

CAN THE FOLLOWING CROPS BE GROWN YEAR-ROUND? IF SO, HOW?

	Temperate Crops	Warm Season Crops	Hot Season Crops
Pacific Northwest	Coast—Yes	Season Extenders	No
	Inland—Season Extenders	Season Extenders	No
Pacific Southwest	Coast—Yes	Season Extenders	No
	Inland—Season Extenders	Season Extenders	No
Southwest	Low altitude—Yes	Season Extenders	No
	High altitude—Season Extenders	Season Extenders	No
Central West	Low altitude—Yes	No	No
	High altitude—Season Extenders	No	No
Great Lakes and Northeast	Season Extenders	No	No
Midwest and Mid-Atlantic	Coast—Yes	No	No
	Inland—Season Extenders	No	No
Upper South	Low altitude—Yes	Season Extenders/Shade	Season Extenders/Shade
	High altitude—Season Extenders	Season Extenders/Shade	Season Extenders/Shade
Lower South	Shade	Shade	Shade

W ← → E

APPENDIX 9

ORGANIC CONTROLS FOR TREE PESTS AND DISEASES

Pest or Disease	Indicator	Organic Control
Gophers	Gopher holes and tunnels near tree. Young tree wilts or dies. Roots are chewed.	Plant trees in gopher cages.
Deer	Leaves are "pruned."	Surround with 6-foot high deer netting.
Birds	Large, fresh-looking holes in fruit.	Cover the tree with bird netting until harvest.
Squirrels	Varmints are seen removing fruit and burying it.	Surround the tree with bird netting until harvest.
Aphids	Ant trails up tree. Seed-size black or green sapsuckers on undersides of curled leaves.	Beneficial ladybug or lacewing beetles and syrphid fly. Use sticky bug paint repeatedly around trunk of tree and ensure no branches are touching a fence or other tree that ants could climb. Spray repeatedly with hard-water spray and/or insecticidal soap. Yearly application of superior oil.
Mites	Pinpoint-size sapsuckers on undersides of curled leaves.	Beneficial ladybug or lacewing beetles and predator mites. Yearly application of dormant oil.
Scale	Ant trail up tree; ⅛-inch to ¼-inch gray to tan bumps on trunk, stems, and leaves with sap sucking insects harbored beneath shell.	Beneficial ladybug or lacewing beetles and syrphid fly. Use sticky bug paint repeatedly around trunk of tree and ensure no branches are touching a fence or other tree that ants could climb. Spray repeatedly with hard-water spray and/or insecticidal soap. Yearly application of superior oil and dormant oil.
Borers	Tiny holes in wood; ½-inch to ¾-inch-long moths.	Remove by hand with a knife.
Coddling moth/ Apple maggot	Find 1-inch-long pale pink larvae and ½-inch gray moths. Larvae holes in fruit.	Beneficial lacewing beetles. Use rotenone.
Fungal and bacterial diseases	See variety of scabs, spots, dieback, fungal powder, lesions, cankers, and galls.	Appropriate application of baking soda spray, bordeaux mixture, sulfur, copper sulfate, or lime sulfur.

APPENDIX 10

SEED SAVING TABLE

Notes:

1. If you want to save seed from a crop that is not listed, match it with the genus and species it is in and use the methods given.

2. With a few exceptions as noted, crops can crossbreed within the same species but not with other species.

3. The "distance" isolation method is not recommended for urban seed savers, since we cannot control what our neighbors are growing.

AMARYLLIDACEAE FAMILY

Genus	Species	Vegetable	Natural Pollination Method	Recommended Isolation Method	Hand Pollination Method	Special Instructions	Contaminating Wild Plants
Allium	ampelo-prasum	Leek	Perfect flowers; outbreeding; self-sterile; blossoms on heads can pollinate each other or other heads. Insect pollinated.	Time	None	Overwinter	
	cepa	Onion	Perfect flowers; inbreeding; insect pollinated	Bagging	Unbag and spread pollen with a brush for two weeks	Store bulbs and replant next season	Wild onions
		Shallot	Same as above	Bagging	Unbag and spread pollen with a brush for two weeks	Store bulbs and replant next season	Wild onions
		Scallion	Same as above	Bagging	Unbag and spread pollen with a brush for two weeks		Wild onions
	fistulosum	Green onions/ Bunching onion	Perfect flowers; inbreeding; insect pollinated	Time	None	Overwinter	
	Schoeno-prasum	Chives	Perfect flowers; outbreeding; insect pollinated	None	None	Perennial. If you wish to save an uncommon variety, use bagging method.	
	Tuberosum	Garlic chives	Same as above	None	None	Same as above	

BRASSICACEAE FAMILY

Genus	Species	Vegetable	Natural Pollination Method	Recommended Isolation Method	Hand Pollination Method	Special Instructions	Contaminating Wild Plants
Brassica	Juncea	Mustard greens	Perfect flowers; self-pollinating; self-compatible	Bagging	None		Wild mustard
	Napus	Rutabaga	Perfect flowers; self-pollinating; self-compatible	Bagging	None	Overwinter	Wild turnips; rape
	Oleracea	Broccoli	Perfect flowers; outbreeding; self-sterile; insect pollinated	Caging	Introduce insects to the cage	Overwinter. Save seed from as many plants as possible.	Wild cabbage (rare)
		Brussels sprouts	Same as above	Same as above	Same as above	Same as above	Same as above
		Cabbage	Same as above	Same as above	Same as above	Same as above	Same as above
		Cauli-flower	Same as above	Same as above	Same as above	Same as above	Same as above
		Collards	Same as above	Same as above	Same as above	Same as above	Same as above
		Kale	Same as above	Same as above	Same as above	Same as above	Same as above
		Kohlrabi	Same as above	Same as above	Same as above	Same as above	Same as above
	Rapa	Turnip	Perfect flowers; outbreeding; self-sterile. insect pollinated	Caging	Introduce insects to the cage	Overwinter. Save seed from as many plants as possible.	
		Broccoli rabe	Same as above	Same as above	Same as above	Same as above	
		Chinese cabbage	Same as above	Same as above	Same as above	Same as above	
		Chinese mustard	Same as above	Same as above	Same as above	Same as above	
Eruca	Sativa	Arugula/ Rocket	Perfect flowers; outbreeding; self-sterile; insect pollinated	Caging	Introduce insects to the cage	Save seeds from as many plants as possible. If purity is not required, use time method.	
Raphanus	Sativus	Radish	Perfect flowers; outbreeding; self-sterile; insect pollinated	Caging	Introduce insects to the cage	Save seed from as many plants as possible	Wild radish

CHENOPODIACEAE FAMILY							
Genus	Species	Vegetable	Natural Pollination Method	Recommended Isolation Method	Hand Pollination Method	Special Instructions	Contaminating Wild Plants
Beta	Vulgaris	Beet	Perfect flowers; outbreeding; self-pollinating; wind pollinated	Bagging	Shake the bag from time to time	Pollen can travel up to five miles. For purity they must be isolated, even if growing only one variety. Over winter	
		Chard	Same as above	Same as above	Same as above	Same as above	
Spinacia	Oleracea	Spinach	Dioecious; outbreeding; self-pollinating; wind pollinated	Caging	Shake the plants within the cage from time to time	Pollen can travel up to five miles. For purity they must be isolated, even if growing only one variety. Over winter. Maintain a ratio of one male to two female plants within each cage.	

COMPOSITAE FAMILY							
Genus	Species	Vegetable	Natural Pollination Method	Recommended Isolation Method	Hand Pollination Method	Special Instructions	Contaminating Wild Plants
Chichorium	Endivia	Endive	Perfect flowers; inbreeding; self-pollinating	Time	If bagging, shake the bag from time to time in the morning	Overwinter. Can't be crossed by chicory but can cross with chicory.	
		Escarole	Same as above	Same as above	Same as above	Same as above	
	Intybus	Chicory	Perfect flowers; outbreeding; self-sterile on same plant; insect pollinated	Time with other chicories, endive, and escarole	If bagging or caging, include multiple plants and hand pollinate by removing flowers from each plant and rubbing them into the flowers of other plants	Overwinter. Can cross with endive and escarole.	
Cynara	Cardun-culus	Cardoon	Perfect flowers; inbreeding; self-sterile within same flower, but not within same flower head	Time		Overwinter. Perennial. Can cross with artichokes.	
Helianthus	Annuus	Sun-flower	Thousands of perfect flowers per head; outbreeding; most varieties self-compatible	Bagging	Remove bags from two plants for ten days and rub heads together		
Lactuca	Sativa	Lettuce	Perfect flowers; inbreeding; self-compatible; self-pollinating	None		Possible small amount of cross-pollination with other lettuces from insects	

				CUCURBITACEAE FAMILY			
Genus	**Species**	**Vegetable**	**Natural Pollination Method**	**Recommended Isolation Method**	**Hand Pollination Method**	**Special Instructions**	**Contaminating Wild Plants**
Citrullus	Lanatus	Water-melon	Imperfect flowers; outbreeding; self-compatible; insect pollinated	Taping	See taping instructions in chapter 14		
Cucumis	Melo	Muskmelon	Same as above	Same as above	Same as above		
		Cantaloupe	Same as above	Same as above	Same as above		
		Honeydew	Same as above	Same as above	Same as above		
		Casaba	Same as above	Same as above	Same as above		
		Armenian cucumber	Same as above	Same as above	Same as above		
	Sativus	Cucumbers	Same as above	Same as above	Same as above		
Cucurbita	Maxima	Banana squash	Same as above	Same as above	Same as above		
		Buttercup squash	Same as above	Same as above	Same as above		
		Hubbard squash	Same as above	Same as above	Same as above		
		Turban squash	Same as above	Same as above	Same as above		
	Mixta	Green striped and white cushaw squash	Same as above	Same as above	Same as above		
	Moschata	Butternut squash	Same as above	Same as above	Same as above		
		Cheese squash	Same as above	Same as above	Same as above		
		Golden cushaw squash	Same as above	Same as above	Same as above		
	Pepo	Acorn squash	Same as above	Same as above	Same as above		
		Crookneck squash	Same as above	Same as above	Same as above		
		Scallop squash	Same as above	Same as above	Same as above		
		Spaghetti squash	Same as above	Same as above	Same as above		
		Zucchini	Same as above	Same as above	Same as above		

LEGUMINOSAE FAMILY

Genus	Species	Vegetable	Natural Pollination Method	Recommended Isolation Method	Hand Pollination Method	Special Instructions	Contaminating Wild Plants
Arachis	Hypogaea	Peanut	Perfect flowers; inbreeding; self-fertile; self-pollinating	None, or Time	None	Some amount of cross-pollination with other same-species legumes via insects. Avoid planting different varieties right next to each other.	
Glycine	Max	Soybean	Same as above	Same as above	Same as above	Same as above	
Paseolus	Coccineus	Runner bean	Same as above	Same as above	Same as above	Same as above	
	Lunatus	Lima bean	Same as above	Same as above	Same as above	Same as above	
	Vulgaris	Common snap beans and dry beans	Same as above	Same as above	Same as above	Same as above	
Pisum	Sativum	Pea	Same as above	Same as above	Same as above	Same as above	
		Edible podded pea	Same as above	Same as above	Same as above	Same as above	
Vicia	Faba	Fava bean	Same as above	Same as above	Same as above	Same as above	
Vigna	Unguiculata Sesquipedalis	Cowpea	Same as above	Same as above	Same as above	Same as above	
		Yard long bean	Perfect flowers; inbreeding; self-fertile; self-pollinating	None, or Time	None	Same as above	

				SOLANACEAE FAMILY			
Genus	**Species**	**Vegetable**	**Natural Pollination Method**	**Recommended Isolation Method**	**Hand Pollination Method**	**Special Instructions**	**Contaminating Wild Plants**
Capsi-cum	Annuum	Sweet peppers	Perfect flowers; inbreeding; self-pollinating; self-compatible	Caging or flower bagging	Remove the cage or bag between 7:00 and 11:00 A.M. and gently thump each flower	Save and mix seeds from at least six plants. Cross-pollination can occur from insects.	
		Chili peppers	Same as above	Same as above	Same as above	Same as above	
Lycoper-sicon	Lycoper-sicum	Tomato	Perfect flowers; inbreeding; self-pollinating; self-compatible	None	None	Save and mix seed from at least six plants. Currant tomatoes, potato-leaved varieties, and double-blossomed beefsteaks can be cross-pollinated by insects.	
Physalis	Ixo-carpa	Tomatillo	Perfect flowers; inbreeding; self-pollinating; self-compatible	None	None	Save and mix seeds from at least six plants. Unknown if *Physalis* species other than tomatillo can cross within their species, so use time method if purity is desired.	
	Peruvi-ana	Cape goose-berry	Same as above	Same as above	Same as above	Same as above	
	Subgla-brata	Ground cherry	Same as above	Same as above	Same as above	Same as above	
Solanum	Melon-gena	Eggplant	Perfect flowers; inbreeding; self-pollinating; self-compatible	Time or flower bagging	None	Save and mix seeds from at least six plants. Cross-pollination can occur via insects.	

UMBELLIFERAE FAMILY							
Genus	Species	Vegetable	Natural Pollination Method	Recommended Isolation Method	Hand Pollination Method	Special Instructions	Contaminating Wild Plants
Apium	Graveolens	Celery	Perfect flowers; outbreedng; self-sterile within same flower, but not with the same umbels; insect pollinated	Bagging	Hand pollinate each day from 7:00 to 11:00 A.M. with a brush by rubbing the surface of the umbels for at least two weeks	Overwinter	
		Celeriac	Same as above	Same as above	Same as above	Same as above	
Anethum	Graveolens	Dill	Same as above	Same as above	Same as above	Same as above	
Corian-drum	Sativum	Cilan-tro/Corian-der	Same as above	Same as above	Same as above	Same as above	
Daucus	Carota	Carrot	Same as above	Same as above	Same as above	Same as above	Queen Anne's lace
Foenicu-lum	Vulgare	Fennel	Same as above	Same as above	Same as above	Same as above	Wild fennel
Pastinaca	Sativa	Parsnip	Same as above	Same as above	Same as above	Same as above	
Petroseli-num	Crispum	Parsley	Same as above	Same as above	Same as above	Same as above	
		Parsley root	Same as above	Same as above	Same as above	Same as above	

GRAMINEAE FAMILY							
Zea	Mays	Corn	Imperfect flowers; outbreeding; wind pollinated	Bagging	See corn-bagging instructions in chapter 14		

LABIATAE FAMILY

Genus	Species	Vegetable	Natural Pollination Method	Recommended Isolation Method	Hand Pollination Method	Special Instructions	Contaminating Wild Plants
Ocimum	Basilicum	Basil	Perfect flowers; inbreeding; insect pollinated	Time	None	Isolation distance of 150 feet reduces the risk that the variety will be spoiled by neighbors' crops	

MALVACEAE FAMILY

Genus	Species	Vegetable	Natural Pollination Method	Recommended Isolation Method	Hand Pollination Method	Special Instructions	Contaminating Wild Plants
Abel-moschus	Escu-lenntus	Okra	Perfect flowers; inbreeding; self-pollinating	Caging or blossom bagging	None		

VALERIANACEAE FAMILY

Genus	Species	Vegetable	Natural Pollination Method	Recommended Isolation Method	Hand Pollination Method	Special Instructions	Contaminating Wild Plants
Valeri-anella	Spp.	Corn salad/Mache	Perfect flowers; outbreeding; insect pollinated	Caging	Hand pollinate with a brush	Its low growth habit makes it hard to bag	Wild corn salad

APPENDIX 11

URBAN FARMING RESOURCES

BEES

Books

Bishop, Holley. *Robbing the Bees: A Biography of Honey—The Sweet Liquid Gold that Seduced the World*. New York: Free Press, 2006.

Conrad, Ross. *Natural Beekeeping; Organic Approaches to Modern Apiculture*. White River Junction, VT: Chelsea Green Publishing, 2007.

Dadant, C. P. *First Lessons in Beekeeping*. Hamilton, Illinois: Dadant and Sons, 1976.

Flottum, Kim. *The Backyard Beekeeper. A Beginners Book*. Gloucester, MA: Quarry Books, 2005.

Graham, Joe M. *The Hive and the Honeybee*. Hamilton, IL: Dadant and Sons, 2003.

Morse, Roger A. ed. *ABC and XYZ of Bee Culture: An Encyclopedia Pertaining to the Scientific and Practical Culture of Honeybees*. 40th edition. Medina, OH: A.I. Root Co., 2007.

Packages and Queens
Check with www.beesource.com/bees-supplies/united-states for a complete listing of suppliers of queens and bee packages.

Supplies

Betterbee, Greenwich, www.betterbee.com, New York

Brushy Mountain Bee Farm, www.brushymountainbeefarm.com, Morvarian Falls, North Carolina

Dadant and Sons, www.dadant.com, Illinois (also many branch locations)

Knight Family Honey, www.knightfamilyhoney.com, Utah

Mann Lake, www.mannlakeltd.com, California and Minnesota

Ruhl Bee Supply, www.ruhlbeesupply.com, Oregon

Shastina Millworks, www.shastinamillwork.com/commercial.asp, Oregon

Walter T. Kelley, www.kelleybes.com, Kentucky

Western Bee Supply, www.westernbee.com, Montana

Top Bar Beekeeping Resources

Chandler, Philip, *The Barefoot Beekeeper*. UK: Lulu.com, 2010

www.biobees.com

BEGINNING FARMER ASSOCIATIONS

www.beginningfarmers.org

www.thegreenhorns.wordpress.com

www.youngfarmers.org

BENEFICIAL INSECTS

From predatory flies that keep barn-fly populations down to ladybugs that eat aphids, beneficial insects are great to have in the garden.

Arbico Organics, www.arbico-organics.com, Arizona

Buglogical, www.buglogical.com, Arizona

BIODYNAMIC GARDENING BOOKS

Stella Natura Biodynamic Planting Calendar. Kimberton, PA: Stella Natura (published yearly, www.stellanatura.com).

Storl, Wolf D. *Culture and Horticulture: A Philosophy of Gardening*. Great Barrington, MA: Steiner Books, 1996.

Wright, Hilary. *Biodynamic Gardening for Health and Taste*. Edinburgh: Floris Books, 2009.

COMPOSTING RESOURCES

Urban Home Composting

www.cityfarmer.org/homecompost4.html

Books

Gershuny, Grace. *Soul of the Soil*. White River Junction, VT: Chelsea Green Publishing Company, 1999.

Gershuny, Grace and Deborah Martin. *The Rodale Book of Composting*. Emmaus, PA: Rodale, 1992.

Worms

Books

Appelhof, Mary. *Worms Eat My Garbage*. Kalamazoo: Flower Press, 1997.

Curriculum for teachers

www.calrecycle.ca.gov/Education/curriculum/worms

How to build a worm bin

Green Gusanito Worm Farm Bin, www.wormswrangler.com

www.seattletilth.org/learn/resources-1/compost/otsbinplans

Worm Factory, www.naturesfootprintinc.com

Where to buy worms

The best practice is to find a local worm farmer—check bait shops and garden centers for a source. If this fails, order from an online farm supply store, such as www.groworganic.com.

CONSTRUCTION AND BUILDING RESOURCES

craigslist.org

Freecycle.com

Alternative Masonry Units

Alternative Masonry Unit Presses, www.aureka.com

Corum, Nathaniel, with Marisha Farnsworth. *Alternative Masonry Unit Systems: A Manual for Village-Scale Block Construction Utilizing a Range of Industrial and Agricultural Byproducts.* San Francisco: Architecture for Humanity, 2010.

Drip Irrigation

Drip Works, www.dripworksusa.com

www.urbanfarmerstore.com

Recycled milk jug lumber

www.axionintl.com

Tool Libraries

There are libraries that will lend tools for building your urban farm from Atlanta to Columbus to Seattle. Instead of listing them we'll just refer you to the wiki page that will be updated as new ones are added: www.en.wikipedia.org/wiki/List_of_tool-lending_libraries

CONTAINER GROWING

www.insideurbangreen.org

FARM PEST PREVENTION

Ants

Tanglefoot. A sticky substance to apply to fruit trees and the legs of beehives to dissuade ants from setting up shop. Widely available at nurseries and online.

Flies

Buglogical predatory flies, www.buglogical.com

Sticky Barn Tape, www.get-revenge.us/flycatchers.html

Rats

Barn Owl Headquarters, www.tommy51.tripod.com

Build an Owl Box, www.carolinaraptorcenter.org/owl_box.php

Peaceful Valley, www.groworganic.com

Rat Terriers, www.ratting.co.uk

Snap traps, www.victorpest.com/store/rodent-control/mouse-traps

Zappers, Traps, and Nontoxic Rat Poison, www.eradibait.com

FARM SUPPLIES

Fedco, www.fedcoseeds.com, Maine

Harmony Farm Supply, www.harmonyfarm.com, California

Naomi's Organic Farm and Garden Supply, Portland

Peaceful Valley Farm Supply, www.groworganic.com, California

FRUIT TREE RESOURCES

Books

Bird, Richard. *Pruning Fruiting Plants.* London: Anness, 2007.

Dave Wilson Nursery. *Taste Results of the Best Fruit.* www.davewilson.com/br40/br40_taste_files/taste_index .html.

Otto, Stella. *Backyard Orchadist: Complete Guide to Growing Fruit Trees in the Home Garden.* Empire, MI: OttoGraphics, 1995.

Reich, Lee. *Uncommon Fruits for Every Garden.* Portland, OR: Timber Press, 2008.

Straub, Jack. *75 Remarkable Fruits for Your Garden.* Salt Lake City, UT: Gibbs Smith, 2008.

Trees and Root Stock

Since most root stocks are used in the wholesale nursery trade, it can be difficult to find small quantities of root stocks. The following companies, one on the West Coast and one on the East Coast, sell split bundles (a bundle contains one hundred root stocks).

Fedco Trees, www.fedcoseeds.com/trees.htm, Maine

One Green World Nursery, www.onegreenworld.com, Oregon

Raintree Nursery, www.raintreenursery.com, Washington

FARMING RESOURCES

Climate Resources

Climate Zones

www.garden.org/zipzone

www.sunset.com/garden/climate-zones

Daylight Predictor

astro.unl.edu/classaction/animations/coordsmotion/daylighthours
explorer.html

Predict the number of daylight hours in your region. You'll need to
know your latitude

Farming Resources

American Community Gardening Association, www.community
garden.org

Appropriate Technology Transfer for Rural Areas (ATTRA), www
.attra.org. ATTRA is *the* go-to resource for organic farming how-to
information.

CFSC Urban Agriculture Committee, www.foodsecurity.org/
ua_home.html

Community Food Security Coalition, www.foodsecurity.org

National Gardening Association, www.gardening.org

Vancouver, Canada's most excellent resource: www.cityfarmer.org

Farming Books

Bartholomew, Mel. *Square Foot Gardening*. Emmaus, PA: Rodale,
2005.

Bubel, Nancy. *The New Seed-Starters Handbook*. Emmaus, PA:
Rodale, 1988.

Chadwick, Alan. *Performance in the Garden: A Collection of Talks on
Biodynamic French Intensive Horticulture*. Mars Hill, NC:
Logosophia, 2008.

Coleman, Eliot. *The Winter Harvest Handbook: Year-Round Vegetable
Production Using Deep-Organic Techniques and Unheated Greenhouses*.
White River Junction, VT: Chelsea Green Publishing Company, 2009.

_____. *The New Organic Grower: A Master's Manual of Tools and Techniques for the Home and Market Gardener.* White River Junction, VT: Chelsea Green Publishing Company, 1995.

Coyne, Kelly, and Erik Knutzen. *Urban Homesteading.* Los Angeles: Process, 2008.

Creasy, Rosalind. *The Complete Book of Edible Landscaping: Home Landscaping with Food-Bearing Plants and Resource-Saving Techniques.* San Francisco: Sierra Club Books, 1982.

Damerow, Gail. *Barnyard in Your Backyard.* North Adams, MA: Storey, 2001.

Emery, Carla. *Encyclopedia of Country Living.* Seattle: Sasquatch Books, 2007.

Flores, H. C. *Food, Not Lawns.* White River Junction, VT: Chelsea Green Publishing, 2005.

Jeavons, John. *How to Grow More Vegetables than You Ever Thought Possible on Less Land than You Can Imagine.* Berkeley: Ten Speed Press, 2002.

Madigan, Carleen. *The Backyard Homestead.* North Adams, MA: Storey, 2009.

Marshall, Fern. *The Organic Gardener's Handbook of Natural Pest and Disease Control.* Emmaus, PA: Rodale, 2009.

Montague, Fred. *Gardening: An Ecological Approach.* Wanship, UT: Mountain Bear Ink, 2009.

Seattle Tilth. *The Maritime Northwest Garden Guide.* Seattle, WA: Seattle Tilth, 2008.

Solomon, Steve. *Growing Vegetables West of the Cascades: The Complete Guide to Natural Gardening.* Seattle, WA: Sasquatch Books, 2000.

Tozer, Frank. *The Organic Gardeners Handbook.* Santa Cruz: Greenman Publishing, 2008.

_____. *The Vegetable Growers Handbook.* Santa Cruz: Greenman Publishing, 2008.

Whitson, Tom. *Weeds of the West.* Laramie, WY, Western Society of Weed Science in Cooperation with the Western United States Land Grant Universities Cooperative Extension Services, 2002.

Wiswall, Richard. *The Organic Farmer's Business Handbook.* White River Junction, VT: Chelsea Green Publishing Company, 2009.

GOATS

Books

Belanger, Jerry. *Storey Guide to Raising Dairy Goats*. North Adams, MA: Storey, 2001.

Hathaway, Margaret. *Living with Goats*. Gullford, CT: Lyons Press, 2009.

Mackenzie, David. *Goat Husbandry*. London: Faber and Faber, 1993.

Cheese making

Glengarry Cheesemaking Supply, glengarrycheesemaking.on.ca

New England Cheesemaking Supply Company, www.cheesemaking.com

Supplies

Caprine, www.caprinesupply.com

Fiasco Farm, http://fiascofarm.com

Hoegger, www.hoeggergoatsupply.com

GREY WATER

Books

Lancaster, Brad. *Rainwater Harvesting for Drylands and Beyond*, Vols. I and II. Tucson: Rainsource Press, 2006.

Ludwig, Art. *The New Create an Oasis with Graywater: Choosing, Building, and Using Graywater Systems*. Santa Barbara, CA: Oasis Design, 2006.

Supplies

www.harvestingrainwater.com

www.oasisdesign.net

POULTRY

Books

Alderson, Dave. *Raising the Home Duck Flock*. North Adams, MA: Storey, 1980.

Brock, Todd, David Zook, Rob Ludlow. *Chicken Coops for Dummies*. Indianapolis, IN: Wiley, 2010.

Damerow, Gail. *Chicken Health Handbook*. North Adams, MA: Storey, 1994.

Kilarski, Barbara. *Keep Chickens!* North Adams, MA: Storey, 2003.

Rossier, Jay. *Living with Chickens*. Gullford, CT: Lyons Press, 2004.

Ordering Chicks/Poults/Ducklings

Ideal Poultry, turkey poults, ducklings, broilers, laying hen chicks, www.ideal-poultry.com, Texas

JM Hatchery, freedom ranger chicks, guinea fowl, muscovy ducklings, www.jmhatchery.com, Pennsylvania

Metzer, goslings, ducklings, and game birds, www.metzerfarms.com, California

Murray McMurray, rare breeds, turkey poults, ducklings, goslings, chicks, www.mcmurrayhatchery.com, Iowa

My Pet chicken, sends small numbers of chicks through the mail, www.mypetchicken.com, Connecticut

Privett Hatchery, chicks, ducklings, goslings, www.privetthatchery.com, New Mexico

Sand Hill Preservation, specializes in heirloom varieties of chicks, www.sandhillpreservation com, Iowa

RABBITS

Bass Rabbit Supply, www.bassequipment.com, Missouri and California

Books

Bennett, Bob. *Storey's Guide to Raising Rabbits*. North Adams, MA: Storey, 2001.

Goodchild, Claude and Alan Thompson. *Keeping Rabbits and Poultry on Scraps.* New York: Penguin, 2008 (reissued from 1941).

ROOFTOP GROWING

www.greenroofgrowers.blogspot.com

SEEDS AND PLANTS

Abundant Life, www.abundantlifeseeds.com, Oregon

Baker Creek, www.rareseeds.com, Missouri

Bountiful Gardens, www.bountifulgardens.org, California

Johnny's Selected Seeds, www.johnnyseeds.com, Maine

High Altitude Gardens/Seeds Trust, www.seedstrust.com /joomla

High Mowing Seeds, www.highmowingseeds.com, Vermont

Kitazawa Seed Company, www.kitazawaseed.com, California

Mountain Valley Seed Company, www.mvseeds.com, Utah

Renee's Garden Seeds, www.reneesgarden.com, California

Seed Saver's Exchange, www.seedsavers.org, Iowa

Seeds of Italy, www.seedsofitaly.com, England

Seeds Trust, www.seedstrust.com, Arizona

Siskyou Seeds, www.siskiyouseeds.com, Oregon

Solana Seeds (French and heirloom varieties), www.solanaseeds .netfirms.com/catalog.html, Quebec

Southern Exposure Seed Exchange, www.southernexposure.com, Arizona

Sustainable Seed Company, www.sustainableseedco.com, California

Territorial Seed, www.territorialseed.com, Washington

Turtle Tree Seed, www.turtletreeseed.com, New York

Uprising Seeds, www.uprisingorganics.com, Washington

Wild Garden Seeds, www.wildgardenseed.com, Oregon

SEED-SAVING BOOKS

Ashworth, Suzanne. *Seed to Seed: Seed Saving and Growing Techniques for Vegetable Gardeners*. Decorah, IA: Seed Savers Exchange, 2002.

Deppe, Carol. *Breed Your Own Vegetable Varieties: The Gardener's and Farmer's Guide to Plant Breeding and Seed Saving*. White River Junction, VT: Chelsea Green Publishing, 2000.

TOOLS

BCS, www.bcsshop.com, Massachusetts

Earth Tools, www.earthtoolsbcs.com, Kentucky

Garden Tool Company, www.gardentoolcompany.com, Colorado

Hida Tool, www.hidatool.com, California

Smith and Speed, www.smithandspeed.com, Washington

URBAN FARMING ORGANIZATIONS, EXTENSION AGENCIES, AND COMMUNITY GARDENS

A State-by-State Listing

ALABAMA

Jones Valley Urban Farm, www.jvuf.org, Birmingham

ALASKA

Anchorage Parks and Rec Community Garden Plots, www.muni
.org/Departments/parks/Pages/GardenPlots.aspx

Fairbanks Community Garden, www.home.gci
.net/~fairbankscommunitygarden/index.html

Juneau Community Garden, www.juneaucommunitygarden.org

ARIZONA

Tucson Village Farm, www.tucsonvillagefarm.org

Urban Farm, www.urbanfarm.org, Phoenix

ARKANSAS

Little Rock Urban Farming, littlerockurbanfarming.blogspot.com

CALIFORNIA

Biofuel Oasis Urban Feedstore, www.biofueloasis.com, Berkeley

Center for Urban Agriculture at Fairview Gardens, www.fairview gardens.org, Goleta

City Slicker Farms, www.cityslickerfarms.org, Oakland

Community Food Security Coalition, www.foodsecurity.org, Los Angeles

Farm Lab, www.Farmlab.org, Los Angeles

Institute for Urban Homesteading, www.iuhoakland.com, Oakland

People's Grocery, www.peoplesgrocery.org, Oakland

Seeds at City Urban Farm, www.seedsatcity.com, San Diego

Soil Born Farms, www.soilborn.org, Sacramento

Spiral Gardens, www.spiralgardens.org, Berkeley

COLORADO

Community Roots Urban Gardens, www.communityrootsboulder .com, Boulder

Denver Urban Gardens, www.dug.org, Denver

The Urban Farm, www.siouxfallsparks.org/community_gardens .aspx, Denver

CONNECTICUT

Hartford Food System, www.hartfordfood.org

DELAWARE

Delaware Center for Horticulture, www.thedch.org, Wilmington

DISTRICT OF COLUMBIA

Common Good City Farm, commongoodcityfarm.org

The Urban Agriculture Network (TUAN), www.cityfarmer.org/TUAN.html

FLORIDA

Earth Learning, www.earth-learning.org, Miami

Florida Organic Gardeners, www.foginfo.org, Gainsville

GEORGIA

Truly Living Well Center, www.trulylivingwell.com, Atlanta

IDAHO

Earthly Delights Farm, www.earthlydelightsfarm.com, Boise

ILLINOIS

Growing Home, www.growinghomeinc.org, Chicago

Growing Power, www.growingpower.org/chicago_projects.htm, Chicago

The Resource Center / City Farm, www.resourcecenterchicago.org, Chicago

INDIANA

Big City Farms, www.bigcityfarmsindy.com, Indianapolis

IOWA

Iowa Network for Community Agriculture, www.growinca.org, Des Moines

KANSAS

Kansas City Center for Agriculture, www.KCCUA.org

KENTUCKY

Louisville Grows, www.louisvillegrows.org

LOUISIANA

New Orleans Food and Farm Network, www.noffn.org

Parkway Partners, www.parkwaypartnersnola.org, New Orleans

Viet Village Urban Farm, www.tulanecitycenter.org/programs/projects/viet-village-urban-farm, New Orleans

MAINE

Cultivating Community, www.cultivatingcommunity.org, Portland

MARYLAND

Baltimore Urban Agriculture, www.baltimoreurbanag.org

MASSACHUSETTS

Allandale Farm, www.allandalefarm.com, Brookline

The Food Project, www.thefoodproject.org, Boston

Our Roots, www.nuestras-raices.org, Boston

Re-Vision Farm, www.vpi.org/Re-VisionFarm, Dorchester

MICHIGAN

Detroit Agriculture Network, www.detroitagriculture.org

Detroit Black Community Food Security Network, detroitblackfoodsecurity.org

Earthworks Urban Farm, www.cskdetroit.org, Detroit,

Georgia Street Garden, www.georgiastreetcc.com, Detroit

Greening of Detroit, www.greeningofdetroit.com

Growing Hope, www.growinghope.net, Detroit

Growing Power, www.growingpower.org. Madison and Milwaukee

MINNESOTA

Eggplant Urban Farm Supply, www.eggplantsupply.com, St. Paul

MISSISSIPPI

The Jackson Inner-city Gardeners, www.jiggarden.org

MISSOURI

Bad Seed, www.badseedfarm.com, Kansas City

Kansas City Community Gardens, www.kccg.org

MONTANA

Garden City Harvest, www.gardencityharvest.org, Missoula

NEBRASKA

City Sprouts, omahasprouts.org, Omaha

Community CROPS, www.communitycrops.org, Lincoln

University of Nebraska–Lincoln, www.food.unl.edu/web/localfoods/urbanag

NEVADA

Mayberry Farm, www.mayberryfarm.info/Mayberry_Farm/Home.html, Reno

NEW HAMPSHIRE

University of New Hampshire Cooperative Extension,
www.extension.unh.edu/HCFG/Map_CommGarden.htm

NEW JERSEY

Garden State Urban Farm, www.gardenstateurbanfarms.com, Newark

NEW MEXICO

Albuquerque Backyard Farms, www.abqbackyardfarms.com

Mother Nature Gardens, www.mothernaturegardens.com,
Albuquerque

Rio Grande Community Farm, www.riograndefarm.org,
Albuquerque

NEW YORK

Added Value, www.added-value.org, Brooklyn

Bronx Green Up, www.nybg.org/green_up

Brooklyn Grange, www.brooklyngrangefarm.com, Queens

East New York Farms, www.eastnewyorkfarms.org, Brooklyn

Green Guerillas, www.greenguerillas.org, Brooklyn

Green Thumb, www.greenthumbnyc.org, Manhattan

Just Food, http://www.justfood.org/jf, Manhattan

La Familia Verde, www.lafamiliaverde.org, Bronx

New York Restoration Project, www.nyrp.org, Manhattan

Oasis Community Garden Map, www.oasisnyc.net/garden/garden
search.aspx

Sustainable Flatbush, www.sustainableflatbush.org, Brooklyn

NORTH CAROLINA

Darko Urban Farm, www.csa.alexissparko.com, Durham

Green Hill Urban Farm, www.greenhillurbanfarm.com, Asheville

SEEDS, www.seedsnc.org, Durham

NORTH DAKOTA

Bismarck Community Gardens, www.bismarckgardens.org, Bismarck

OHIO

Ohio City Farm, www.ohiocityfarm.com, Cleveland

Ohio State University Extension, cuyahoga.osu.edu/topics/agriculture-and-natural-resources/urban-agriculture-program

Urban Growth Farms, www.urbangrowthfarms.com, Cleveland

OKLAHOMA

Closer to Earth, www.closertoearthokc.blogspot.com, Oklahoma City

OREGON

Growing Gardens, www.growing-gardens.org, Portland

Naomi's Organic Farm Supply, www.naomisorganic.com, Portland

Portland Fruit Tree Project, www.portlandfruit.org

Portland State University, Learning Gardens Laboratory, www.pdx.edu/elp/learning-gardens-laboratory

Sauvie Island Center, www.sauvieislandcenter.org, Portland

Slow Hand Farms, www.slowhandfarm.com, Portland

Urban Farm Store, www.urbanfarmstore.com, Portland

Urban Farm University of Oregon, www.landarch.uoregon.edu/resources/urbanfarm, Eugene

Your Backyard Farmer, www.yourbackyardfarmer.com, Portland

Zenger Farm, www.zengerfarm.org, Portland

PENNSYLVANIA

American Community Gardening Association, www.communitygarden.org, Philadelphia

Emerald Street Farm, www.emeraldstreeturbanfarm.wordpress.com, Philadelphia

Green's Grow, www.greensgrow.org, Philadelphia

Grow Pittsburgh, www.growpittsburgh.org

Mill Creek Farm, www.millcreekurbanfarm.org, Philadelphia

Weaver's Way Coop, www.weaversway.coop, Philadelphia

RHODE ISLAND

Southside Community Land Trust, www.southsideclt.org, Providence

SOUTH CAROLINA

City Roots, www.cityroots.org, Columbia

SOUTH DAKOTA

Sioux Falls Community Gardens, www.siouxfallsparks.org/community_gardens.aspx

TENNESSEE

Urban Farm, www.bdcmemphis.org, Memphis

TEXAS

Austin Urban Farming, www.austinurbanfarming.com

Boggy Creek Farm, www.boggycreekfarm.com, Austin

East Dallas Community Garden and Market, www.gardendallas.org/east_dallas_community_garden.htm

North Haven Gardens, www.nhg.com, Dallas

Sustainable Food Center, www.sustainablefoodcenter.org, Austin

Urban Harvest, www.urbanharvest.org, Houston

UTAH

Red Butte Garden, www.redbuttegarden.org, Salt Lake City

Wasatch Gardens, www.wasatchgardens.org, Salt Lake City

VERMONT

Intervale Center, www.intervale.org, Burlington

VIRGINIA

Lynchburg Grows, www.lynchburggrows.org

Tricycle Gardens, www.tricyclegardens.org, Richmond

WASHINGTON

Alley Cat Acres Urban Farm Collective, www.alleycatacres.com, Seattle

Goat Justice League, www.goatjusticeleague.org, Seattle

King County Community Gardens, www.kingcounty.gov/recreation/parks/rentals/communitygardens.aspx, Seattle

P-Patch, www.seattle.gov/neighborhoods/ppatch, Seattle

Seattle Farm Cooperative, www.seattlefarmcoop.com

Seattle Tilth, www.seattletilth.org

Seattle Youth Garden Works, www.sygw.org/home

Urban Eden Farm, www.urbanedenfarm.com, Spokane

WEST VIRGINIA

West Virginia University Extension Service, www.ext.wvu.edu/lawn_garden/home_gardening

WISCONSIN

Growing Power, www.growingpower.org, Madison and Milwaukee

Milwaukee Urban Gardens, www.Milwaukeeurbangardens.org

STARTING AN URBAN FARMING BUSINESS

After a season or two of successful urban farming, it's almost inevitable: You want to sell your surplus and try to make a little money. Maybe it's just an American impulse to become an entrepreneur, but we've noticed many urban farmers (including ourselves) take the plunge to make it pay—as an extra source of income, that is. You won't be very happy with your business if you expect it to provide all your income, but an urban farming business can certainly cover the costs of your hobby and provide a little extra cash. There are two different ways to go about making money off your farm: underground sales or a legitimate business.

UNDERGROUND

The underground farmer often starts by selling one product: chicken eggs or duck eggs to coworkers, for example. Many times these farmers also sell to underground or secret restaurants—things like salad mix, a few apples, or a brace of rabbits. Sometimes the sales occur through the Internet on sites like Craigslist.com or through blogs. Underground urban farm sales give the farmer a sense of pride, some needed cash, and a way to offset a glut in the garden or barnyard.

Sounds perfect, right: cash economy, untaxed income, happy customers? It can be good—and we don't want to discourage sales, but there are pitfalls. One is that you may sell someone a contaminated product and they become sick or worse. Not only will you feel terrible, you may be sued. Avoid this problem by being incredibly careful with everything you sell. Another possible problem is the tax man—at the city, state, or federal level. If you have an obvious business presence and no business license displayed, the city may come after you. If you are making a sizable income from your urban farm, you may get a call from the IRS. Finally, all manner of city, county, and state health department inspectors and/or U.S. Department of Agriculture agents may come to your house and try to bust you.

GOING LEGIT

If you can't take the heat, start a legitimate business. Consider starting a community-supported agriculture subscription (a weekly produce box), vending at farmers markets, selling to restaurants, or opening your own farm stand. Rules vary from city to city, but in most places you'll need county agriculture department certification, a city business license, liability insurance, and a way to track your sales for tax purposes in order to start a business. The downside of going legitimate is that the licenses and insurance will cost a pretty penny, eating into your profit margin.

If you want to start a legit business, first call your county agriculture department and ask for the paperwork necessary to receive a county producer's certificate. Once granted, this certificate will allow you to legally engage in direct sales at farmers' markets or farm stands, or to retail businesses on the county level.

Next, go to the city department that deals with business licenses and ask if they will license your type. Depending on where your farm is located, it may not be allowed at present. You will need to find out the zoning designation of your location. If it's zoned for mixed use, meaning that both businesses and residences are allowed, chances are you may be able to get a license. If you are in a zone established for residential use, however, it may not be allowed. Check what is allowed under the "home-based business" license category. If you have a friendly city worker look over your application, you may be in luck. If not, you may need to get involved in advocating for new city legislation.

Find an insurance broker who specializes in small businesses and work out what type of liability insurance you will need. Since agricultural businesses are less common in cities, you may have to do some research to help your agent. Call a rural agent if you are stumped. In addition to a policy covering the business, you may also want to purchase a personal umbrella one to cover you in case of a major lawsuit. Your business insurance policy holder may find a loophole negating your coverage. If you have assets this is important.

Finally, you will need to begin tracking expenses and income for tax purposes. This will also help you determine if you are making a profit. Set up a separate business account as soon as possible (the feds don't like to see combined accounts). The easiest way to set up your bookkeeping is by using a computer program for businesses. You may need to employ a bookkeeper or accountant once a year to organize and file your tax returns for you.

If you go legit, we highly recommend writing a business plan. Doing this will help you make money from your farming rather than paying money just to keep it going. What are you going to sell and for how much? Who is going to buy your product? Set up Excel spreadsheets that predict how much you will spend bringing the crop to market, what you should charge, and how much profit you need to make on it. Don't forget to account for your time working!

Looking at a budget will help you gauge what products make a profit and which don't. From our experience, honey, salad greens, and value-added products (which require the use of a commercial kitchen) show the best profit margin. Selling organic eggs yields small profit. But perhaps you have a niche, so don't be afraid to experiment.

Remember too that it's okay to charge more for urban-farmed products. Just being grown on an urban farm gives the product a niche, an appeal for city people who want extremely fresh and truly local food. We've noticed that it helps to have a logo and a Web site, so people can find you and recognize your brand. You can also make some extra money selling T-shirts with your logo. For help in writing a business plan, contact your local U.S. Small Business Administration office and ask about free or affordable classes and mentoring in your area.

Because you may be producing so little that you are unable to satisfy even one customer's demand for product, you may want to create a producers' cooperative. This is an agricultural business set up collectively by small-scale growers so that they can compete with larger producers. By increasing the quantity of product available under the cooperative's brand, it can market to customers an individual producer wouldn't be able to serve. For instance, if your local restaurant uses fifty pounds of salad mix a day and all you can supply is fifty pounds a week, you can't sell to them. If, however, you gather together with other small producers, the demand can be met, and each producer will have a consistent buyer for their product.

Another benefit of operating as a producers' cooperative is that capital expenditures for materials and equipment can be shared. This will allow you to share equipment you need to use only occasionally and to purchase materials in larger quantities, making them less expensive. You can collectivize aspects of production as well, such as propagating seedlings or packaging produce for the market. Contact your local cooperative extension service for instructions on how to set up a producers' cooperative.

WORKING WITH BUSINESS PARTNERS

If you decide to start an urban farm business with a friend, proceed with caution. In executing group projects, beautiful visions can sometimes end with strained nerves and tenuous relationships. In theory everything sounds great, but getting down to the practicalities can reveal deep differences in goals, strategies, work styles, and philosophies.

We think you should carefully consider the implications of working in a partnership as opposed to being sole proprietor. You'll still have to apply for business licenses and file tax forms, but there's another layer of power dynamics. There's nothing wrong with being the person making the decisions in a business that involves others as long as you are up front about that. It can certainly save a lot of time, and there are meaningful ways to include friends and community members in your urban farm short of considering them equal decision makers.

Those who want to run a business should be excited about doing a lot of hard work, some of it thankless. Over time we've developed a motto: "Your degree of decision-making power is determined by your degree of responsibility for business-sustaining work, even when it might be personally inconvenient." This doesn't preclude collaboration, and it doesn't give people unearned rights or excessive power over others. While some feedback and ideas can be a godsend, some people have an unrelenting flow of new ideas but very little will to carry anything out. If you genuinely think a new idea would help the business, a very positive response is "What a great idea! Why don't you go ahead and do that and let me know how it works out." Otherwise, new ideas may constantly get in the way of carrying out your core goals.

If you do decide to work in partnership, we urge you to address some key questions at the outset, and come up with a list of initial agreements. Just discussing them will bring out issues that could turn into major problems later on. People who were enthusiastic about the project may voluntarily opt out when they realize the commitments involved, and people who really mean business will become even more enthusiastic with clarity of expectations. Later, in the event of broken commitments, you can return to your record of initial agreements to help resolve conflicts in a businesslike manner. It's much simpler to point out that if someone wants to stay involved they must satisfy the commitments they agreed to rather than to fume for months in the belief that you are unjustly shouldering more than your share of the responsibilities.

Suggested questions:

1. What are the mission and goals of the business?

2. Who can be considered business partners, and what are the criteria for new partners to be admitted, if at all?

3. What commitments in terms of time, money, and other resources must partners meet, and over what time period (each month, quarter, year, etc.)?

4. What are the various types of decisions to be made, and who gets to make them?

5. How are decisions made and how are meetings conducted (vote, consensus, rotating facilitation, etc.)?

6. What would be the ideal way for business partners to leave the project if necessary?

Issues of authority and hierarchy also come up in the day-to-day execution of tasks. Planning a project, gathering the materials, and doing the necessary preparatory work collaboratively, with a group of people, is a setup for inefficiency and conflict. As much time will be spent simply figuring out who's doing what as it would have taken one person to do what's required. People love to feel like they're contributing, not like they're trapped in an endless discussion over whether screws or bolts would be best for the planter boxes. For any project that will be executed by a number of people, it will be most efficient if one partner sets up the job and leads the others through the work. There will be no confusion over who was supposed to bring the extension cord, and everyone will feel the pride of accomplishment when the project is completed on time. Partners can, of course, rotate this responsibility.

ACKNOWLEDGMENTS

Without the support of our families this book would have never been written. Willow thanks her dad Arthur, with his gardens full of summer bounty, for inspiring her; her husband Lew Summer for his heartfelt encouragement; her sisters Mira and Zoe for their support; and her mom Claire and stepmom Nancy for their encouragement in all her endeavors. Novella learned to farm at a young age from her mom Patricia who kept a wonderful homestead in the early years and is still a kick-ass gardener; from her dad who encouraged communion with animals; and from her sister Riana who has been an enormous support even across the ocean. Novella's Billy Jacobs kept the trucks running.

One of the joys of being a farmer is that the learning never ends. Before we could call ourselves farmers, we studied at the knee of a number of generous mentors. Willow gratefully acknowledges Anna Ransom, her first farming teacher, as well as Peggy Kass and Helen Krayenhoff of Oakland's treasured Kassenhoff Growers. Doug Gosling of the Occidental Arts and Ecology Center has inspired Willow since her teen years. Willow learned the joys and sorrows of urban farming along with a core team of collaborators in the early years of City Slicker Farms and she gratefully acknowledges their camaraderie. The City Slicker Farms apprentices and staff tested and contributed to many of the methods in this book. Willow is truly thankful that Barbara Finnin, executive director, and Kelly Saturno, board president, took the reins at City Slicker Farms, giving her the time to write this book. Novella wants to thank the women of the BioFuel Oasis, and their students at the urban farm store who were unwitting guinea pigs for the livestock curriculum. For both of us, the community of West Oakland has truly been our learning ground and we thank our neighbors for their collaboration and wisdom.

A few notable people have served to inspire many, including us: Will Allen of Growing Power; Allen Chadwick and John Jeavons, who brought intensive farming techniques to the United States and inspired a generation of farmers; and Alice Waters, whose Edible Schoolyard Project captured our imaginations.

This book has been a community effort; we gratefully acknowledge the expertise of the following technical advisers: Celia Bell of Salt Lake City; Steve Calanog of San Francisco Environmental Protection Agency; Patrick Crouch of Detroit; Shirley DeLeon; Hannah Farley of New Haven; Nathan McClintock, Jim Montgomery, and Steve Norwick of Sonoma State University; Ann Ralph; Jordan Shay of New Orleans; Donna Smith, Robyn Streeter, and Joshua Volk of Portland; and Adam Wolf at Princeton University. Despite their close readings, errors may have occurred—all of these are strictly the fault of the authors.

Beyond technical support, there are also people in our lives who have inspired us by their words, wisdom, or acts of unbridled urban farming support. They include: Dan and Brook Salvaggio at Bad Seed Farm; Maurice Cavness, Joy Moore, Kim Allen, Daniel Miller, Jason Mark, and Antonio Alcala of Alemany Farm; OBUGS; People's Grocery; Brooke Budner and Caitlyn at Little City Gardens; Max Cadji at Phat Beets; Rich Pederson; Esperanza Pallana at Pluck and Feather; Ashley Willihite at NYRP; Owen Taylor at Just Food; Brandon Hoy, Annie Novak, and Ben Flanner of Brooklyn; Food First and the Oakland Food Policy Council; and the newly revived Berkeley Food Council.

Words are flimsy without images. We were so lucky to work with the talented Bronwyn Barry, whose delightful line drawings grace these pages; Arianna Rosenthal (who happens to be Willow's sister) who effortlessly produced technically accurate yet whimsical illustrations, and Sophia Wang who meticulously created photographic images when an illustration just wouldn't do.

At Penguin, we would like to thank the following team: production editor Jennifer Tait, copy editor Rachel Burd, proofreader Andrea Polselli, designer Sabrina Bowers, editor Janie Fleming who started our project, and the diligent Lindsay Whalen who brought our project to completion and has been a source of delight. Our agent, Richard Morris, is a peach.

W — E

INDEX